HETEROCHROMATIN
Molecular and structural aspects

HETEROCHROMATIN

Molecular and structural aspects

Edited by
RAM S. VERMA
Division of Genetics, Long Island College Hospital
and
Department of Medicine, State University of New York,
Health Science Center at Brooklyn

CAMBRIDGE UNIVERSITY PRESS
CAMBRIDGE
NEW YORK NEW ROCHELLE MELBOURNE SYDNEY

Published by the Press Syndicate of the University of Cambridge
The Pitt Building, Trumpington Street, Cambridge CB2 1RP
32 East 57th Street, New York, NY 10022, USA
10 Stamford Road, Oakleigh, Melbourne 3166, Australia

First published 1988

Printed in the United States of America

Library of Congress Cataloging-in-Publication Data
Heterochromatin : molecular and structural aspects / edited by Ram S. Verma.
p. cm.
Includes bibliographies and index.
ISBN 0-521-33480-2
1. Heterochromatin. I. Verma, Ram S. (Ram Sagar)
QH599.H48 1988
574.87′322–dc19 87-22510

British Library Cataloguing in Publication Data
Heterochromatin : molecular and structural aspects.
1. Heterochromatin
I. Verma, Ram S.
574.87′32 QH599
ISBN 0-521-33480-2

To
Herbert A. Lubs, M.D.
and
David B. Walden, Ph.D.

Contents

Contributors

ARVIND BABU, Division of Medical Genetics, Department of Pediatrics, Beth Israel Medical Center–Mount Sinai School of Medicine, New York, NY 10003, U.S.A.

BERNARD JOHN, Research School of Biological Sciences, The Australian National University, Canberra City, A.C.T. 2601, Australia

CHYI-CHYANG LIN, Departments of Pathology and Laboratory Medicine, University of Alberta, Edmonton, Alberta, Canada T6G 2R7

ALLAN LOHE, Division of Entomology, CSIRO, Black Mountain, Canberra, A.C.T. 2601, Australia

CLAUDIO NICOLINI, Department of Biophysics, School of Medicine, University of Genoa, Viale Benedetto XV, Genoa 2 Italy

DANIEL A. PEPPER, Department of Cell Biology, Boehringer Mannheim Research and Development, 9115 Hague Road, Indianapolis, IN 46250

JEROME B. RATTNER, Departments of Medical Biochemistry and Anatomy, The University of Calgary and Health Science Center, 3330 Hospital Drive N.W., Calgary, Alberta, Canada T2N 4N1

PAUL ROBERTS, Department of Zoology, Oregon State University, Corvallis, OR 97331, U.S.A.

RAM S. VERMA, Division of Genetics, Long Island College Hospital, Brooklyn, NY, and Department of Medicine, State University of New York Health Science Center, Brooklyn, NY 11201, U.S.A.

Preface

Understanding of the structure and function of chromosomes has advanced significantly during the last three decades, largely due to the proliferation of newer staining techniques, use of electron microscopy, *in situ* hybridization, and most recently the application of recombinant DNA technology. These technical advances have led to the exploration of the components of chromosomes. Today eukaryotic chromosomes are primarily described according to the structural and functional aspects of euchromatin and heterochromatin. New information regarding heterochromatin has accumulated at such an accelerated rate that old dogma is no longer relevant with respect to chromosomal structure. Therefore the need has arisen for a volume containing current information on the organization of chromosomal heterochromatin.

Each chapter provides a comprehensive review and critical discussion of the recent advances within each subtopic. The first and most elaborative chapter, written by Bernard John, covers the old and new biological enigmas of heterochromatin. Certain portions of eukaryotic chromosomes show remarkably different physical properties when compared to the bulk of DNA. These fractions, called satellite DNA, are highly conserved, a topic discussed in Chapter 2. The unique heterochromatic structure of kinetochores is described in Chapter 3, while in Chapter 4 the organization of the centromere and centromeric heterochromatin is discussed. Despite significant coverage regarding the functional and structural aspects of heterochromatin in various chapters, the basic question regarding its three-dimensional organization would have been neglected without Chapter 5. Often there is no direct relationship between a heteromorphism (variant) identified by one staining technique and that identified by others. This intriguing heterogeneity is explained in Chapters 6 and 7. Interesting and eminently readable reviews dealing with meticulous details of this fascinating subject are listed in the final chapter.

It is my sincere hope that the contributions I have selected for this volume are highly informative and contain the most recent information available. To my knowledge, there is no comprehensive contemporary work on this very exciting subject. I have deliberately attempted to choose topics of special concern to graduate students who wish to pursue advanced careers in cell biology. Because it is difficult to keep informed of every possible relevant development, seasoned scientists will also find this book an invaluable reference source.

Ram S. Verma

Brooklyn, New York
November 10, 1987

Acknowledgments

My thanks to the distinguished contributors for revising the chapters after their initial conception; special gratitude to those who completed their chapters early. I thank those scientists and publishers who generously contributed figures or granted permission for figure reproduction. The publisher, scientific editors, and many staff members of Cambridge University Press deserve much of the credit.

I would like to express my sincerest appreciation to Arvind Babu, who suggested the names of scientists to be considered for this volume. I owe a debt of thanks to him for bringing the most recent advances to my attention and for numerous instances of discussion, diversion, and technical assistance. Thanks and profound gratitude are also extended to Michael J. Macera for editorial suggestions and proofreading. Furthermore, the financial support of Labrepco is greatfully acknowledged. I am also thankful to many secretaries for typing the manuscript. Lastly, this project would not have been possible without the understanding and patience of my wife Shakuntala, who graciously provided moral support and encouragement during the most trying times of this project; to her I will always be thankful.

R. S. V.

I

The biology of heterochromatin

BERNARD JOHN

"Most new ideas are wrong. Most old ideas are wrong also except that having stood the test of time an old idea is judged better than a new idea. The real trouble is that most new ideas are old. They can fail twice."

Art Seidenbaum

1. Defining heterochromatin

"Those who cannot remember the past are condemned to repeat it."

George Santayana

1.1. Modes of chromocenter formation

The term *heterochromatin* has proven to be a major biological enigma ever since it was first introduced into the genetic vocabulary by Heitz in 1928. It was conceived as a modification of an earlier word, *heterochromosome,* used by cytologists before Heitz's time to distinguish between sex chromosomes, which commonly undergo differential condensation at male meiosis, and euchromosomes, or as we now refer to them, *autosomes,* which do not. Heitz invented his new term to describe and denote chromosome segments, or in some cases entire chromosomes, that maintain a condensed state during the interphase of the mitotic cell cycle and therefore appear in the resting nucleus as a chromocenter. By carefully following chromosomes through successive division cycles, he was able to demonstrate that structures which formed chromocenters at interphase were simply specific chromosome regions that did not decondense at telophase of the mitotic division.

The definition proposed by Heitz was thus purely morphological. It identified regions of mitotic chromosomes that retained a compact structure during interphase, displayed a differential condensation, or *heteropycnosis,* at prophase and, unlike euchromatin, did not unravel at telophase of the mitotic cycle. In all of these respects, heterochromatin was

My thanks to Elizabeth Robertson for typing the manuscript and to Gary Brown for his assistance with the illustrations.

out of step, or *allocyclic,* in its condensation behavior relative to euchromatin, which was assumed to be dispersed during interphase and maximally contracted only at metaphase. Since the time of Heitz, the term *chromocenter* has come to be used in two rather different ways: first, to describe individual condensed regions within interphase nuclei; and second, to describe aggregations of condensed regions. At the extreme, all condensed segments may fuse into a single structure, a collective chromocenter. Alternatively there may be several, distinct, compound chromocenters. The interactions responsible for the production of collective or compound chromocenters in fact vary with cell type and stage of development, both within and between species (Schmid, 1967).

Implicit in the original definition proposed by Heitz was the assumption that all condensed intermitotic chromatin belonged to a unitary class. This, as we now know, was incorrect. In general terms, the DNA–protein complex which underlies chromosome organization may exist in the interphase state in two distinct conformations, diffuse or condensed. This difference in state reflects different levels of condensation resulting from the interaction of DNA with chromosomal or other proteins (Weith, 1985). Autoradiographic studies using tritium-labeled uridine ([^{3}H]UdR) provides convincing evidence that RNA synthesis is possible only in the diffuse or dispersed state, whereas condensed chromatin is transcriptionally inactive (Ris and Korenberg, 1979). A chromosome region may, however, be inactive for one of two quite different reasons:

1. Because it is in a repressed state where, despite the fact that it includes coding DNA, that DNA is temporarily and reversibly inactivated.
2. Because it is composed of noncoding DNA and so is permanently incapable of transcription.

Relative to these two quite distinct forms of inactivity, it is possible to distinguish three different kinds of condensed chromatin within interphase nuclei.

1.1.1. Tissue-specific condensed euchromatin

In animal development, although the various cell types produced during somatic cell differentiation contain the same coding DNA sequences, only a part of the available information within a genome is expressed in any given cell type. This selective transcription depends upon specific sections of the euchromatic component of that genome becoming differentially inactivated in specific tissues following their loss of mitotic activity. In general, enbryonic cells and permanently proliferating stem cells exhibit a diffuse chromatin structure, whereas particular portions of euchromatin condense at differentiation as part of a tissue-specific transcriptional control system. The pattern of condensed euchromatin so formed is highly variable, both qualitatively and quantitatively, from one cell type to another within a given species. It is seen in its most pronounced form in the sperm nuclei of animals where the entire haploid genome is in a condensed state.

In this type of repression, chromosome condensation involves a specifically inactivated chromatin fraction within a given cell generation. This inactivation is often reversible, at least under experimental conditions. Thus, during phytohemaglutinin (PHA) -induced blastic transformation of peripheral lymphocytes in mammals, the nucleus enlarges some two or three fold and the condensed chromatin decondenses. Associated with this marked hypertrophy of the nucleus there is an extensive new synthesis of both ribosomal (r) and messenger (m) RNA (Frenster, 1974). Even so, it is important to recognize that not all dispersed chromatin automatically adopts an active state so that one cannot simply equate euchromatin with transcriptionally active chromatin. Although decondensation is most certainly a prerequisite for gene activity in the case of euchromatin, it is not necessarily sufficient for the onset of transcription.

Condensed euchromatin can be readily distinguished from heterochromatin in the light microscope; however, it shows the same density as heterochromatin in the electron microscope and, for this reason, has sometimes been inaccurately referred to as heterochromatin (see, for example, Frenster, 1974).

In contrast to the situation in animals, where the pattern of condensed euchromatin appears to be development- and tissue-specific, within a given species of plant the amount of condensed chromatin is much more species-specific (Nagl, 1979, 1985). This can be correlated with the fact that many plant cells remain totipotent whereas animal cells become irreversibly determined early in development.

1.1.2. The facultative heterochromatinization of euchromatin

This is a long-term form of euchromatin repression in which chromosome segments, entire chromosomes, or sometimes even chromosome sets, become inactivated and condensed during early development and then remain inactive over many cell generations in all or many somatic tissues. It is also associated with a process of chromosome imprinting in which only one of two genetically homologous chromosomes, as in the case of the single repressed X of female (XX) mammals, or sets of chromosomes, as in the case of the repressed paternal set of male coccid bugs, is affected, the homologous counterpart in the same nucleus continuing to code actively for proteins. The material that is heterochromatinized has a normal complement of genes, and these may be active both in the preceding and the succeeding generations. Such regions may also sometimes deheterochromatinize and resume genetic information during the heterochromatinization process.

1.1.3. Constitutive heterochromatic regions

These are the regions to which Heitz originally gave the name heterochromatin. They differ from euchromatin in being composed predominantly of noncoding, largely repetitive, DNA. They are thus permanently nontranscribable rather than simply repressed, and so represent a distinct structural kind of chromatin rather than a state of chromatin, as is the case both for tissue-specific condensed euchromatin and facultatively heterochromatinized euchromatin. There was a time when it could be argued that heterochromatin differed from euchromatin in its behavior but not in its fundamental structure (Baker and Callan, 1950; Dyer, 1964; Brown, 1966). That time is now past. Constitutive heterochromatin is indeed composed of DNA sequences with distinctive characteristics. Moreover, there are not two classes of heterochromatin, constitutive and facultative, as is still commonly claimed. The use of the noun *heterochromatin* to describe euchromatin that is facultatively inactivated is both misleading and unnecessary. The facultative heterochromatinization of euchromatin, in principle, has more in common with the tissue-specific condensation of euchromatin, though it obviously takes place on a more extended scale, both temporally and spatially, within the individual organism and is usually restricted to only one sex.

Although Heitz used chromocenter formation as his basis for defining heterochromatin, it is now clear that chromocenters do not form in all cell types. This appears to be true both for constitutive heterochromatin and for facultatively heterochromatinized euchromatin. In *Drosophila melanogaster,* for example, all three autosome pairs carry large blocks of pericentromeric heterochromatin. Additionally, the entire Y, and a substantial part of the X, chromosome is heterochromatic, too. The net result is that some 29% of the female (XX) genome and 35% of the male (XY) genome is composed of constitutive heterochromatin (Gatti, Pimpinelli, and Santini, 1976). However, no chromocenters are present in embryonic cleavage nuclei (Huettner, 1933). Heteropycnotic regions first appear during the prolonged interphase which accompanies the delayed cytokinesis of the syncytial blastoderm. Moreover, when a 160-minute lateral blastoderm nucleus, with well-developed chromocenters, is injected into an unfertilized egg, there is a loss of chromocenters (Illmen-

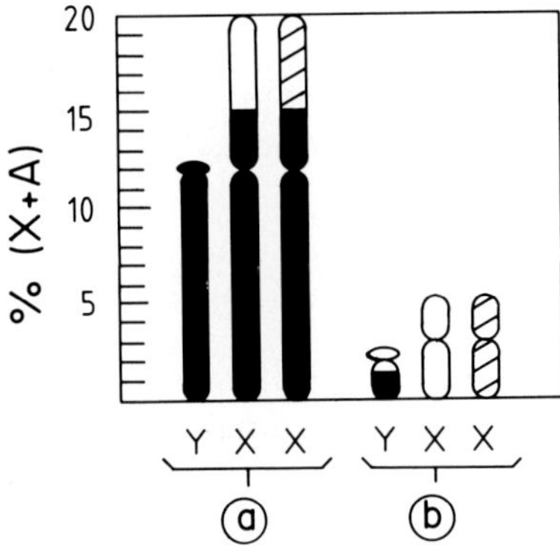

Figure 1.1. Structure of the sex chromosomes of *Microtus agrestis* (a) compared to those of *Homo sapiens* (b). Constitutive heterochromatin is depicted as solid (filled) areas, facultatively heterochromatinized regions as hatched areas, and euchromatin as clear areas. (After Wolf, Flinspach, Böhm, and Ohno, 1965.)

see, 1972). Evidently, the capacity of constitutive heterochromatin to form condensed chromocenters is readily reversible in this species.

The same appears to be true of many mammals. Fraccaro et al. (1969) observed blocks of condensed constitutive heterochromatin at all stages of preimplantation mouse embryos, beginning as early as the four-cell stage, though they note a remarkable degree of asynchrony between cells within the same embryo. At later stages in the mouse, constitutive heterochromatin appears fully condensed in brain cells but is not discernible in a majority of fibroblasts, whereas in liver it shows a more diffuse pattern (Schmid, 1967). Likewise, in the kangaroo rat *Dipodomys ordii,* which has some 50% of its genome in the form of constitutive heterochromatin, the appearance of this heterochromatin varies considerably in different kinds of somatic cells. In interphase nuclei from the kidney, it is highly condensed and localized at the nuclear membrane. In brain cells it is similarly condensed, though centrally located in smaller nuclei; however, larger nuclei, which are fewer in number, are devoid of chromocenters (Comings and Okada, 1976).

In the European field vole, *Microtus agrestis,* heterochromatin is localized primarily within two giant sex chromosomes (Figure 1.1). About a quarter of the total length of the X is composed of euchromatin, whereas the remainder of the X and the entire Y are composed of constitutive heterochromatin. In the female vole, as in all female placental mammals, the euchromatic portion of one of the two X chromosomes is facultatively heterochromatinized. Consequently, when the constitutive heterochromatin is condensed at interphase in female nuclei, the sex chromatin body formed by the single facultatively heterochromatinized X segment lies within one of the two larger chromocenters formed by the constitutive heterochromatin. Although there are certainly differences in the degree of compactness of the two X chromosomes, they are condensed at interphase in many cell types, including the zygote (Lee and Yunis, 1971; Yunis and Yasmineh, 1971). The variation they exhibit in their degree of compactness appears to be related to the extent of mitotic activity shown by a given tissue. Nondividing tissues, such as the brain, have large compact chromocenters in both sexes. On the other hand, male fibroblasts have no chromocenters of any kind whereas some 30–50% of female fibroblasts have a small sex chromatin body, the remainder again showing no heterochromatic structures of any kind. No RNA synthesis is detectable in the nuclear areas occupied by the sex chromosomes, even when they do not form microscopically visible chromocenters (Sieger, Pera, and Schwarzacher, 1970). Evidently, the transcriptional inactivity of the heterochromatic areas is not governed by heteropycnosis.

A majority (85%) of inactive cells in the thyroid gland of *Microtus agrestis* contain nuclei with two large chromocenters. By contrast, in particularly active regions of the same gland, the nuclei of the follicular epithelium are devoid of large chromocenters, and here 92% of female nuclei contain only the small sex chromatin body. Treatment with thyroid stimulating hormone (TSH), or with the thyrostatic methylthiouracil (MTU), leads to an enlargement of the thyroid gland; accompanying this is a pronounced decrease in the number of nuclei with large chromocenters. The facultatively heterochromatinized sex chromatin body, however, is not affected by these changes (Schneider, Heukamp, and Pera, 1973).

In plant nuclei, variation in the amount of condensed chromatin is only partly due to variation in heterochromatin content. Here, basic nuclear DNA content has been held to play a definitive role in determining the species-specific pattern of nuclear structure. Low 2C values lead to a diffuse euchromatin and the clear resolution of constitutive heterochromatin. Values higher than 9 pg, on the other hand, lead to a more complex organization in which heterochromatic chromocenters are coupled with condensed euchromatin (Nagle and Fusenig, 1979).

It should now be clear that the term *condensed chromatin* covers a variety of chromatin moieties with different molecular and biological characteristics. It should also be clear that the extent and type of condensation visible with the light microscope at interphase represent a rather labile property and that the production of chromocenters is not, in itself, an adequate basis for defining heterochromatin. It is equally clear that the extension of Heitz's original definition to include any pattern of condensation that differs from that of euchromatin at any mitotic or meiotic stage is even less justified. Perhaps the most blatant example of this relates to the use of the term *negative heteropycnosis.* As has already been mentioned, heteropycnosis describes the differential condensation of chromosomes, or chromosome segments, during mitosis or meiosis, though the term was, in fact, first coined to describe the differential condensation of certain chromosomes at meiosis (Wilson, 1925). In a number of orthopterans the univalent X chromosome of XO males is lighter staining, and seemingly undercondensed, during the early spermatogonial divisions. White (1940) referred to this as negative heteropycnosis, in contrast to the positive heteropycnosis displayed by this same chromosome at first meiotic prophase, with the clear implication that negative, as well as positive, heteropycnosis could be used to define heterochromatin. We now know that this was incorrect. The most compelling evidence for this conclusion comes from the work of Church (1979), who analyzed the behavior of the negatively heteropycnotic X chromosome of the male grasshopper *Brachystola magna* by using the techniques of autoradiography and low-power electron microscopy. In this species, the X occupies a separate compartment from the autosomes at premeiotic interphase and so is easily identified at that time. When tritiated uridine ([^{3}H]UdR) was used as a probe, in conjunction with autoradiography, no label could be detected over the X when it was negatively heteropycnotic during division stages so that it was certainly not transcriptionally active at this time as postulated by White (1970). Furthermore, electron microscopic observations on the negatively heteropycnotic X indicated that the condition results from an intermingling of X chromatin with persistent membrane material and is not the result of less condensed chromatin per se. Thus, while heterochromatic regions are necessarily heteropycnotic, heteropycnotic regions are not necessarily heterochromatic.

1.2. Facultative heterochromatinization

Two rather different systems of facultative heterochromatinization are known. Both operate in only one sex but in two quite different modes.

1.2.1. Sex-limited inactivation of the paternal set in coccids

The coccids are a family of homopteran insects which are closely related to aphids. Three different, but clearly related, cytogenetic systems (the Lecanoid, Comstockiella, and Diaspidid systems) are known within the group. In the first two, one or more entire sets of chromosomes become facultatively heterochromatinized during early embryogeny in prospective males (Nur, 1980). Sex determination in coccids is not associated with morphologically distinguishable sex chromosomes. Rather, it is regulated in some unknown way by the egg. In the cleavage divisions immediately following fertilization, all chromosomes appear euchromatic. During the blastula stage, however, the entire paternally derived set becomes facultatively heterochromatinized, but only in embryos destined to become males, and at interphase these chromosomes condense to form a conspicuous chromocenter. Facultative heterochromatinization in coccids is thus closely related to the mechanism of sex determination.

In male mealy bugs, one of the most thoroughly understood of the Lecanoid systems (Brown, 1969), heterochromatinization of the paternal set occurs at the coenocytic stage between the fifth and sixth cleavage divisions and shortly after the migration of cleavage nuclei to the egg periphery (Sabour, 1972). This coincides with the onset of RNA synthesis in the embryo. In some tissues the paternal set remains permanently heterochromatinized and genetically inactive, whereas in others it reverts to a euchromatic state and, correlated with this, regains its transcriptive capacity. The phenomenon occurs in the Malpighian tubules, the cyst wall cells of the testes, some cells of the skeletal muscles, and the intestinal wall (Mur, 1967). Alternatively, in some gut tissues that become endopolyploid, the heterochromatinized (H) set may fail to replicate during the truncated endomitotic sequences responsible for the formation of enlarged nuclei. Consequently, whereas the euchromatic (E) maternal set may be represented at a variety of multiple levels, a single paternal H set is maintained in a majority of endopolyploid nuclei (Table 1.1).

Interspecific hybridization between mealy bugs confirms that the paternal set of one species can be facultatively heterochromatinized in the egg of another, even though the two species may belong to different genera. Additionally, F_1 hybrid embryo tissues in which the H set deheterochromatinizes show gross developmental disturbances, whereas tissues like the hypodermis and its derivatives, which show no H set reversal and no resumption of genetic activity, also show no abnormalities (Chandra, 1971).

During the first and most of the second instar stage, female and male mealy bugs are morphologically indistinguishable and approximately the same size, despite the fact that two gene sets are active in the female whereas only the maternal set is capable of transcription in many of the male tissues. However, male embryos have on average 1.45 times as many cells as females and 1.47 times as much DNA. The precise ratio of cell numbers varies from 0.99 in the case of the Malpighian tubules, where deheterochromatinization of the paternal set occurs, to 4.46 in the case of brain cells, where the paternal set remains inactive (Table 1.2). The ratio of 1.47:1 corresponds to the estimate that 47% of all cells in the male contain a heterochromatinized paternal set. Thus, in organs where there is neither deheterochromatinization nor endopolyploidization,

Table 1.1. The number of H and E chromosomes in 33 uninucleate oenocytes from a third instar male of *Planococcus citri* ($2n = 5H + 5E$)

No. of H chromosomes	No. of cells	No. of E chromosomes									
		20E	30E	40E	60E	80E	100E	120E	140E	160E	200E
5H	24	6	2	9	2	2	—	2	—	1	—
10H	9	9	—	—	—	2	—	—	1	5	1

Note: The presence of 10H is due to a single replication of the H set. The nongeometric numbers of 30, 60, and 120E chromosomes are due to differential replication of the E and H sets coupled with nuclear fusion.
Source: Nur, 1966.

Table 1.2. Comparative cell numbers in males and females of the mealy bug *Planococcus citri*

Tissue and stage	Sex	No.	Average no. cells	Ratio	Cell type
Thoracic ganglion					
2nd instar	♂	10	6480 ± 87	1.47	diploid
	♀	10	4395 ± 98		
Brain					
290-μm embryo	♂	3	3290 ± 93	1.45	diploid
	♀	3	2246 ± 92		
Early 2nd instar	♂	5	5924 ± 440	3.95	
	♀	2	1498 ± 126		
Late 2nd instar	♂	5	10005 ± 605	4.46	
	♀	7	2239 ± 72		
Malpighian tubules	♂	10	60.0 ± 0.8	0.99	polyploid and deheterochromatinized
2nd instar	♀	10	61.0 ± 0.5		

Source: Berlowitz, Loewus, and Pallotta, 1968.

males appear to increase their amount of genetically active material by increasing cell number, a distinctive form of dosage compensation (Berlowitz, Loewus, and Pallotta, 1968).

In mealy bugs, the paternal set remains heterochromatinized in the male germ line. Consequently the H and E chromosomes do not pair at the first meiotic division, and both sets divide equationally. The H and E chromosomes than segregate at the second division. Following meiosis, the four spermatids produced by each meiocyte fuse to form a quadrinucleate syncytium, but within this structure only the two nuclei of E origin elongate to form functional sperm. The two H derivatives degenerate (Figure 1.2). Consequently, the facultatively heterochromatinized paternal set is discarded by postmeiotic elimination and so has to be formed anew in each generation. The paternal chromosomes transferred by the E sperm must then, in some sense, be conditioned to becoming imprinted by the egg cytoplasm before they combine with the maternal set.

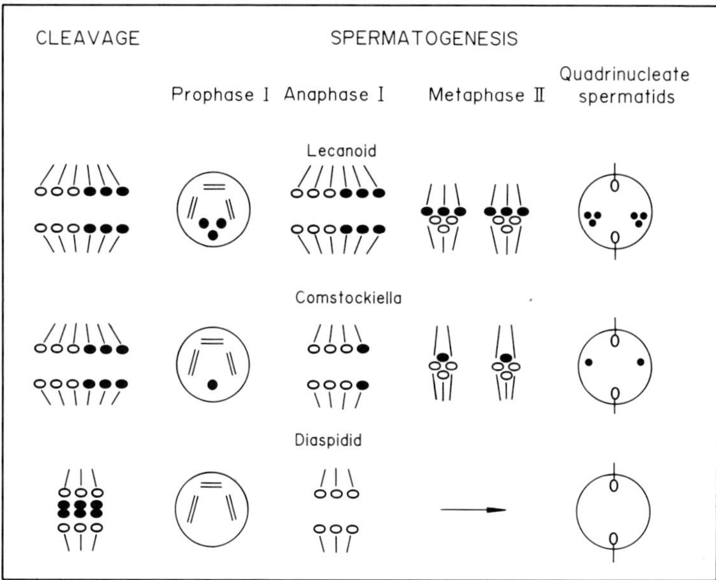

Figure 1.2. Early cleavage and spermatogenesis in the three coccid systems involving either facultative heterochromatinization (Lecanoid and Comstockiella systems) or else elimination (Diaspidid system) of the paternal chromosome set (shown solid). All three are characterized by an inverse meiosis in which the first anaphase division is equational. (After Nur, 1980.)

1.2.2. X chromosome repression in female mammals

A very different system of facultative heterochromatinization is found in female mammals where it takes rather different forms in the three major mammalian orders. In placental mammals (Eutheria), all but one of the X chromosomes is functionally repressed and transcriptionally inactive in all somatic cells. Such X inactivation depends on the number of X chromosomes present in the cell rather than simply on gender, although, under normal circumstances, it affects females and not males since in a majority of placentals the female is XX and the male XY in constitution.

In placentals, each repressed X forms a distinct condensed sex chromatin body in the interphase nucleus. The decision relating to which X chromosome remains active is made early in embryonic life so that the timing of heterochromatinization is comparable to that of coccids in the sense that it takes place while the embryo is still undifferentiated. In the very early female embryo, both X chromosomes are euchromatic. X inactivation occurs first in the trophectoderm, then in the primitive endo-

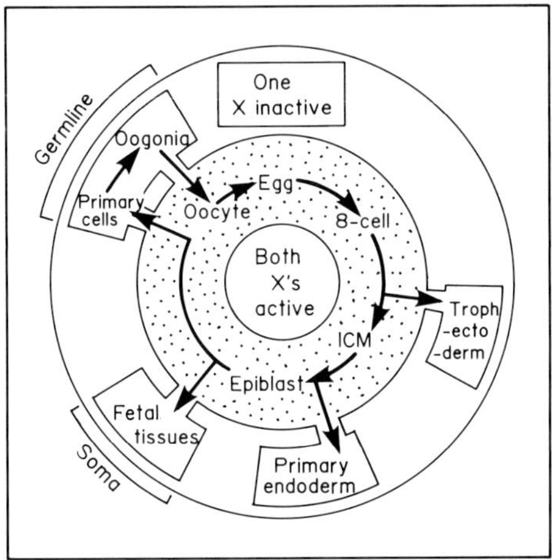

Figure 1.3. X chromosome inactivation and reactivation in female placental mammals. (After McLaren, 1983.)

derm, and finally in the embryonic ectoderm (Gartler and Riggs, 1983). Each of these three inactivation events correlates with an early step in tissue differentiations, which suggests that they involve a common element. X inactivation in nonembryonic tissues is nonrandom, with the paternal X being preferentially affected. However, there is a later and random pattern of X inactivation in the cells of the inner mass, the embryonic component destined to form the adult somatic cells. Female germ cells, too, are derived from a common inactivated embryonic cell pool, but they undergo a reactivation (deheterochromatinization) event about the time of entry into meiosis which in placentals also occurs early in embryonic life (Figure 1.3). In somatic tissues, on the other hand, once the inactivation event has taken place it becomes fixed in the subsequent somatic heredity of the cell and is maintained with complete fidelity for more than 50 somatic cell generations. Indeed, the stability of the inactivated state is maintained even following long-term in vitro culture though not all nondividing cells in such a culture actually include a sex chromatin body. Likewise, female interphase nuclei that lack sex chromatin have been found in various tissues of human and other mammals (Moore and Barr, 1954). Even when decondensed, however, the X still remains inactive. Condensation is thus, in some sense, secondary to inactivation (Comings, 1972).

The chromosome changes involved in X inactivation in placentals include genetic silencing, facultative heterochromatinization, and the formation of a sex chromocenter. These events involve a change in chromosome conformation which appears to be initiated by a single inactivation center whose influence then spreads throughout most of the X, possibly by functioning as a nucleation center (Therman and Sarto, 1983). In humans, it has long been clear that not all of the repressed X is inactivated and that the distal portion of the short (p) arm, which functions in humans as the pairing segment and includes the Xg, Sts, and Mic2 loci, remains active (Craig and Tolley, 1986). It is presumably partly for this reason the XO, XXY, and multiple-X individuals are abnormal in phenotype (Therman, 1983).

It is generally agreed that X inactivation functions to bring about dosage compensation since it ensures that the same balance between one functional X chromosome and two sets of autosomes obtain in both sexes. Inactivation also compensates for the presence of any additional X chromosomes generated as a result of meiotic or early mitotic errors. Chandra (1985, 1986) has suggested that X inactivation originated as a means of sex determination; that is, sex determination was first achieved by inactivation of the relevant genes in one sex. This inactivation might then have provided a basis from which dosage compensation evolved secondarily as the chromosomes became heteromorphic. Dosage compensation is not, in fact, a universal accompaniment of sex chromosome differentiation in vertebrates. It is absent in birds, in which a ZW female–ZZ male sex chromosome mechanism operates, and both Z chromosomes remain active in the male (Baverstock, Adams, Polkinghorne, and Gelder, 1982).

Although the molecular mechanism of X inactivation is not understood, several lines of evidence link inactivation to DNA methylation (Gartler and Riggs, 1983). Individual markers on the inactive X can, for example, be reactivated following 5-azacytidine (5azaC)-induced hypomethylation, which also induces a shift in the time of replication of the inactive X (Schäfer and Priest, 1984). Transcriptionally active genes are hypersensitive to digestion with DNase I. This sensitivity is present before the onset of transcription and is still demonstrable after RNA synthesis has ceased. Indeed, it is maintained at mitotic metaphase and so can be visualized in fixed chromosomes by autoradiography following controlled DNase I treatment and nick translation with DNA polymerase I in the presence of labeled deoxynucleoside triphosphates. This approach has been used to distinguish between active and inactive X chromosomes in female cells of *Gerbillus gerbillus* on the basis of their differential sensitivity to DNase I (Kerem et al., 1983, 1984). More recently, Jablonka, Goitein, Marcus, and Cedar (1985) have shown that 5azaC treatment of the inactive X of *Gerbillus* causes a dramatic and coordinated change in both its DNase I sensitivity and its time of replication, converting it into an early replicating and DNase I-sensitive structure, equivalent in both respects to the active X of this species. Although 5azaC treatment altered the time of initiation of replication, it did not affect the order of replication within the X. Even so, the correlation between a shift in early replication and the conversion to a DNase I-sensitive conformation suggests that a link between gene inactivity and the time of onset of replication might be of more general importance in eukaryotes.

In marsupials (Metatheria), by contrast with placentals, it is always the paternal X chromosome that is inactivated (Cooper, Johnston, Sharman, and Vandeberg, 1977) though, interestingly, it does not give rise to a sex chromocenter at interphase (Hayman and Martin, 1974). Added to this, the level of inactivation varies between loci and even between tissues for the same locus. In the embryo, it is completed only in blood cells. Elsewhere, it is incomplete and tissue specific (Vandeberg, 1983). In the extraembryonic membranes, X inactivation when present, as in the Dasyuridae, also involves the paternal X, but it may be absent, as in the Macropodidae (Johnston and Robinson, 1985).

In the monotremes (Prototheria), the sex chromosomes are only minimally differentiated. Thus the platypus, *Orinthorhynchus anatinus* has an XY system in which the long arms of the two chromosomes are entirely homologous, with differentiation affecting only part of the short arm. Though the echidna, *Tachyglossus aculeatus,* has an X_1X_2Y / $X_1X_1X_2X_2$ system, the X_1 is identical in G banding with the X of the platypus and thus constitutes the basic X chromosome of a system that must have arisen from a Y autosome translocation. In both these cases, therefore, X chromosome inactivation should, if present, be confined to the region of the short arm over which the X and Y are unpaired. Wrigley and Marshal Graves (in press), who made these observations, have sought evidence of inactivity in terms of late replication, which is a consistent feature of the inactive X in both the Metatheria and the Eutheria. While the X chromosomes of female fibroblasts from both species show no evidence of asynchrony, lymphocytes from female echidna show asynchronous replication of the short arm of the two X_1 chromosomes. From this, they conclude that asynchrony is tissue specific and is, as expected, confined to the region of the X that is nonhomologous with the Y. They therefore argue that the evolution of X inactivation in mammals has been from a paternal, incomplete, and tissue-specific mechanism to one with superimposed additional control mechanisms leading through the marsupial system, where it is still paternal and incomplete but more extreme, to that of placentals where it is complete but random. Accompanying these changes is a reduction in the size of the Y chromosome and a parallel spread of inactivation throughout the X as it becomes increasingly differential.

In summary, it is clear that facultative heterochromatinization is an epigenetic phenomenon, a consequence of some form of chromosome imprinting that temporarily and reversibly inactivates a section of coding DNA. In coccids, the chromosomes of the sperm must be imprinted by the egg cytoplasm before the fusion of the egg and sperm nuclei. Male embryos result from eggs in which the paternal chromosomes become imprinted, whereas female embryos are the products of eggs in which no imprinting has occurred. What little evidence there is on imprinting of the X chromosome in placental mammals comes from ovarian teratomas. These are benign tumors that simulate sexually produced embryos in the sense that they contain disorganized assemblages of differentiated tissues. They have an XX constitution and are postmeiotic products of maternal origin, probably resulting from the fusion of a second polar body with the haploid egg nucleus (Linder, McCaw, and Hecht, 1975). The presence of an inactive X in these teratomas indicates that, as in coccids, the factors that regulate X chromosome inactivation reside in the cytoplasm of the egg.

1.3. Constitutive heterochromatin

Heitz's original observations were based on a comparative study of telophase, interphase, and prophase of the mitotic cell cycle since euchromatin and heterochromatin are equally condensed at metaphase. Over the years, numerous attempts have been made to explain in precise biochemical terms the morphological definition of heterochromatin provided by Heitz. These attempts have depended largely on the development of three novel techniques for probing chromosome structure.

1.3.1. Differential staining

A number of techniques are now known which produce a differential response in the isopycnotic euchromatic and heterochromatic regions of condensed metaphase chromosomes. The simplest of these is the C-banding procedure. When fixed metaphase chromosomes are treated with hot alkali, followed by warm saline, regions that include constitutive heterochromatin give a positive staining reaction with a variety of dyes including, most commonly, Giemsa. The sensitivity of this method lies in the order of 1×10^{-2} pg 10^7 base pairs (bp) of DNA (Schweizer, 1980). Interphase chromocenters formed from constitutively heterochromatic regions behave in a similar fashion (Figure 1.4). Facultatively heterochromatinized chromosomes, however, do not give positive C-banding. Consequently, the advent of C-banding has made it possible to differentially stain constitutive heterochromatin at a cell-cycle stage when it can-

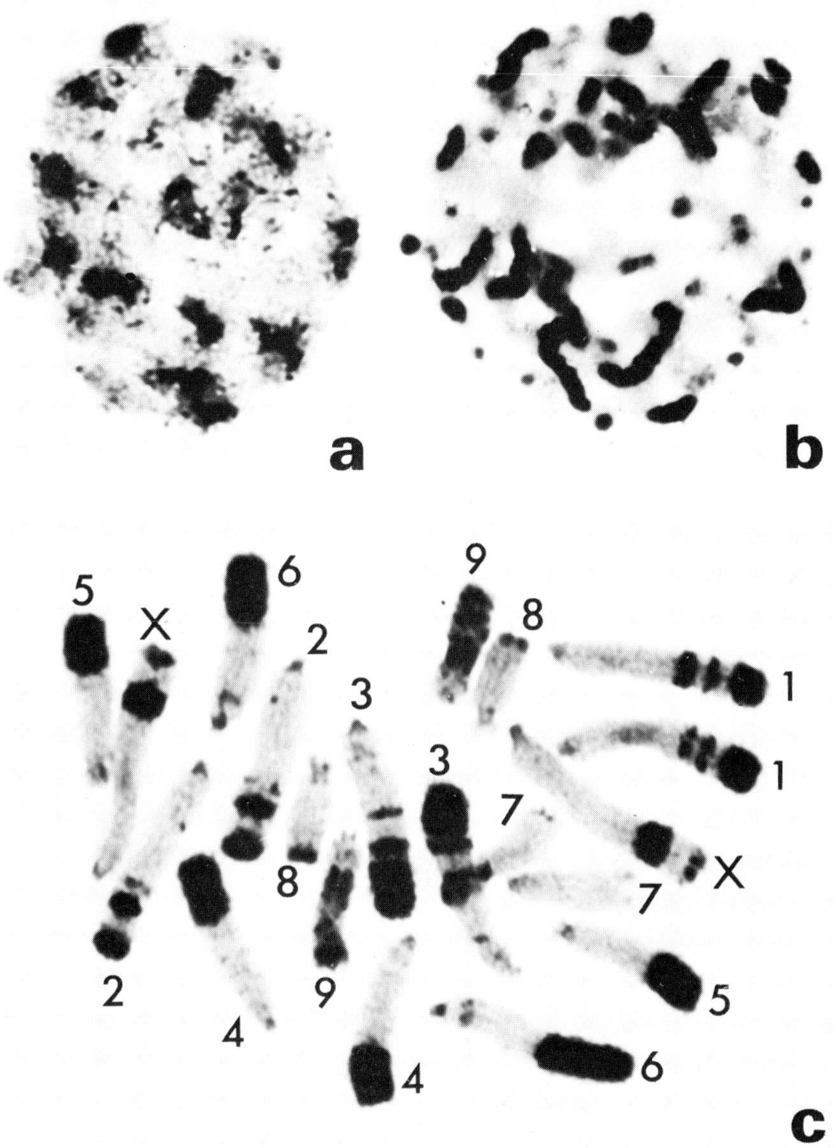

Figure 1.4. C-band somatic interphase (a), early prophase (b), and metaphase (c) of neuroblast nuclei from the grasshopper *Atractomorpha similis.*

Table 1.3. Dye binding properties of different chromatin fractions

Type of chromatin	Banding technique[a]						Base composition of chromatin
	C	Q	DAPI	H33258	MM	CMA	
Constitutive heterochromatin	+	+	+	+	−	−	AT rich
	+	−	−	−	+	+	GC rich
	+	−	−	−	−	−	AT+GC rich (?)
Facultative heterochromatin	−	−	−	−	−	−	
Euchromatin	−	−	−	−	−	−	

[a]C, C-banding; CMA, chromomycin A_3; DAPI, 4′,6-diamidino-2-phenylindole; H33258, Hoechst 33258; MM, mithramycin; Q, quinacrine.
Source: Schmid, 1980a.

not be distinguished from euchromatin by conventional staining. Additionally, the application of the C-banding technique to the giant polytene chromosomes of dipteran flies has led in some instances to the identification of C-band regions that are beyond the limits of resolution in conventional mitotic chromosomes. At mitosis in the black fly *Simulium ornatipes,* for example, there are three chromosome pairs, all of which are characterized by centric C-bands as well as by a faint interstitial C-band in two of the pairs. In polytene preparations, all centromere and telomere regions give a positive C-band reaction as also does the nucleolus organizing region (NOR) on chromosome I. Additionally, however, in all chromosome arms, numerous individual polytene bands are evident which show positive C-banding (Bedo, 1975), a result that was confirmed in chironomids by Hägele (1977).

It is also possible to obtain differential banding of the constitutive heterochromatin in metaphase chromosomes by use of a number of fluorescent dyes, or fluorochromes, several of which are known to be specific for particular nucleotides, showing either AT or GC specificity (Table 1.3). The AT-specific fluorochromes include quinacrine (Q), Hoechst (H) 33258, and 4′,6-diamidino-2-phenylindole (DAPI) whereas the GC-specific fluorochromes include chromomycin A_3 (CMA) and mithramycin (MM). While the brightness of chromosome segments stained with fluorochromes is related to the AT richness of that segment, the length of uninterrupted T sequences is also an important factor regulating the reaction. The chromosome banding produced with certain of the fluorescent dyes can be enhanced by counterstaining with DNA-binding, base-specific, peptide antibiotics. Thus, counterstaining with the nonfluorescent,

AT-specific, oligonucleotide antibiotic distamycin A_1 (DA) heightens DAPI fluorescence (Schweizer, 1981).

1.3.2. Replication behavior

The introduction of chromosome autoradiography using tritium-labeled thymidine ([^{3}H]TdR) as a probe provided the first means of monitoring chromosome replication patterns. From a survey of the earliest studies using this technique, Lima de Faria and Jaworska (1968) concluded that heterochromatic regions were late replicating, though it was accepted that not all late replicating regions were necessarily heterochromatic.

The very first demonstration of such late replication came, in fact, from the study of the univalent X chromosome of the grasshopper *Melanoplus differentialis* at first prophase of male meiosis. This chromosome is condensed at this time but not in spermatagonia or in somatic nuclei. Neither, as we now know, is it C-band positive except for a small procentric block of constitutive heterochromatin. Thus it behaves in a frankly facultative fashion though, since it is hemizygous in the male and shows distinctive behavior only in the germ line, it clearly differs from the conventional categories of facultatively heterochromatinized euchromatin. Moreover, in the female, where two X chromosomes are present, both are euchromatic. For these reasons, Comings (1972) proposed that the male X chromosome be considered as semifacultative. A simpler solution, however, would be to broaden the concept of facultative heterochromatinization to cover this case, and others like it, where a chromosome is condensed in the germ line but not in the soma. Thus, there are at least two other categories of chromosomes which show consistent positive heteropycnosis at first prophase of male meiosis and which are either not C-band positive at all or else only partially banded. These are the megameric chromosomes of orthopteran insects (King and John, 1980) and the B chromosomes of some eukaryotes (Jones and Rees, 1982). As will become apparent later, the Y chromosome of some mammals and flies pose a not too dissimilar problem.

In their summary paper of 1968, Lima de Faria and Jaworska formulated the rule that "heterochromatic regions replicate their DNA later than the euchromatic segments" since all the cases they listed did indeed behave this way. This same generalization was echoed by Back (1976) who wrote: "Late DNA synthesis is the only constant character of heterochromatizable chromatin of somatic cells known thus far." Like several of the other statements that have been made concerning heterochromatin, this has proven to be an overgeneralization, as hinted at by Hsu, Schmid, and Stubblefield as early as 1964. Subsequently, Ray Chaudhuri and Singh (1972) found that the sex chromosome of female (ZW) snakes,

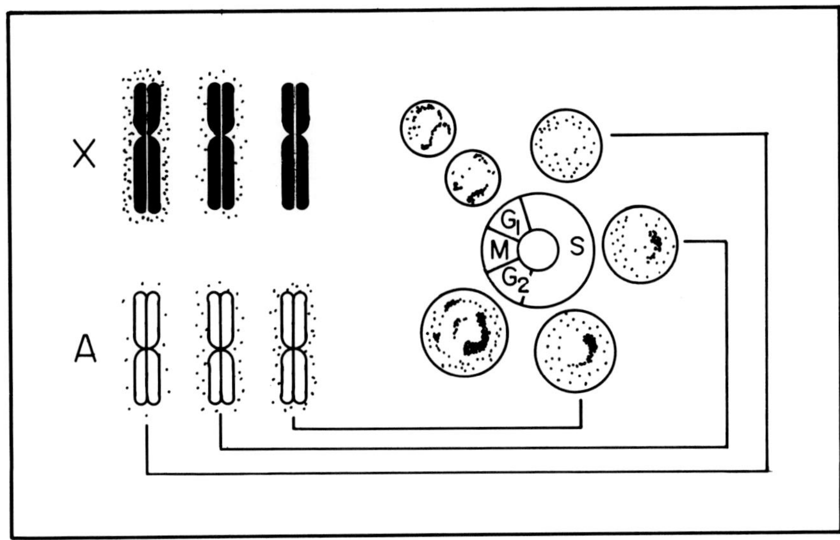

Figure 1.5. Replication behavior and labeling pattern of the heterochromatic X chromosome (shown solid) compared to the euchromatic autosomes (shown clear) of the liverwort *Pallavicinia longispina.* (After Masubuchi, 1976.)

which invariably forms a heteropycnotic sex chromatin body at somatic interphase, was early replicating in *Boiga forsteini, B. Trigonata, Naja naja naja,* and *Vipera ruselli russelli,* even though it was late replicating in *Bungarus coerulens* and *Naja naja kaouthia.* A number of plants also show early replicating constitutive heterochromatin. These include the angiosperm *Spiranthes sinensis* (Tanaka, 1965) and several species of liverworts (Masubuchi, 1976; and also Figure 1.5) including the genus *Pellia* in which Heitz (1928) first identified heterochromatin and which recent C-banding studies have shown to be constitutive in character (Newton, 1985). Thus late replication is not a universal feature of heterochromatin. Rather, as Masubuchi (1976) points out, the most common characteristic of heterochromatin replication is its tendency to be significantly shorter in duration than that of euchromatin. This applies whether it is late or early replicating.

The timing of DNA replication in constitutive heterochromatin may also be related to the time during interphase when it disperses and condenses. For example, in a number of liverworts, including *Pellia,* the heterochromatin is condensed through G_1 but then becomes diffuse at the onset of the S phase when it starts to replicate its DNA. By mid S, it has completed its replication and then recondenses and persists in this form through G_2 and into the next prophase. In the liverwort *Calobryum rotundifolium,* however, the heterochromatin does not decondense until middle to late S and its replication is correspondingly delayed (Masubuchi, 1976). There is, in fact, a stage during interphase in many plants where chromocenters regularly disperse (Nagl, 1977). This Z, or *Zertaübungsstadium,* phase was again first observed, and named, by Heitz, though he mistakenly believed it to be in early prophase. Nuclei with Z-phase morphology have, however, been shown to be engaged in DNA synthesis (Barlow, 1978).

1.3.3. DNA composition

The occurrence of different biochemical types and fractions of DNA within the genome has made it possible to construct specific biochemical probes from these different fractions. This, when coupled with in situ hybridization of radioactively labeled copies of these probes, provides a simple visual method for locating particular kinds of DNA within specific chromosome regions. In this method, fixed chromosomes are first treated with hot alkali to denature the DNA and then incubated with tritium-labeled cDNA or cRNA under conditions that allow for renaturation. The chromosome segments involved in hybridization can then be visualized by autoradiography. Among the earliest probes to be used in this way were the satellite DNA sequences, the first repeated sequences to be discovered within the genome (Kit, 1961; Sueoka, 1961). The characteristic organizational feature of these satellites is the tandem repetition of a unit DNA sequence, though the repeat unit may vary from a few to several thousands base pairs (Southern, 1984). The resolving power of this technique has, however, been considerably enhanced since the introduction of highly specific probes obtained by digestion of particular DNA fractions with restriction enzymes. These divide up the DNA into a series of fragments whose size depends on the spacing between successive enzyme recognition sites. Molecular cloning then provides a means of purifying and amplifying the DNA segments generated by restriction endonuclease digestion.

The use of both satellite and restriction probes has indicated that within chromosomes, the distribution of constitutive heterochromatin, as defined by C-banding or by fluorescence, and highly repeated (hr) DNA as defined by in situ hybridization, is often, though certainly not always, coincident (Table 1.4). This does not mean that constitutive heterochromatin is made up only of hr sequences. Nor does it mean that these sequences are located only in regions of constitutive heterochromatin. It is clear, however, that the greater bulk of hr DNA occurs at constitutively heterochromatic sites when these are present within a genome. Cases are known where satellite or other hr DNA is not the only,

Table 1.4. Molecular characterization of C-band heterochromatin

Species	C-bands		hr DNA			Monomer units
	% Genome	Location	% Genome	Type	Location	
(A) Proximal C-bands only						
(1) *Mus musculus*	6	Centric (except Y)		Major satellite	Centric (except Y)	234 bp comprising 8 shorter unrelated tandem repeats
				Minor satellite	Some centromeres	?
(2) *Macropus rufogriseus*	—	Centric (except Y)		Major satellite	Centric (except Y)	2500 bp with internal repeats of 300 bp
(B) Proximal and distal C-bands						
(3) *Scilla siberica*	16–20	Small centric and interstitial; larger terminal	19	GC-rich satellite	Interstitial and terminal	Dominant palindromic fragment of 35 bp
(4) *Secale cereale*	35			480-bp fragment	Predominantly terminal, all 7 pairs	
				610-bp fragment	Exclusively terminal, 6 pairs	
				630-bp fragment	Exclusively terminal, 5 pairs	
(5) *Allium cepa*	4			Cryptic satellite	Terminal only	375 bp
(6) *Atractomorpha similis*	13–44	Centric, short arm, subterminal, and terminal		Cryptic major satellite	Subterminal and Terminal	179 bp

Sources: (1) Singer, 1982; (2) Dennis, Dunsmuir, and Peacock, 1980; (3) Deumling, 1981; (4) Appels, 1982; (5) Barnes, James, and Jamieson, 1985; (6) John, Appels, and Contreras, 1986.

or even the most common, component of some categories of constitutive heterochromatin. This is specially clear in humans: Here C-banding indicates an average of 13% constitutive heterochromatin within the genome (Brown 1983); however, only half of these C-band sites actually contain satellite sequences, and the relative proportions of these sequences in autosomes 1, 9, and 16 are not related to the size of the C-bands (Miklos and John, 1979).

1.3.4. Heterochromatin heterogeneity

It should now be clear that the term *heterochromatin,* as used in the past, covers a number of quite distinct behavioral systems. This is especially true for the class we have distinguished as constitutive heterochromatin. Thus, despite the fact that C-banding provides the most consistent marker for identifying constitutive heterochromatin, it sometimes requires considerable modification of the conventional procedure to obtain positive results. Hedgehogs, for example, possess substantial blocks of autosomal heterochromatin which are distributed differently in different species. These form large chromocenters at somatic interphase and are heteropycnotic at first meiotic prophase (Gropp and Citoler, 1969). They are also late replicating in somatic cells (Citoler, Gropp, and Natarajan, 1972). Despite this, they fail to stain by conventional C-banding methodology and do not show bright fluorescence after quinacrine mustard (QM) staining. When preparations are subjected to extreme conditions of denaturation and reassociation, however, they do form C-bands (Gropp and Natarajan, 1972). Likewise the heterochromatic segments present at the centromeres of cattle, goats, and sheep autosomes are unstained by the conventional C-banding technique. By shortening the incubation times in SSC or phosphate buffer to 20 minutes, or by omitting these steps altogether, the otherwise unstained centric heterochromatin can be made to stain deeply (Schnedl and Czaker, 1974).

Among other notable exceptions to conventional behavior, four are worth detailed comment:

1. While carnivores, in general, possess only small amounts of constitutive heterochromatin, some few species have relatively large blocks of autosomal heterochromatin (Table 1.5). These, however, do not C-band, though they are late replicating.
2. The flesh fly, *Sarcophaga bullata,* has a genome some five times that of *Drosophila melanogaster* (O.61 pg, vs. 0.12 pg), but contains only small amounts of hr (9%) and moderately repeated (mr) DNA (6%), with 81% of the genome in the form of single-copy sequences. Using cRNAs for all three sequence classes, Samols and Swift (1979)

Table 1.5. Carnivores with unusually large quantities of constitutive heterochromatin (het.)

Species	$2n$	% Constitutive het. in ♂ haploid set (X+A)	No. chromosomes with het.[a]
Herpestes edwardsi (Indian mongoose)	35/36	3.6–5.0 (polymorphic)	2, both in S.A.
Mustela putorius (ferret)	40	6.5	3, 2 in S.A. 1 in L.A.
Mustela rixosa (pygmy weasel)	42	11.0	7, all in S.A.

[a]S.A., short arms; L.A., long arms.
Source: Fredga and Mandahl, 1972.

showed by in situ hybridization that hr probes were restricted to the centric heterochromatin of the C and E chromosomes although all five autosomes in the haploid set have regions that are C-band positive (Figure 1.6). Moreover, no mr or hr sequences were present in either of the sex chromosomes although both were totally C-band positive. The C-banded sex chromosomes of the Chinese hamster, *Cricetulus griseus* likewise do not contain particular concentrations of satellite or other hr DNA (Arrighi, Hsu, Pathak, and Sawada, 1974). The same is true of the Y chromosome of the mouse, which lacks the satellites present in the autosomes; and the X, which does not give a positive C-band reaction.

3. The Y chromosome of *Drosophila hydei* which shows clear C-bands, also does not contain hr DNA and is built predominantly from mr sequences. Renkawitz (1978b) isolated two hr DNA fractions in this species. The 1.696 fraction, which comprises 13% of the genome, is restricted to the X heterochromatin. The 1.714 hr fraction, which constitutes 4% of the genome, hybridizes to all four acrocentric autosomes but to neither the X nor the Y. Two mr nuclear fractions, with densities of 1.707 and 1.697, respectively, are also located in specific heterochromatic positions (Renkawitz, 1978a). The former is found in all four autosomes and in the middle of the Y, while the latter is present in the NOR sites on the X and the Y (Figure 1.7).
4. In the domestic pig, *Sus scrofa* ($2n = 38$), there is a particularly striking difference between the composition of the centromeric blocks of constitutive heterochromatin between the 12 acrocentric members of the diploid set, which have the same satellite DNA and exhibit a DA/DAPI-positive fluorescence, and the 26 biarmed chromosomes,

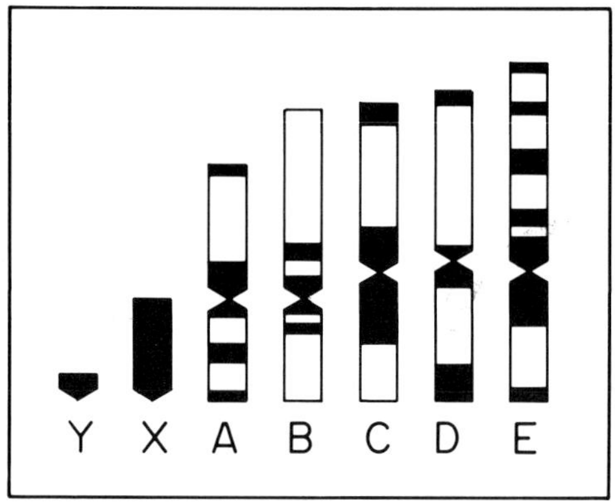

Figure 1.6. C-banded mitotic karyotype of the flesh fly *Sarcophaga bullata*. (After Samols and Swift, 1979.)

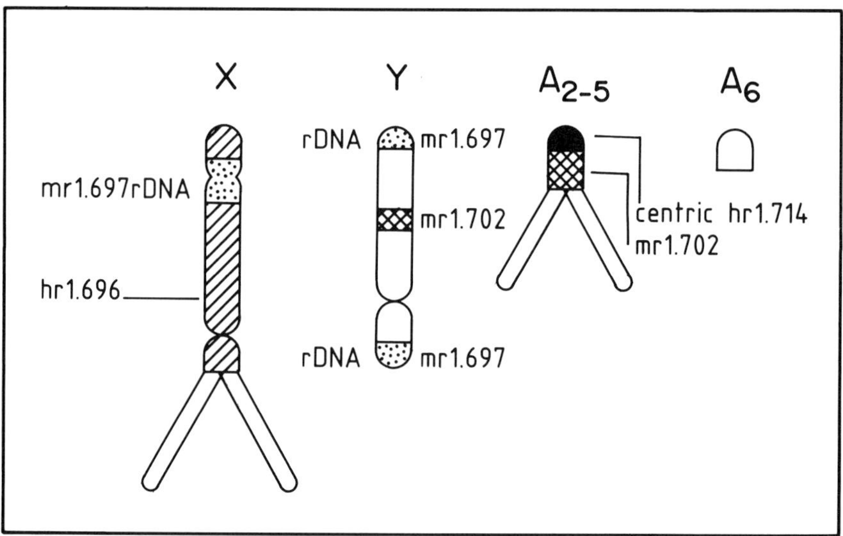

Figure 1.7. Composition and arrangement of mr and hr sequences in the genome of *Drosophila hydei*. (After Renkawitz, 1978b.)

which have chromomycin A_3-positive C-bands but whose satellites appear to be chromosome specific (Schwarzacher, Mayr, and Schweizer, 1984; Schweizer and Loidl, in press). Moreover, while the DA/DAPI-positive C-bands form one or two large chromocenters at pachytene of male meiosis, though not at lymphocyte interphase, the chromomycin A_3-positive, bright C-bands are well scattered in both cell types. A more general difference in the composition of constitutively heterochromatic segments is that which occurs when proximally and distally located C-bands are both represented in the same genome. Here, particular repeat families are commonly restricted to specific heterochromatic regions, either proximally or distally but not both (John, Appels, and Contreras, 1986).

These cases, and others like them (John and Miklos, 1979) emphasize that the DNA composition of constitutively heterochromatic regions does vary, both within and between species, and that this variation can influence the properties of these regions. It is abundantly clear, therefore, that constitutive heterochromatin is heterogeneous in its sequence composition and cannot be uniquely characterized by any particular base composition. Neither is it the sole repository of repeated sequences, though DNA of extremely repeated base composition is most likely to be confined to heterochromatin.

Even so, in very general terms, there is a good one-to-one relationship among hr DNA, C-band-positive regions, and constitutive heterochromatin within a genome. That is, in most instances, constitutive heterochromatin corresponds to chromatin containing hr DNA and staining positively after C-banding. This carries the rider that the inertness of constitutive heterochromatin, in terms of its transcriptional inability, is a consequence of its distinctive DNA structure and not of its condensed nature, which may itself be a secondary consequence of its peculiar DNA sequence organization.

1.4. The euchromatin–heterochromatin boundary

It has been assumed by a number of authors that the normal boundary zones between euchromatin and heterochromatin have special properties because the chromatin at such junctions exists in a state of flux between condensed and noncondensed forms. On this basis, it has been argued that the access of repair enzymes in junction regions might be impaired, thus facilitating exchanges between non partner DNA strands (Hatch and Mazrimas, 1977). Erroneous rejoining might then lead to sister chromatid exchange, to inversion (homologous exchange), to translocation or fusion (Heterologous exchange), or to the accumulation of heterochromatic DNA by unequal exchange. Kato (1979), for example, noted the preferential occurrence of sister chromatid exchanges (SCEs) between euchromatin heterochromatin junctions in wallaby and rat hamster chromosomes (Table 1.6). The X chromosome of the Indian muntjac, *Muntiacus Muntjak vaginalis* [$2n = 6$ (female), 7 (male)], is characterized by

Table 1.6. Distribution of sister chromatid exchanges (SCEs) in BrdU-treated cells of two mammals[a]

Species	N	Region	Relative length	No. SCEs: Exp. by length	No. SCEs: Observed
Macropus parma (white-throated wallaby)	70	Euchromatin (E)	0.672	498.6	484
		C-band (C)	0.328	243.4	114
		H–E junction	—	0	114
E C E		Totals	1.000	742.0	742
Cricetulus triton (ratlike hamster)	75	Euchromatin (E)	0.679	450.2	468
		C-band (C_1)	0.116	76.9	17
		(C_2)	0.205	153.9	52
		H–E junction	—	0	79
		C_1–C_2 junction	—	0	47
					t
E C^1 C^2		Totals	1.000	663.0	663

[a]BrdU, 5-bromodeoryuridine.
Source: Kato, 1979.

an unusually long and C-band-positive centromeric neck which represents a compound kinetochore region (Brinkley, Valdivia, Tousson, and Brenner, 1984). This constitutively heterochromatic neck separates a short chromosome arm of X origin from a long arm of autosomal origin. Within the euchromatin of both the X and other members of the diploid set, SCEs are distributed according to the amount of DNA present. Within the X chromosome, however, SCEs occur at very low frequency in the neck region itself but at exceptionally high frequencies at the interface between euchromatin and heterochromatin (Carrano and Wolff, 1975). Bostock and Christie (1976) likewise report that relatively few SCEs occur within C-band regions of *Dipodomys ordii,* but that a high frequency of such exchanges occur at the junction between C-band and non–C-band regions.

The molecular organization of transition zones between euchromatin and heterochromatin is still largely unknown. Only in one case has it been possible to examine a junction region of this kind in some detail. This is in the polytene chromosomes of larval *Drosophila.* These chromosomes are giant, multistranded units produced by a series of successive curtailed endoreduplication cycles in which the conventional sequences of cell division are bypassed, with the products of replication remaining in association. This process, coupled with the retention of somatic pairing between homologous chromosomes, results in the many hundreds or thousands of homologous chromatids associating as single multistranded chromosomes. Two features that regularly accompany polytenization are increasing length and width and the production of a system of transverse chromatic bands, separated by achromatic interbands, in euchromatic regions of the chromosomes. Thus, from one polytenic level to the next, Hartmann Goldstein, and Goldstein (1979) reported that the chromosomes of *D. melanogaster* increase by about 10% in length and 40% in width. The length of the major heterochromatic regions in the polytene chromosomes is, however, relatively much shorter than in the somatic chromosomes. About one-third of the length of the X is heterochromatic in mitotic chromosomes, whereas the same region forms less than one-twentieth of the X in salivary gland nuclei. The Y, which is even longer than the X in the mitotic set, is even further reduced in the polytene set.

During the process of polytenization in many species of *Drosophila,* the heterochromatic regions associate to form a single chromocenter. The chromosomes themselves are in a permanent prophase in which the individual arms are arranged in a polarized Rabl orientation, despite the fact that these nuclei actually completed their last mitotic division early in embryogenesis some eight to ten replication cycles earlier.

On the basis of the staining reaction, Heitz (1934) suggested that the chromocenter of polytene nuclei was made up of two distinct regions. He distinguished between a small compact central region surrounded by a much larger, loosely compacted and granular region. He also proposed that these two morphologically distinct regions represented two distinct types of heterochromatin, which he termed *alpha* (α) and *beta* (β). The α heterochromatin, he argued, could not be uncoiled and so formed a dense compact core to the chromocenter. The β heterochromatin, however, he believed to be partially uncoiled, thus giving rise to a diffuse network. Normally this exhibited only faint traces of the banding found in the euchromatic arms. Thus the junction between β heterochromatin and the euchromatic portion of a chromosome arm was marked by a transition of the apparently disorganized fibrils of the β heterochromatin into the regularly arranged patterns of bands and interbands.

Heitz also suggested that the α heterochromatin corresponded to the bulk of constitutive heterochromatic segments so evident in the conventional mitotic chromosomes, since, in his opinion, not all the polytene arms had associated β heterochromatin whereas all of them showed mitotic heterochromatin. The photographic maps of *Drosophila melanogaster* provided by Lefevre (1976) indicates that, in this species at least, β heterochromatin exists in sections 20 (X), 40 (2L), 41 (2R), 80 (3L), 81 (3R), and 101 (4), though there are certainly differences in amount in these different regions. While there is no simple relationship between the amount of α and β heterochromatin in a given chromosome of *D. melanogaster,* it is clear that, as pointed out by Heitz, the β heterochromatin must represent only a very small part of the conventional mitotic chromosome and that it does indeed lie at the junction between euchromatin and α heterochromatin.

Whereas the chromocenters formed by all species of the *virilis* group of *Drosophila* include distinct α and β regions (Gall, Cohen, and Polan, 1971; Cohen and Kaplan, 1982; see also Figure 1.8), this is not the case in *Drosophila melanogaster.* Here, distinct α and β regions cannot be distinguished in conventional squash preparations. In sections of unsquashed nuclei, however, two distinct types of chromatin organization, consisting of a centrally located single mass of homogeneous dense chromatin and a surrounding network of smaller dense blocks, can be identified with both light microscopy and electron microscopy (Lakhotia and Jacob, 1974). This is reminiscent of the situation identified by Heitz in *D. virilis.*

We now know that the two types of heterochromatin which Heitz distinguished differ both in sequence content and in replication behavior. The α heterochromatin contains hr DNA sequences that undergo few, if any, rounds of replication during polytenization. Compelling evidence for this comes from the observation that when labeled cRNA, synthesized from the satellite DNA of *D. virilis,* is hybridized in situ on the same slide

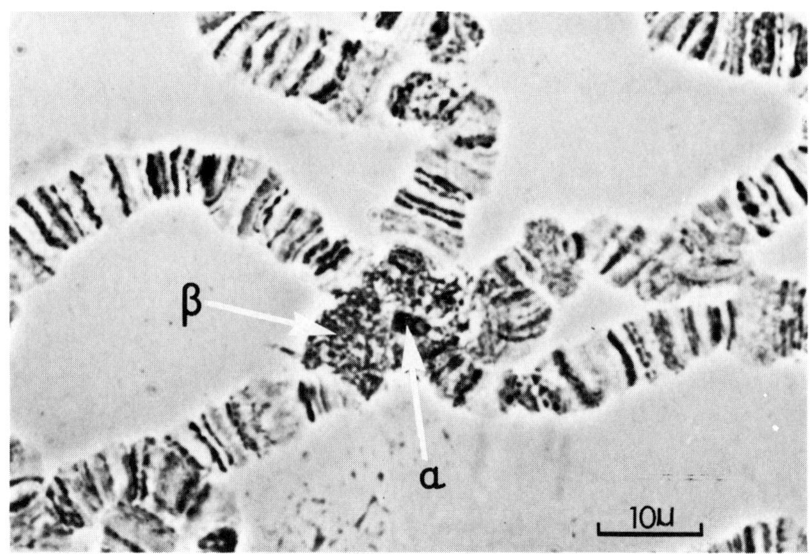

Figure 1.8. The polytene chromocenter of a salivary gland nucleus from *Drosophila virilis.* The centrally located, compact α heterochromatin is clearly distinguished from the surrounding granular β region. (Photograph courtesy of Professor J. G. Gall.)

to both diploid and polytene nuclei, they bind comparable amounts of the labeled probe (Gall, 1973). Since the satellite DNA in question is restricted to regions of α heterochromatin it is clear that, despite the 1000-fold difference in total DNA content, the two types of nuclei have identical amounts of heterochromatin.

Less is known about β heterochromatin. It does not contain satellite or other hr sequences, and it replicates at least partially during polytenization since it is known to incorporate [^{3}H]TdR (Plaut, 1963). Moreover, following in vitro pulse labeling of salivary glands of *D. melanogaster* with [^{3}H]UdR, the majority of silver grains lie over the β heterochromatin whereas no grains occur over the α heterochromatin (Table 1.7). Indeed, in terms of average grain density, the β heterochromatin appears to be as active as euchromatin.

Constitutive heterochromatin, as we have seen, is considered to be transcriptionally inactive, although the NOR that is sequestered within the secondary constriction located in the basal heterochromatin of the X certainly functions in the polytene nucleus. The notion that β heterochromatin is not inert is consistent with the fact that poly(A)$^+$ mRNA, transcribed by cultured cells of *D. melanogaster,* binds strongly to the β het-

Table 1.7. Labeling patterns observed in 30 EM autoradiographs of the polytene chromosomes of *Drosophila melanogaster* following incubation of excised salivary glands from late third instar larvae in [^{3}H]UdR

Region	Total area (μm^3)	Total no. grains	Grains per 100 μm^2
α Heterochromatin	26.48	0	0
β Heterochromatin	1,639.84	774	47.20
Euchromatin	1,301.24	596	45.80
α/β Border	33.84	12	35.46

Source: Lakhotia and Jacob, 1974.

erochromatin of the polytene chromocenter as well as to some 50 euchromatic bands (Spradling, Penman, and Pardue, 1975). Coupled with the evidence for active transcription within the β heterochromatin (Lakhotia and Jacob, 1974), this finding raises the question of whether it should indeed be regarded as heterochromatin in the sense we have defined it. Spofford (1976) uses the term *quasi-heterochromatin* to describe β heterochromatic regions with the obvious implication that they are not equivalent to constitutive heterochromatin.

While the X chromosome of the nematoceran fly *Phryne cincta* includes some characteristic α heterochromatic sites, the remainder has been described as β heterochromatin (Wolf, 1970). The β heterochromatin of *Phryne,* however, differs from that of *D. melanogaster* in two respects: It is not late replicating, and it is not restricted to the proximity of α heterochromatin. The X chromosome of *Phryne* also shows an amazingly variable appearance in polytene nuclei. In some cells, and especially in larvae raised at low temperatures, it is long, thin, and clearly banded, presenting a typical euchromatic appearance. In others, however, and especially in larvae raised at high temperatures, it is short and thick, and lacks visible bands. By contrast, the entire Y is α heterochromatic and is underreplicated in polytene tissues where it appears as a small compact, unbanded structure. Wolf (1970) has commented that so-called β heterochromatin should be considered as euchromatin that under certain conditions displays modified properties. This implies that it represents a form of facultative heterochromatinization – one that is confined to the polytene state and operates in both sexes.

The observations of Bedo (1982) on the behavior of the X chromosome of *Lucilia cuprina* in polytene trichogen cells reinforces this opinion. In mitotic cells, the X is darkly stained for most of its length following C-banding, though there is a lighter-staining distal segment on the long arm (Bedo, 1980). In both male (XY) and female (XX) individuals, the X in trichogen cells is a short, thick granular element lacking any sign of polytene banding. In experimentally constructed XXY strains of this species, however, a short, uniquely banded, polytene chromosome is present together with the conventional granular X (Figure 1.9).

In what sense, then, can the β heterochromatic region of *D. melanogaster* be regarded as a trainsitional zone between euchromatin and heterochromatin? The varying width of the base of each chromosome arm, which is evident when individual chromosomes are dislodged from the chromocenter by squashing, indicates that in terms of its precise level of replication the β heterochromatic region certainly forms a transitional zone between the totally unreplicated and heterochromatin and the fully replicated euchromatin. Indeed, because of its proximity to α heterochromatin, it cannot avoid coming under the influence of a domain of nonreplication. This, as Laird (1973) first suggested, might generate a hierarchical multifork transition zone, composed of varying levels of replication, which bridges the completely unreplicated α heterochromatin and the fully replicated base of the euchromatic arm (Figure 1.10).

The best-studied β heterochromatic region is that at the base of the X chromosome of *D. melanogaster* (Miklos et al., 1984). Here, sections 19 and 20 lie at the interface between two very distinct chromatin landscapes (Figure 1.11). This junction is not, however, abrupt in a molecular sense since, as Schalet and Lefevre (1973) have shown, there are at least 11 major gene loci within the β heterochromatin of the X at or after 20 D. In situ hybridization indicates that this region certainly contains middle repetitive DNA sequences. Indeed the labeling of mobile sequence probes to the chromocenter is most likely to result from their presence in the β, rather than the α, heterochromatin. Unlike α heterochromatin, however, the β region does not contain satellite or other hr DNA sequences, so the two regions differ in sequence content, in replicative behavior, and in transcriptional ability.

2. Quantitative changes in heterochromatin content

"Any in vitro system is a contrivance. We can never be certain that the results obtained with such a system are relevant to the in vivo state."

Charles G. Kurland

2.1. In vivo changes

There are two types of in vivo changes in heterochromatin content. The one involves a partial or a complete loss of heterochromatin from the soma. The other involves an under- or nonreplication of heterochromatin in endoploid somatic tissues.

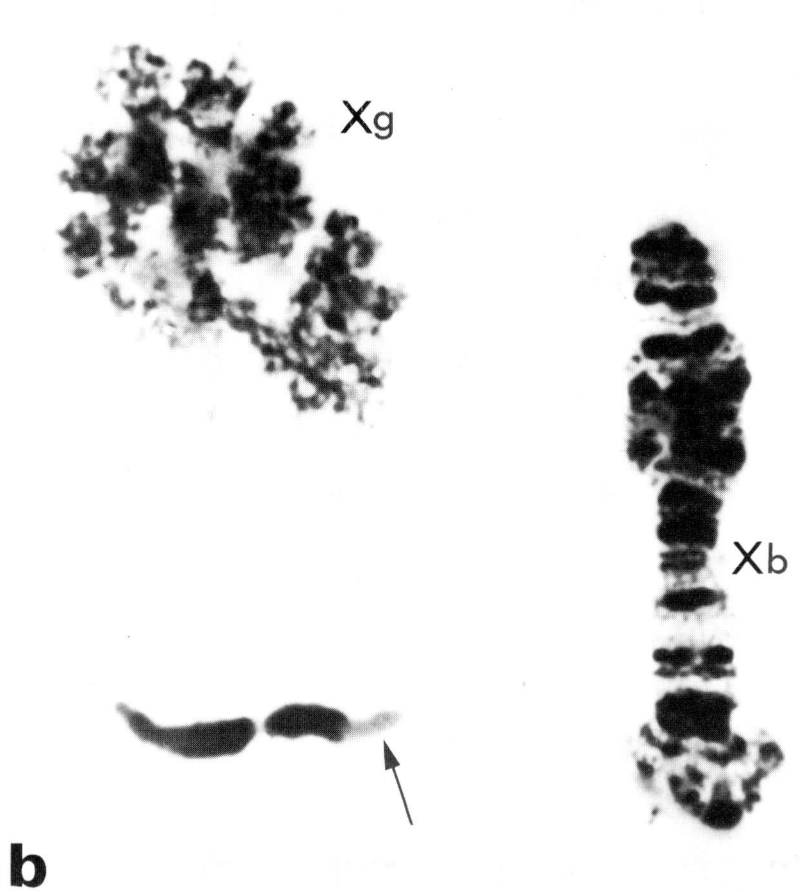

Figure 1.9. Behavior of the X chromosome of *Lucilia cuprina* in polytene (a) and C-banded mitotic (b) nuclei. At mitosis, the X has two C-band-positive arms, one of which is terminated by a paler segment (arrow). In polytene nuclei from trichogen cells, the X is usually granular (Xg) but in experimentally constructed XXY males one of the two X chromosomes (Xb) now has a characteristic, though distinctive, pattern of bands and interbands. (Photographs courtesy of Dr. D. G. Bedo.)

2.1.1. Presomatic elimination

In two groups of invertebrates, there is a regular and complete loss of constitutive heterochromatin from all presomatic cell lineages during early ontogeny so that, unlike the germ line, the soma contains only euchromatin. The classic case of this occurs in nematode worms and

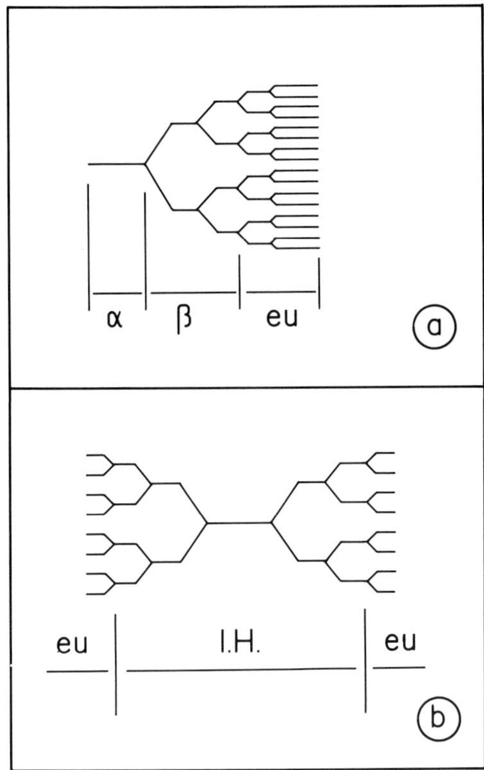

Figure 1.10. Diagrammatic representation of the structure of the α and β regions (a), and of an intercalary heterochromatic region (I.H.) (b), of the polytene nucleus of *Drosophila melanogaster,* as compared to that of euchromatin (eu). This diagram is based on the frozen replication-fork model of Laird (1973). The α region is nonpolytenized, while the β region and the I.H. region are partly polytenized.

involves chromosome fragmentation, followed by the elimination of all heterochromatin, in presomatic cells. The cytological details are best known in *Parascaris univalens* ($2n = 2$) and *P. equorum* ($2n = 4$). These two species differ with respect to the location of heterochromatin within the chromosomes. It is distal in *P. univalens* but distal and intercalary in *P. equorum.* In both cases all the heterochromatic regions give positive C-banding, but there are qualitative differences between the two species with respect to their reaction with fluorochromes (Goday and Pimpinelli, 1986). In *P. univalens,* where the central region of the germ-line chromosomes is entirely euchromatic, and shows only a weak fluorescence with H33258 and Q, the heterochromatic blocks that terminate both ends of each chromosome fluoresce strongly with both dyes though, in each

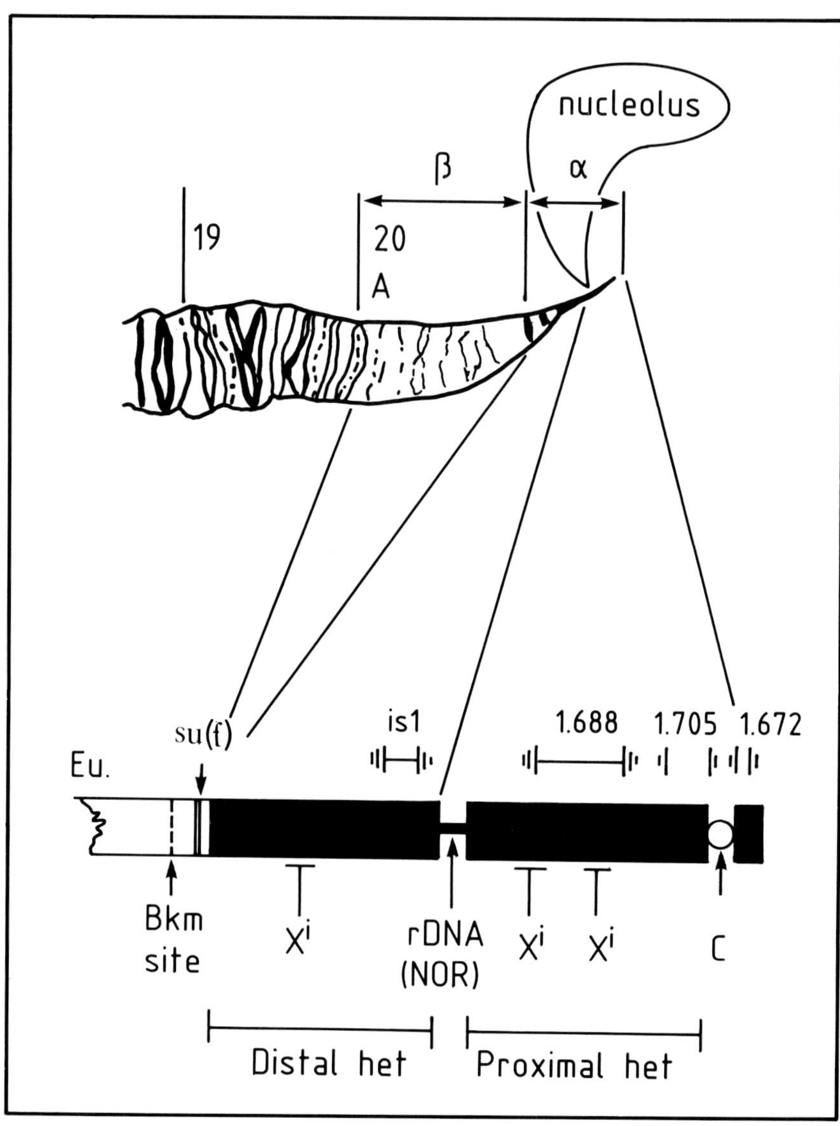

Figure 1.11. Structure of the euchromatin (Eu.)–heterochromatin (het) boundary region of the polytene X chromosome of *Drosophila melanogaster* (upper) relative to the mitotic organization of the same region (lower). Also shown for subsequent reference are the lcoations of the Bkm (banded krait minor, see Sec. 4.11) sequences, the mutation suppressor of forked [su(f)] three of the satellite sequences (1.688, 1.705, and 1.672), the type 1 insertion sequences (is1) and the three presumptive X inactivation sites (X^i). NOR, nucleolus organizer region; C, centromere.

case, there is a terminal, moderately fluorescing subregion and a proximal, brightly fluorescing subregion. CMA reverses this pattern and also gives bright fluorescence in the central euchromatic region.

In *P. equorum,* both the terminal and intercalary heterochromatic regions give bright fluorescence with Q, so that a series of brightly fluorescing segments alternate with weakly fluorescing regions interstitially. Bot H33258 and CMA give weak fluorescence for the terminal heterochromatic segments but stronger fluorescence for the interstitial region, though some of the small interestitial segments that are nonfluorescent with H33258 fluoresce brightly with CMA. Thus, whereas there are at least two categories of heterochromatin in *P. univalens,* there are three in *P. equorum* and no type is common to both species. There are also differences in the behavior of the euchromatin in the two species. These individual qualities are maintained in F_1 hybrid embryos, though such embryos always possess abnormal karyotypes which lead to their lethality.

In both species, elimination affects all the heterochromatic segments, regardless of their location, and takes place after the four-cell stage, partly during the third cleavage mitosis and partly during the fourth. By the 16-cell stage, therefore, only two cells within the embryo retain heterochromatin. In the fifth cleavage, one of these undergoes the last elimination division whereas the other gives rise to the two pregerminal cells without elimination. Elimination, in both species, leads to the production of an elevated number of small somatic chromosomes with a "diploid" count of $2n = 60$.

Electron microscopic (EM) observations indicate that there is a single uninterrupted kinetochore profile in the germ-line chromosomes of both species (Goday, Ciofi Luzzatto, and Pimpinelli, 1985). If, as this evidence suggests, kinetochores are present in both intercalary and terminal heterochromatic segments, these kinetochores are certainly not active in the presomatic lineage at the time of elimination.

The detailed molecular composition of the DNA sequences that are eliminated in *Parascaris* are not known, though *P. univalens* contains two satellite DNAs, both AT rich (Moritz and Roth, 1976). In *Ascaris lumbricoides* var. *suum* (= *A. suum*), some 22% of the genome (1C = 0.32 pg) consists of germ-line-limited DNA in the form of an hr satellite (Table 1.8). Restriction enzyme analysis indicates that this satellite is composed of two families of tandemly repeated sequences. One repeat unit is 125 bp long, whereas the other is 131 bp long (Roth and Moritz, 1981). At least 99.5%, but not quite all, of these satellite copies are expelled from the presumptive somatic cells so that somatic DNA consists, at most, of about 5000 copies of the satellite sequence. Many of the somatic nuclei in *Ascaris,* however, are known to increase their DNA content following

Table 1.8. Chromatin diminution in ascarids and copepods

Species	DNA content (pg)		% DNA eliminated
	Germ line	Soma	
1. *Pascaris univalens*	1.2–2.1 (polymorphic)	0.25	80%
2. *Ascaris suum*	0.32	0.25	22%
3. *Cyclops strenuus*	2.2	0.9	59%
4. *Cyclops furcifer*	2.9	1.4	52%
5. *Cyclops divulsus*	3.1	1.8	42%

Sources: (1, 2) Davis, Kidd, and Carter, 1979; Roth and Moritz, 1981. (3–5) Beermann, 1977.

elimination. Whereas muscle and nerve ganglia appear to retain the postelimination "diploid" level of DNA, intestinal nuclei are in the "tetraploid" range; the uterine epithelium, the pharyngeal glands, and the excretory cells all have a high degree of endoploidy, reaching 150 ploid in some nuclei of the uterine epithelium (Swartz, Henry, and Floyd, 1967).

A remarkably similar system of elimination is found in the copepod *Cyclops*. Here, segments that comply with Heitz's definition of heterochromatin are present in the meiotic prophase of female oocytes as well as in the fifth to sixth cleavage mitoses, though their precise molecular composition has not been determined. No chromocenters are present at interphase in the first three to four cleavage mitoses. Since no nucleoli are evident also during this period, it is probable that the chromosomes are not transcriptionally active at this time despite their apparent euchromatic appearance (Beermann, 1977). This phase is followed in the subsequent cleavages by the presomatic elimination of all the heterochromatic segments which now become apparent. As in the case of *Parascaris,* these segments may be solely terminal and present at both ends of each chromosome *(C. divulsus);* alternatively, there may be additional interstitial blocks present in single *(C. furcifer)* or multiple *(C. strenuus)* copies. Elimination again results in the loss of a large fraction of the genome from the presomatic cell lineage (Table 1.8).

An even more bizarre behavior is known in the case of the supernumerary L, or limited, chromosomes of *Sciara* (Figure 1.12). These are initially retained within the germ line but are eliminated from the presumptive soma following their arrest in anaphase movement at the fifth cleavage division when they are left at the equatorial plate and subsequently degenerate. These L chromosomes are characterized by their

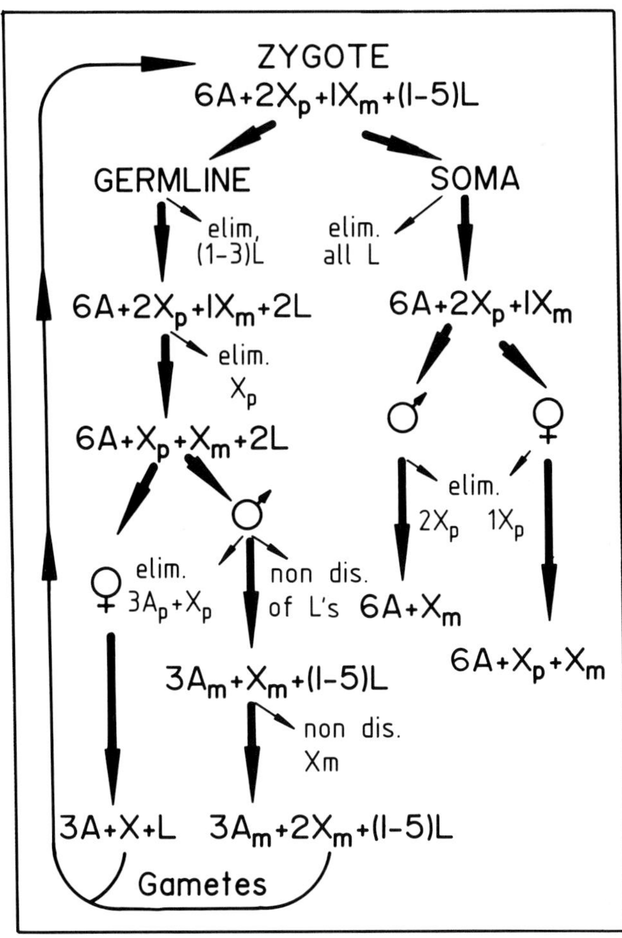

Figure 1.12. Behavior of the X and L chromosomes in the life cycle of *Sciara coprophila*.

large size and by their condensed nature in resting nuclei, though they have not been C-banded, and their molecular composition is unknown. Their number varies from embryo to embryo, the range being 1–5. All but two of them are also eliminated from each germ cell nucleus so that, according to the constitution of the embryo, one to three L chromosomes are lost from the developing germ line, too. This elimination, however, takes place at a considerably later stage of embryonic development, namely immediately after the cluster of 300 germ cells has migrated from the pole plasm, and is said to occur by the migration of whole chromo-

Table 1.9. Intraindividual L chromosome variation in *Sciara coprophila*[a]

	% Constitution					
Stage	0L	1L	2L	3L	4L	5L
Zygote	—	—	18	56	21	3
Spermatogonia	—	—	100	—	—	—
Spermatocytes	0.3	11.3	78	9.5	0.9	—
Oocytes	—	—	100	—	—	—

[a]Compare this table with Figure 1.10.
Source: Crouse, Brown, and Mumford, 1971.

somes through the nuclear membrane (Crouse, Brown, and Mumford, 1971).

During the six gonial divisions that precede the primary spermatocyte, the L chromosomes become variable in number, presumably as a result of mitotic nondisjunction; thus by the time of formation of the primary spermatocytes, only 78% of the cells still have two L chromosomes, and males may transmit from one to five of them. By contrast, all oocytes retain two L chromosomes and, in a majority of cases, these form a bivalent at female meiosis so that each egg transmits one L except for a low number of meiocytes where L univalents are present at first metaphase and so give rise to a low frequency of 0L eggs (Table 1.9).

A quite distinct form of elimination occurs in the Diaspidid coccid system. Here the entire paternal set is eliminated, from each male embryo during early cleavage, and at a time when the paternal set becomes facultatively heterochromatinized in males of the Lecanoid and Comstockiella systems. Similarly, somatic X chromosome elimination replaces X inactivation as a mechanism for dosage compensation in the female of *Microtus oregoni* (Ohno, Stenius, and Christian, 1964). Although there is no actual loss of heterochromatin in either of these cases, they clearly derive secondarily from systems of facultative inactivation with accompanying heterochromatinization.

There is also an elimination of sex chromosomes from some, though not all, somatic tissues in several members of the marsupial family Peramelidae (Hayman, Martin, and Waller 1969). The chromosomes involved are the Y of the male and one of the two X chromosomes of the female. However, a full complement is present in the germ line and, in some species, in the corneal epithelium as well. This sex chromosome elimination takes place at much later stages of morphogenesis than does the appearance of the sex chromatin body in cells of eutherian mammals

Table 1.10. Combinations of eu- and heterochromatin classes in brain ganglion nuclei of *Drosophila hydei*

Euchromatin value	Heterochromatin value	Total value	No. times observed
2C = 11	2C = 5	16 (diploid)	24
4C = 22	4C = 10	32 (tetraploid)	10
	8C = 20	42	3
8C = 44	2C = 5	49	3
	4C = 10	54	10
	8C = 20	64 (octoploid)	3
	16C = 40	84	4
16C = 88	4C = 10	98	1
	8C = 20	108	4
	16C = 40	128 (16 ploid)	1
32C = 176	4C = 10	186	2
	8C = 20	196	1

Source: Berendes and Keyl, 1967.

or does the elimination of the X from the female soma in the case of *Microtus oregoni*. Whether this implies that sex chromosome inactivation also occurs at a much later stage in bandicoots is not clear. In a second family of marsupials, the Petauridae, Murray and McKay (1979) found that sex chromosome elimination is restricted to the Y chromosome of males, and, as in the Peramelidae, there are timing differences between different tissues.

2.1.2. Differential somatic replication

We have already referred to the fact that the constitutive α heterochromatin of *D. melanogaster* and *D. virilis* is not replicated during polytenization. Satellite DNA nonreplication or underreplication is in fact a common phenomenon in both polytene and endoploid insect tissues, so that constitutive heterochromatin is often underrepresented in such tissues relative to euchromatin. The brain ganglion of *D. hydei* consists mainly of diploid cells but also includes a number of nuclei with a low level of polyteny. In both classes of nuclei, the heterochromatin is represented as a single chromocenter, consisting predominantly of the heterochromatic arm of the X in females, and this same section of the X together with the entire Y in males. The proportion of euchromatin and heterochromatin in polytene nuclei is, however, different from that in diploid nuclei (Table 1.10). Most commonly, polytene nuclei show a substantial underrepresentation of heterochromatin, though in some the heterochromatin has evidently completed one replication step more than the euchromatin (Berendes and Keyl, 1967). The correlated increase in nucleolar size and nuclear size indicates that, as in the salivary gland cells of this species which attain much higher levels of polyteny, the NOR regions of the two sex chromosomes do not follow the behavior of the chromocentral heterochromatin.

In *Drosophila nasuta,* all the large chromosomes carry substantial blocks of pericentromeric α heterochromatin which come together as a chromocenter both in interphase nuclei and polytene nuclei. These heterochromatic regions share similar AT-rich DNA sequences and so fluoresce very brightly with QM and H33258. Using cytofluorimetry, Lakhotia (1984) found that the amount of flourescing material was similar in diploid imaginal disk nuclei and in the highly polytenized nuclei of both the Malpighian tubules and the salivary glands. His measurements indicated that the α heterochromatin remained at the 2C level in Malpighian tubules, whereas in salivary glands it varied between 2C and 4C. In *D. nasuta* as in *D. melanogaster,* this underreplication affects all the constitutive heterochromatin, but in other cases specific chromosomes or chromosome segments are excluded from replication. Perhaps the most spectacular of these is found in *D. nasutoides.* Here, the diploid (2C) DNA content, calculated from brain metaphases, is 0.79 pg, and the individual members of the complement constitute, respectively, 9% (X), 16% (2), 13% (3), and 62% (4) of the genome (Zacharias, 1986). Some 60% of the genome is composed of satellite DNA, the greater bulk of which occupies a pair of giant metacentric isochromosomes (No. 4) in the mitotic karyotype. These, like the Y chromosome of this species, stain darkly after C-banding, though not homogeneously so, which is indicative of their heterochromatic structure. In polytene nuclei, however, this pair of giant chromosomes is replaced by a dot chromosome which is not represented as such in the diploid set. Thus, in the larval salivary gland, which has undergone 11 endoreplication cycles, chromosome 4 now comprises only 1.5% of the total genome (Zacharias, 1986). As a result, there is no detectable satellite DNA in the salivary gland cells (Cordeiro et al., 1975; Cordeiro Stone and Lee, 1976; Wheeler, Arrighi, Cordeiro Stone, and Lee, 1978).

Similarly, studies of DNA clones from the Y of *Drosophila hydei* indicate that all sequences other than the rRNA genes, which undergo partial replication, are completely excluded from polytenization (Vogt and Hennig, 1983). This chromosome, as we have seen earlier, is built up predominantly out of mr sequences, although it is C-band positive, so that underreplication is not the sole prerogative of hr DNA. Indeed underreplication is not confined to repetitive DNA. Thus, in *Sarcophaga bullata,* where the X and Y are both totally C-band positive, no hr or mr DNA is present in either sex chromosome, yet both are totally absent from polytene nuclei (Samols and Swift, 1979).

The precise factors that determine the replicative behavior of particular heterochromatic segments remain to be defined. The centric block of heterochromatin on chromosome 3 of *Anopheles stephani,* to which the satellite 1 DNA of this species hybridizes both in mitotic and polytene nuclei, certainly replicates during polytenization, whereas sequences complementary to this same satellite on the X do not (Redfern, 1981). A parallel situation is evident in polytene trichogen cells of the fly *Lucilia cuprina,* where both the X and the Y, which are predominantly heterochromatic and C-band positive, are represented by very much smaller, unbanded, elements (Bedo, 1982). The mitotic Y of this species carries a single paracentromeric and Q-positive band; in trichogen cells this is disproportionately represented and may reach up to half of the total Y. Thus, whereas the bulk of the Y is underreplicated, the Q-band is not. In the Medfly, *Ceratitis capitata,* the X and Y chromosomes of the polytene male orbital bristle cells are represented not by banded elements but by a heterochromatic network. Quinacrine staining suggests that these sex chromosomes, like those of *Lucilia,* are also present differentially underreplicated (Bedo, 1986).

The presence of greatly reduced amounts of constitutive heterochromatin in the polytene chromosomes, relative to that evident in somatic metaphase chromosomes of *Sciara coprophila,* likewise indicates that most of the DNA of these heterochromatic regions is also underreplicated during polytenization. However, AT- and GC-rich sequences that are not detectable in somatic metaphase chromosomes are demonstrable in the constitutive heterochromatin of polytene chromosomes; this implies that here, too, there are DNA fractions within the heterochromatin that may replicate preferentially during polytenization (Eastman, Goodman, Erlanger, and Miller, 1980).

Yet a further variant form of behavior is found in the gnat *Phryne cincta.* In this species, the Y chromosome is represented in male polytene nuclei as a small, unreplicated, heterochromatic body as also are the one to seven supernumerary elements that may occur in natural populations. Moreover, when present in any number, these fuse with one another, as well as with the Y, to form a noticeable chromocenter in polytene nuclei although no chromocenter forms in the absence of supernumeraries (Wolf, 1961).

What applies to the polytene tissues of some larval dipterans applies also to the endoploid adult tissues which replace those of the larvae following their histolysis at pupation. Adult midgut, hindgut, thoracic musculature, Malpighian tubules, and salivary glands all include nuclei with DNA contents ranging from diploid to high degrees of polyploidy. Endow and Gall (1975) showed that in these tissues the amounts of the three satellite DNAs that are represented in the genome of *Drosophila virilis,* and are confined to the constitutively heterochromatic regions of the karyotype, were lower than in diploid brain used as a control. The actual deficits ranged from a 28% decrease in thoracic muscle to nondetectable amounts in the Malpighian tubules. Midgut and hindgut showed intermediate, though different amounts, and the salivary glands showed the least reduction (Table 1.11).

In contrast to the α heterochromatin of higher flies, that of chironomids coreplicates with euchromatin during polytenization. This has been confirmed for *Glyptotendipes barbipes* (Walter, 1973; Schmidt, 1980), *Chironomus melanotus* (Steinemann, 1978) and *Chironomus thummi thummi* (Schmidt and Keyl, 1981). Significantly, in these cases there is no chromocenter and no β heterochromatin.

2.2. In vitro changes

In *Microtus agrestis,* the constitutive heterochromatin of the giant sex chromosomes can be readily deleted or translocated from the sex chromosomes of cells in culture. Thus, in cell line Ma/26, 93% of the sex heterochromatin has been lost, yet cells of this line survive and propagate in vitro (Cooper, 1977a). Likewise, bone marrow cells carrying a wide spectrum of radiation-induced deletions and translocations involving the constitutive heterochromatin of *M. agrestis* persist for more than a year after induction (Cooper, 1977b). Even so, it must be emphasized that no cells in nonirradiated individuals of *M. agrestis* have even been found to show any variation in the amount of constitutive heterochromatin; nor do individuals of this species show any polymorphisms with respect to constitutive heterochromatin. This contrasts markedly with the situation in vitro where the heterochromatin is remarkably labile. In vitro behavior is thus not necessarily an indicator of in vivo potential.

A similar situation applies to *Drosophila,* in the sense that it is possible to construct flies with a variety of heterochromatic deletions without producing any overt effect on either viability or fertility (Yamomoto and Miklos, 1978).

A distinct form of in vitro change is that involving the amplification of satellite sequences with the parallel formation of homogeneously staining regions (HSRs). For example, in a number of mouse cell lines that have been selected for resistance to methotrexate, the dihydrofolate reductase activity has been increased as a result of amplification of the gene responsible for the production of this enzyme. Resistant cells sometimes contain large marker chromosomes with distinctive HSRs. After C-banding, these can be seen to be composed of many fine bands linked

Table 1.11. Differential distribution of satellite *(Sat.)* DNA in adult tissues of *Drosophila virilis*[a]

	% Total genome			% Difference relative to brain		
	Main peak	Sats. I + II	Sat. III	Sats. I + II	Sat. III	Cell type
Brain	53.0	38.4	9.4			Predominantly diploid
Hindgut	65.3	27.5	7.3	−40.8	−38.9	Mixed diploid/polyploid
Thoracic muscle	64.6	32.5	2.9	−28.2	−72.2	Mixed diploid/polyploid
Midgut	73.5	20.5	6.1	−60.6	−55.6	Mixed diploid/polyploid
Salivary gland	76.6	20.3	3.1	−62.0	−77.8	Mixed diploid/polyploid
Malpighian tubule	100.0	—	—	−100.0		Polytene
Whole adult	62.2	33.1	4.7	−23.9	−55.6	

[a]In those tissues containing a mixture of diploid and endopolyploid cells, it is not possible to determine the true extent of satellite DNA replication.
Source: Endow and Gall, 1975.

closely together. The in situ hybridization of purified mouse liver satellite DNA labeled with iodine-125 indicates that these HSRs contain a high proportion of sequences that cross-hybridize with mouse satellite DNA. This case is important in demonstrating that heterochromatic segments can arise in response to artificial selection in vitro, though it is not clear whether the satellite sequences of the repeating unit have played a positive role in promoting the amplification process or whether they have been amplified simply by chance along with the dihydrofolate gene (Bostock and Clark, 1980).

The chromosomes of 500γB mouse cells include a characteristic marker chromosome with several interstitial C-bands, all of which, on the basis of autoradiography, appear to contain satellite DNA which is normally confined to centric regions of mouse chromosomes (White, Pasztor and Hu, 1975). Likewise, the 440B and p/51 markers, which also carry interstitial C-bands, occur with very high frequency (>85%) in their respective cell lines. While saltatory replication of sequences already present at levels too low for detection by in situ hybridization may be involved, these C bands may also result from structural rearrangements leading to subsequent amplification. A chromosome with an interstitial C-band has certainly been introduced in a stock of *Mus musculus* following a translocation involving a large block of centric heterochromatin (Dev, Miller, Charen, and Miller, 1974) though, in this case, there was no evidence of subsequent amplification.

3. Effects and functions of heterochromatin

"The functions of heterochromatin must be involved in the differences between somatic and germinal cells."

Richard Goldschmidt

3.1. Somatic effects

3.1.1. Ectopic pairing

The heterochromatin of nonhomologous chromosomes sometimes shows a tendency to associate in somatic cells. The formation of the polytene chromocenter in *Drosophila melanogaster* is one instance of such an association, which involves the pairing of the proximal portions of sections 20, 40, 41, 80, 81, and 101. However, not all species with centric heterochromatin form chromocenters. They are lacking in many chironomids and black flies, although there are substantial procentric blocks of α heterochromatin in these organisms. In *Rhynchosciara angelae,* a polytene chromocenter forms in the Malpighian tubules and in the seminal vesicles but not in the salivary glands, despite the fact that equivalent

Table 1.12. The 49 sections of the polytene chromosomes of *Drosophila melanogaster* most frequently involved in ectopic pairing

Chromosome no. (sections)	
I (1–20)	1A, 3C, 4C, 7ABC, 8B, 9AB, 11A, 12EF, 16D, 18F/19A, 19E.
II (21–60)	21A, 22AB, 25A, 25E, 33A, 35A, 35F, 36D, 39A, 39E, 42B, 56F, 57AB, 59D, 60F
III (61–100)	61A, 64C, 65B, 67A, 67D, 70AB, 70C, 71C, 73F/74A, 75C, 80ABC, 81F, 83D, 84A, 84D, 86D, 89E, 94A, 94D, 98C, 100A.
IV (101–102)	101F, 102F

Telomere regions are indicated by underscoring.
Source: Zhimulev, Semeshin, Kulichkov, and Belyaeva, 1982.

amounts of heterochromatin are present at the same sites within the chromosomes (Pavan and Breuer, 1952). In stocks of *D. melanogaster* carrying deletions of up to 82% of the heterochromatin of the X, on the other hand, this chromosome is still able to participate regularly in chromocenter formation (Yamamoto and Miklos, 1978), so that the actual amount of heterochromatin appears not to be a necessary requirement.

In black flies, Rothfels and Freeman (1977) distinguish between "a genuine glassy chromocenter and ectopic pairing, no matter how tight." In the latter case the unit is composed of the conjoined centromeres themselves, whereas in the former the centromeres are inserted into an extra, Feulgen-positive, chromocenter. Additionally, Bedo (1975) has commented that, whereas the heterochromatic regions of the centromeres of *Simulium melatum* form a loose pseudochromocenter, those of *S. ornatipes* never associate; this, despite the fact that in *S. melatum* the centromere regions consist of an unstained central region flanked on both sides by intense C-banding, whereas in *S. ornatipes* there is a darkly stained C-band-positive centromere with unstained pericentric regions.

It has been known since the pioneer work of Bridges (1935) that certain nonhomologous bands in the euchromatic polytene arms of *D. melanogaster,* both in the same and in different chromosomes, may associate with one another and with telomeric heterochromatin. They may also associate with the chromocenter and so simulate the behavior of the centromere-associated α heterochromatin. This phenomenon, known as *ectopic pairing,* was used by Kaufmann (1939) to define a category of intercalary heterochromatin (IH) which Beermann (1965) regarded as α heterochromatin. In *D. melanogaster,* some 350 bands in 260 different chromosome regions have been found to be capable of ectopic pairing; Table 1.12 lists the 49 regions most frequently involved. Despite earlier suggestions that the peculiarities of IH regions might be attributed to a high DNA content, Bolshakov, Zharkikh, and Zhimulev (1985) have shown that this is not so.

Using fluorescent dyes to mark specific polytene bands, Barr and Ellison (1972) concluded that ectopic pairing results from a chemical similarity of the DNA in the bands concerned. Thus, ectopic pairing between the Q bright spots at 83 D,E and 81 F involves a strong similarity, if not actual homology, between the DNA sequences in these bands since the brightness is associated with the presence of poly dAT. In *Samoaia leonensis,* two satellite DNAs are present within the genome. One of these consists entirely of AT sequences, comprises 75% of the total satellite DNA, and is represented on all of the chromosomes. The other is 60% AT rich. In interphase nuclei from the larval brain, a single large chromocenter is present. When formalin-fixed nuclei are stained with coriphosphine-*o* after removal of RNA and histones, two distinct regions in this chromocenter are identified. The larger, which comprises 78% of the total mass, corresponds to the most AT-rich satellite (Table 1.13). There is an equivalent correspondence between the volume of the three components of the interphase chromocenter of *D. virilis* and the three satellites present in the genome of this species (Mayfield and Ellison, 1975). Thus, as with ectopic pairing in polytene systems, the production of interphase chromocenters in dipterans may also result from a mutual association of chromatin containing similar DNA sequences.

Apart from ectopic pairing, regions of intercalary heterochromatin show a number of other characteristic properties not all of which are equally expressed in every band. Cytologically they may appear deeply stained and often strongly vacuolated. Some form obvious weak points at which the polytene chromosomes are drawn out into a thin thread, which suggests that these regions are underreplicated, leading to incomplete polytenization of the band. Significantly, in *D. melanogaster* the number and distribution of late replicating regions is closely correlated with the occurrence of ectopic pairing (Table 1.14).

When labeled RNA preparations, constructed either from total newly synthesized RNA or from stable cytoplasmic RNA isolated from a cell culture of *D. melanogaster,* were hybridized in situ to polytene chromosomes of the same species, labeling was observed over the chromocenter, the nucleolus, and the IH regions, and strong hybridization also occurred over fibers formed by ectopic pairing (Gvozdev, Ananiev, Kotelyanskaya, and Zhimulev, 1980). Intercalary heterochromatin was also labeled with cloned DNA fragments coding for abundant classes of mRNA produced by cell cultures of *D. melanogaster.* Gvozdev et al. (1980) thus concluded that IH regions represent nests of different types of genes that are involved in active transcription, at least in cell culture.

Table 1.13. Structure of the composite heterochromatic (het.) masses present in interphase nuclei of brain cells in two dipterans

Species	Satellite no.	% AT	% Total satellite	Heterochromatic mass	Q fluorescence	% Total vol. het. mass
S. leonensis	1	65	25	small	slight	23 ± 3
	2	100	75	large	intense	78 ± 3
D. virilis	1	71	59	large	slight	78 ± 6
	2	86	17	small	moderate	11 ± 6
	3	86	24	small	moderate	11 ± 6

Source: Mayfield and Ellison, 1975.

Table 1.14. Correlation coefficients between different behavioral properties of intercalary heterochromatin in *Drosophila melanogaster*

Chromosome arm	Property	Ectopic pairing	Late replication	Chromosome rearrangement
X	Weak points	0.77	0.61	0.47
	Ectopic pairing		0.65	0.55
	Late replication			0.34
2L	Weak points	0.72	0.54	0.21
	Ectopic pairing		0.62	0.27
	Late replication			0.24
2R	Weak points	0.54	0.40	0.08[a]
	Ectopic pairing		0.67	0.17[a]
	Late replication			0.32
3L	Weak points	0.83	0.62	0.24
	Ectopic pairing		0.59	0.35
	Late replication			0.20
3R	Weak points	0.78	0.58	0.06[a]
	Ectopic pairing		0.60	0.16[a]
	Late replication			0.25

[a]Not statistically significant.
Source: Zhimulev, Semeshin, Kulichkov, and Belyaeva, 1982.

As Biessmann, Kruger, Schröpfer, and Spindler (1981) have pointed out, however, strong hybridization of this kind does not necessarily mean that the RNA is actually transcribed by every site to which it hybridizes. Thus, RNA transcripts produced by genes or sequences that occur in multiple copies scattered throughout a genome would be expected to hybridize to copies of these sequences, even if they were not themselves transcriptionally active. In fact, IH regions, when exposed to [^{3}H]UdR, do not show intense uridine incorporation (Gvozdev et al., 1980). If IH regions are indeed α heterochromatic, they are not expected to transcribe, though there remains the possibility that they may sometimes contain multiple or mobile sequences that are capable of transcription but are not an essential part of normal heterochromatin structure. For example, salivary gland nuclei of *D. melanogaster* may contain nucleolar-like structures, which vary in number and in size, attached over 80 euchromatic sites, in addition to the principal nucleolus connected by a chromatin strand to section 20 of the X. All of these structures hybridize in situ with a cloned *Drosophila* DNA fragment containing the 26*S* ribosomal gene. Most of these nucleoli are connected to IH regions, often by a discrete chromatin fiber (Ananiev, Barsky, Ilyin, and Churikov, 1981).

Ananiev et al. (1978) had earlier proposed that mobile elements are predominantly localized in IH regions and are a diagnostic feature of them. More recently, Belyaeva, Ananiev, and Gvozdev (1984) considered that the mobile element *mdg-1* might be an indicator of IH regions. Bolshakov, Zharkikh, and Zhimulev (1985), however, discount both these claims. This is not to say that mobile elements are not represented in IH regions, though it is clear that they are not present in all of them and their frequency of occurrence varies in different strains of *D. melanogaster*. This does not exclude the possibility that IH regions consist of a complex set of specific mr sequences which differ both in composition and in frequency at different sites.

Telomeric heterochromatin, which is also regarded as α heterochromatin (Beermann, 1965), is included in the list of IH regions. Rubin (1978) isolated a cloned segment of DNA from *D. melanogaster*, containing a 3-kilobase (kb) repeated sequence, that hybridized to all the telomere regions of the five major arms of the polytene chromosomes. Young et al. (1983) subsequently isolated several clones from a 12-kb fragment of DNA from this same species which also hybridized specifically to the six free polytene telomeres and additionally to the β heterochromatin of the chromocenter. Although these authors believe that hybridization to the chromocenter indicates that telomeric sequences are also represented in pericentric heterochromatin, there is no good evidence to support this argument. Thus, as they themselves point out, the chromocenter of *D. melanogaster* must include the telomeres of the right arm of the X and the left arm of autosome 4 as well as both of the Y telomeres though, because the Y is not detectable in male polytene nuclei, it is presumably un- or underreplicated.

The cloned sequences described by Young et al. (1983) appear to hybridize to about the same degree to all the telomeres and occur, at least in part, as tandem repeats. Renkawitz Pohl and Bialojan (1984), on the other hand, have identified a further DNA sequence (8–19T) of 2.3 kb, which was differentially distributed to the ends of the polytene chromosomes. It was absent from the telomere of 3L and was represented at the telomeres of 2L, 3R, 2R, and X^L with a relative abundance of 3.4:2.7:1.9:1.0, implying that different copy numbers were present at individual telomeres. It was also evident that 8–19T was not a simple sequence, nor was it composed of tandem repeats. Restriction enzyme analysis indicated that 8–19T sequences were interspersed with other sequences. Finally, apart from the labeling of the free telomeres, the sequence was also present at one internal site, band 44D, and in the chromocenter. Since 8–19T sequences were not visible following in situ hybridization to mitotic chromosomes, it follows that they cannot be contained with the α heterochromatin; that is, they are present in the β heterochromatin and as such cannot be pericentric in the sense implied by Young et al. (1983).

Bedo (1986) reports that in the Medfly, *Ceratitis capitata,* the extent of ectopic pairing varies among different tissues. Most polytene nuclei in this species are characterized by exceptionally high levels of ectopic pairing, and, for this reason, it is difficult to separate the individual chromosomes in a spread. Trichogen cells, however, do not show such pairing and give well-spread, though thin, polytene chromosomes. This may well be a consequence of a lower level of overall replication in trichogen cells since in *Drosophila* males it is known that the single X is never involved in ectopic pairing.

Polytene chromosomes are also represented in the Collembola (spring tails), a group of small wingless insects. Deharveng and Lee (1984) analyzed ectopic pairing in a population of the collembolan *Bilobella aurantica* ($2n = 14$) from Sainte Baume in southern France. In this population, no regular chromocenter ever forms, although each of the polytene chromosomes carries a large heterochromatic band in the center of the swollen centromeric region. Additionally, 12 of the 14 telomeres are heterochromatic, as also are nine of the arm regions immediately adjacent to the swollen centromeres. Two of the short arms and two of the long arms also carry IH sites (Figure 1.13).

Synapsis of homologous chromosomes is rare in the polytene nuclei of *B. aurantica,* but ectopic pairing is extraordinarily common. In five nuclei from one individual, 27 cases were observed (Table 1.15) and 24 of these involved sites on nonhomologous chromosomes.

3.1.2. Position effects

In *D. melanogaster,* experimentally induced transpositions in which a section of euchromatin is relocated next to a heterochromatic break point lead to an alteration in the morphology of the juxtaposed euchromatin in a proportion of polytene nuclei. For example, in the T(1;4)$w^{m258-21}$ rearrangement, the white Notch region of X euchromatin is translocated to the immediate vicinity of the centric heterochromatin of autosome 4 as a result of breaks at 3E5/6 and 101F. In polytene chromosomes that include this translocation, there is a variable heterochromatinization of region 3C–E in terms of both appearance and extent. In some cases the heterochromatinized euchromatin adopts a more homogeneous and compact form and is incorporated within the chromocenter. More frequently, there is an evident underreplication so that the entire region is represented by a thin thread (Reuter, Werner, and Hoffmann, 1982). By comparing each subdivision in nontranslocated, heterochromatinized,

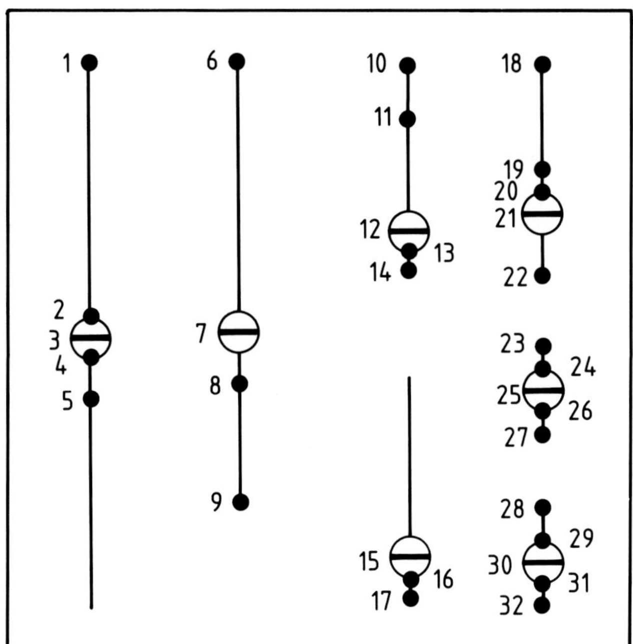

Figure 1.13. The distribution of the (1–32) heterochromatin sites present in the seven chromosomes of the haploid polytene set of the collembolan *Bilobella aurantica.* The chromosomes are all characterized by swollen centromeric bulbs, and seven of the heterochromatic sites (3, 7, 12, 15, 21, 25, and 30) are located within these regions. (After Deharveng and Lee, 1984.)

Table 1.15. Heterochromatic sites involved in ectopic pairing in five polytene nuclei from an individual of *Bilobella aurantica*

Location:	Terminal								Proximal	
Site No.:	1	6	9	10	14	17	23	32	16	29
No. times observed	2	2	2	5	2	5	1	5	1	2
Totals	24								3	
	27									

Note: Compare this table with Figure 1.11.
Source: Deharveng and Lee, 1984.

Table 1.16. Labeling frequencies of polytene chromosomes following exposure of excised salivary glands of *Drosophila melanogaster* to [^{3}H]TdR in standard (S) and heterochromatinized (het) and euchromatic (eu) subdivisions of translocated (T) X chromosomes from a T(1;4)w$^{258-21}$ strain

Subdivision	S	T(het)	T(eu)
3A	0.42	1.0	0.38
3B	0.25	0.75	0.26
3C1–4	0.56	0.76	0.59
3C5–7	0.58	0.83	0.56
3C8–11	0.30	0.72[a]	0.42
3D	0.35	0.59[a]	0.43
BE	0.32	0.61[a]	0.41

Note: Compare with Figure 1.12.
[a]Significant difference.
Source: Wargent and Hartmann Goldstein, 1976.

and euchromatic regions of the translocated homologue after [^{3}H]TdR application, Wargent and Hartmann Goldstein (1976) found an overall higher labeling frequency in the three subdivisions nearest to the heterochromatic breakpoint (Table 1.16). If labeling frequency provides an objective measure of the duration of replication, then these observations suggest that the heterochromatinized region takes longer to replicate. Additionally, relocated regions in this system, whose morphology remains unaffected may, nevertheless, undergo a change in DNA content. Thus, in microdensitometric studies on w$^{m258-21}$ heterozygotes in which the rearranged X section retained a euchromatic morphology (Cowell and Hartmann Goldstein, 1980), region 3D1–E2 immediately adjacent to the breakpoint contained less DNA than the homologous nontranslocated region, whereas 3C1–10, which includes an IH band at 3C, contained more DNA (Figure 1.14).

In experiments using T(1;4)w$^{m258-21}$ larvae, Kornher and Kauffman (1986) showed that at 17°C the *Sgs4* locus was polytenized to approximately one-third its level at 29°C. Likewise compaction of this locus was enhanced at 17°C and reduced at 29°C. On the other hand, Henikoff (1981), in a study of variegation at a heat-shock (HS) puff, found that the degree of polytenization at the site of inactivation was not significantly different between variegating and nonvariegating nuclei. Translocation T(Y;3)A78, with breakpoints in the long arm of the Y and at 87B in chro-

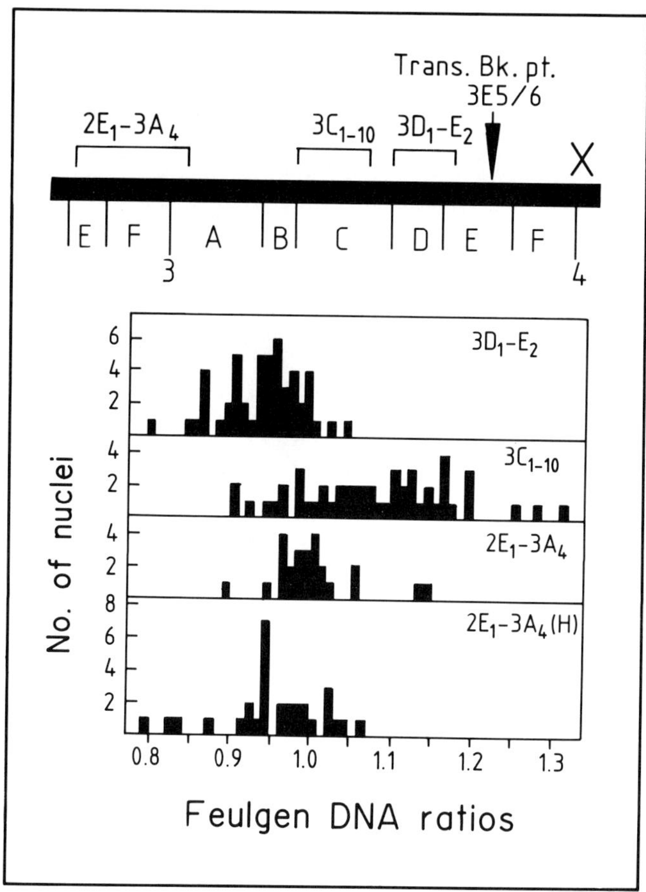

Figure 1.14. Frequency distributions of DNA ratios from translocated and non-translocated chromosomes in three euchromatic regions proximal to the break-point (3E 5/6) in the X chromosome of T(1;4)$w^{m258-21}$ mutants of *Drosophila melanogaster.* In $2E_1$–$3A_4$(H), the position-effect heterochromatinization has spread as far as band 3Cl. (After Cowell and Hartmann Goldstein, 1980.)

mosome 3, places the 87 HS puff locus next to Y heterochromatin. In this location, region 87C fails both to puff and to accumulate RNA in some nuclei after heat shock treatment, though it puffs normally in other nuclei from the same individual. In situ hybridization indicated that the level of polyteny was not significantly affected at the inactive gene site, despite the fact that in polytene nuclei the Y heterochromatin is unreplicated. This implies that a reduction in gene dosage is not responsible for gene inactivation, though it must be admitted that a difference in DNA

Table 1.17. Long-distance position-effect variegation in *Drosophila melanogaster*

Rearrangement	No. polytene bands
1. In(1)N^{264-51}	50 (4C9/D3 to 3C5)
2. Dp(1;4)N^{264-85}	35 (5B to 6A2)
3. T(1;4)$w^{m258-21}$	67 (2B14 to 3E5)
4. Dp(1;f)R	80 (1B1/2 to 3A)

Sources: (1, 2) Demerec, 1940; (3) Hartmann Goldstein, 1967; (4) Spofford, 1976.

amount of the level detected by Cowell and Hartmann Goldstein (1980) in the case of $w^{m258-21}$ – that is, 10–15% – would certainly not have been detected by the method employed by Henikoff.

Correlated with these variable alterations in euchromatin transposed to the vicinity of a heterochromatic breakpoint, there is an associated position-effect variegation of euchromatic genes placed cis and close to the breakpoint. This effect is evident, however, only when a recessive allele of the same locus is present in a normal partner homologue. Moreover, the effect is polarized since genes located closer to the breakpoint are more likely to be affected than are genes distant to it, though there are a number of cases known where the variegation effect operates over long distances (Table 1.17). When the altered gene function involves cell autonomous expression, the position effect is visualized in the form of a variegated phenotype giving rise to a visible mosaicism within the affected tissue. When it involves nonautonomous cell functions, it is expressed as a reduced activity of that gene within the affected individual.

Despite the fact that genes relocated adjacent to a heterochromatic breakpoint may be functionally altered, loci like uncoordinated (19E8) are regularly located adjacent to heterochromatin and still function normally. Others, like light *(lt),* which is located within the basal heterochromatin of 2L, variegate only if moved to the distal three-fourths of the euchromatin in the X, 2L, 2R, 3L, or 3R arms (Hessler, 1958).

Hazelrigg, Levis, and Rubin (1984) have used P-element-mediated transformation to generate transformants carrying segments of DNA from the white locus of *D. melanogaster.* Two such transformants, $A^R4.3$ introduced at 39E–40F (2L) and $A^R4.4$ introduced at 100F (3R), showed a dramatic alteration in gene expression simulating a variegation effect. In the case of $A^R4.3$ the white locus was inserted into β heterochromatin, whereas in $R^R4.4$ the variegating transduced gene lay in the telomeric α heterochromatin at the tip of chromosome 3. This is the only case so far described of variegation resulting from the transposition of a gene next to a telomere.

There is clear evidence that not all heterochromatic regions are equally effective in inducing variegation (Spofford, 1976). Nor do all regions in the same block of heterochromatin lead to the same degree of variegation. Four inversions in *D. melanogaster* – namely sc^{v2}, sc^{8}, sc^{L8}, and sc^{s1} – are known to move the NOR of the X chromosome from its normal position in the basal heterochromatin to the euchromatic tip of the X (Figure 1.15a). Baker (1971) showed that XO males carrying an inverted X whose rightmost breakpoint was between the NOR and the centromere have decreased viability compared with XO males with an inversion breakpoint distal to the NOR. Thus sc^{4} males without a Y are just as viable as those with one, whereas sc^{v2}, sc^{8}, sc^{L8}, and sc^{s1} males that lack a Y have viabilities ranging from 38% to less than 1% (Table 1.18). The qualitative differences between the viabilities when region 4 of the proximal block heterochromatin is present and when it is absent are especially striking. Baker therefore concluded that the lethality of XO males in the presence of In(1)sc^{L8} and In(1)sc^{s1} results from position-effect suppression of the rRNA genes following a break between the NOR and the centromere. The extent of variegated suppression is, however, ameliorated when region 4 of the basal heterochromatin is intact. Nix (1973) subsequently provided biochemical data to show that the lethality of sc^{s1}/0 larvae was due not to a reduction in the number of 18*S* and 28*S* rRNA genes, but rather to a suppression of the transcription of these genes. This leads to an overall reduction of approximately 15% in the amount of rRNA produced by sc^{s1}/0 larvae as compared to control levels. Lethality is thus related to an inability to produce enough ribosomes.

Three variegating white mutants of *D. melanogaster,* namely w^{m4}, w^{mMc}, and w^{m51b} have their euchromatic breakpoints clustered some 25 kb downstream from the white structural gene and flanked by a 1-kb sequence that is homologous to the 1.688 satellite DNA present proximally in the basal heterochromatin of the X. In all three, the heterochromatic breakpoints occur within sequences that have the characteristics of mobile elements (Tartof, Hobbs, and Jones, 1984). In the case of w^{m4} and w^{mMc}, the flanking heterochromatin in the variegated rearrangements is homologous to the type 1 inserts (is1) found in some portions of the rDNA repeats and present as a tandem array distal to the rRNA genes in the X. Thus, the most proximal breakpoint of w^{m4} is located just within this tandem array of is1 sequences (Figure 1.15b). In the case of w^{m61b}, the flanking mobile element has not been characterized. By using x-ray mutagenesis, Tartof et al. (1984) succeeeded in obtaining three w^{m4} revertants in which the white locus was moved by inversion from the centric heterochromatin to a euchromatic site at locations 1F, 4A, and 6B, respec-

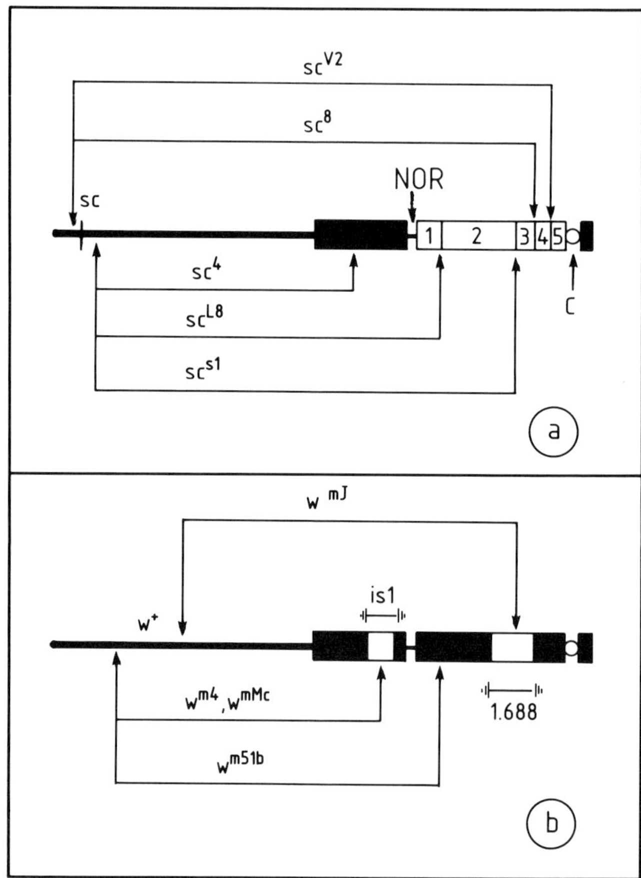

Figure 1.15. Breakpoints of the X chromosome inversions in *Drosophila melanogaster* studied by Baker [1971; scute mutants (a)] and by Tartof, Hobbs, and Jones [1984; white mutants (b)]. In the scute inversions, the right-hand breakpoints delimit five segments (1–5) within the proximal heterochromatin between the nucleolus organizer region (NOR) and the centromere (C). In the white inversions the location of the right-hand breakpoints relative to the type 1 insertion sequences (is1) distal to the NOR, and the 1.688 satellite sequences of the proximal heterochromatic block are indicated.

tively. In all three revertants, it was shown that the heterochromatic breakpoint was distal to the rDNA cluster but proximal to the w^+ heterochromatic junction in w^{m4} so that at least some of is1 sequences must have remained adjoined to white. It was argued, therefore, that these mobile elements could not themselves have produced variegation; otherwise, it would not have been possible to obtain reversion of position-

Table 1.18. Viability indices of *Drosophila melanogaster* males carrying X chromosomes with acute inversions

Paternal genotype	Maternal genotype	
	sc^i/Y	sc^i/O
sc^4/Y	100	95
sc^{v2}/Y	100	38
sc^8/Y	100	14
		<1
sc^{L8}/Y	100	<1
sc^{s1}/Y	100	<1

Note: See also Figure 1.13.
Source: Baker, 1971.

effect variegation. This, if true, would imply that variegation need not be initiated at the actual heterochromatin–euchromatin junction. Rather, like X inactivation, it might spread from an initiation site within the heterochromatin but not located at the junction itself.

A variant, intermediate variegating line of w^{m4h} gives an eye phenotype consisting of red patches on a white background. By either x-ray or ethyl methane sulfonate (EMS) mutagenesis, Reuter, Wolff, and Freide (1985) obtained 51 revertants of w^{m4h}. Of these, 19 were reinversions; 14 others were also inversions but with different euchromatic breakpoints, 12 of which were in IH regions; 4 were translocations; and the remaining 14 showed no accompanying cytological change and so must have been due to a mutational event linked to the heterochromatic breakpoint of w^{m4}. About a half of these revertants were clearly not completely w^+ in function, but had reverted to a very weak mottled phenotype. By using strong enhancer mutations, Reuter et al. were able to demonstrate variegation in 48 out of the 51 revertants. Contrary to the conclusion of Tartof et al. (1984), they argued that the heterochromatic sequences immediately flanking the white gene are essential for inactivation of the latter in w^{m4} revertants. They also concluded that the different phenotypic classes of revertants identified according to their interaction with enhancers of position-effect variegation may well reflect the presence of different numbers of flanking sequences at the heterochromatic junction.

Notions concerning the mechanism responsible for position-effect variegation have used the term *heterochromatinization* (meaning "becoming like heterochromatin") to describe what happens to euchromatic segments brought into proximity with breaks in heterochromatin. Position-effect variegation is then assumed to result from a local change in chromatin structure. (i.e., heterochromatinization) which is in some way causally connected with the inactivation of euchromatic genes within the effected region. The facts of position-effect variegation, and especially the capacity for reversion of variegation, clearly exclude any change in DNA sequence that might convert a euchromatic region into a heterochromatic state. Both increased compaction and underreplication have been associated with position-effect variegation and gene inactivation in polytene tissues. As Kornher and Kauffman (1986) point out, this need not imply that similar events operate in nonpolytene tissues whose nuclei are not amenable to direct study. Much, therefore, depends on the precise cell types involved in variegating tissues. Several monoclonal antibodies, prepared from major nonhistone chromosomal proteins, have been shown to be restricted to centromeric and intercalary heterochromatin in *Drosophila melanogaster* (James, Elgin, and Eissenberg, 1968). This includes the tip of 3R which, it will be recalled, was associated with the inactivation of a transduced white gene (Hazelrigg et al., 1984). Additionally, the cDNA recovered from the screening of a λgt11 expression library with one of these antibodies hybridized in situ to 29A, which is near a cluster of EMS-induced dominant suppressors of position-effect variegation (Sinclair, Mottus, and Grigliatti, 1983).

In male meally bugs, inactivation of the paternally derived chromosomes is known to be associated with an enriched histone content (Berlowitz, 1965). With this in mind, Moor, Sinclair, and Grigliatti (1983) tested whether position-effect variegation is in any way influenced by a reduction in histone gene multiplicity.

In *D. melanogaster,* the histone genes are reiterated about 100 times in the haploid genome and are located on a segment of chromosome 2 corresponding to polytene bands 39D2–3 to E1–2. In their study, Moor et al. (1983) used a series of small deficiencies located close to the histone gene complex at 39D–E (Figure 1.16). Four of these – Df(2L)DS8, Df(2L)DS9, Df(2L)1, and Df(2L)12 – lie at the distal edge of, but do not intrude into, the complex. Four others – Df(2L)DS5, Df(2L)DS6, Df(2L)65, and Df(2L)161 – remove the entire complex. One, Df(2L)84, is a partial deficiency of the histone complex. None of the deficiencies that cumulatively delete euchromatic segments from the distal edge of the complex (39D2 to 37F) have any effect on the white^{+} variegation associated with IN(1)wm^{4}. Those deletions that remove the entire histone complex all cause marked suppression of variegation. Df(2L)84, which deletes only the distal 40–50% of the complex, also causes a level of suppression comparable to that of the deletions that remove the entire

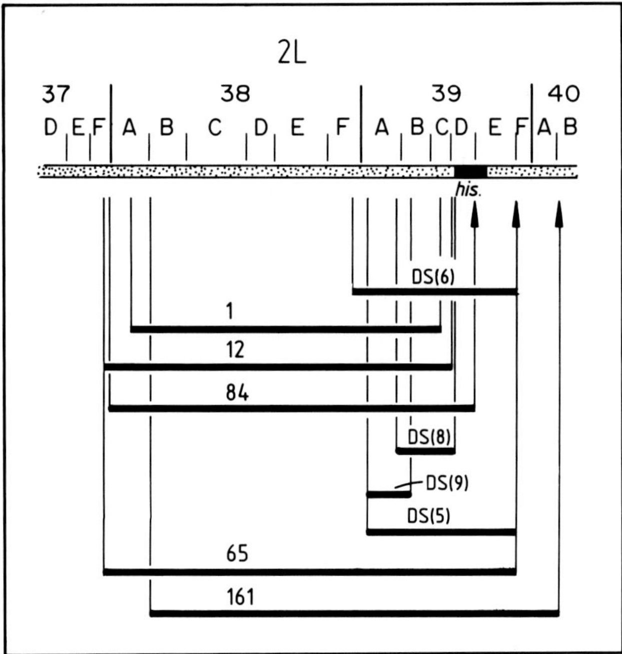

Figure 1.16. Proximal region of the 2L polytene arm of *Drosophila melanogaster* to show the extent of the long [Df(2L)] and short [Df(2L)DS] deficiencies studied by Moore, Sinclair, and Grigliatti (1973) and their relationship to the histone gene locus *(his)*. Right-hand breakpoints that delete all or part of this locus are shown with a solid arrowhead. (After Moore, Sinclair, and Grigliatti, 1973.)

complex. The conclusion seems inescapable that either the histone genes themselves, or else nonhistone genes that map within the histone gene cluster, act as agents of suppression.

The chemical *n*-butyrate inhibits histone deacetylase enzymes and, since no acetylated histones bind preferentially to inactive chromatin, Mottus, Reeves, and Grigliatti (1980) also tested the capacity of *n*-butyrate to suppress variegation of the w^+ locus in the IN(1)w^{m4} system. They reported positive results and concluded that position-effect variegation could be suppressed by influencing the affinity of histone for DNA. This implies that position-effect variegation depends primarily on histone-mediated condensation of DNA.

More than 50 dominant suppressors, as well as 3 dominant enhancers, of position-effect variegation have been isolated in *D. melanogaster* (Sinclair et al, 1983). One of these suppressors, Su-var(2)1^{01}, displays recessive butyrate sensitivity and results in a significant hyperacetylation of histone H_4 (Dorn, Heymann, Lindigkeit, and Reuter, 1986). This suggests that the wild-type product of this locus might be involved in H_4 deacetylation. Since the accessibility of chromatin to exogenous nucleases is significantly increased in larvae containing this suppressor mutation, it can also be classified as a chromatin condensation mutant.

However, Rushlow, Bender, and Chovnick (1984) have emphasized the need for caution in interpreting experiments involving variegated morphological phenotypes since, as they point out, such experiments often provide data with no adequate controls. Thus, in contrast to the findings of Moore et al. (1983) and Mottus et al. (1983), they failed to find any effect of either deficiency Df(2L)84 or of butyrate feeding on the expression of position-effect variegation at the rosy locus, though they were able to confirm the earlier results on the white locus. Their studies with the position-affected rosy allele, however, provided clear evidence that there was no reduction in the amount of rosy locus DNA in polytene tissues. Additionally, they were able to show that heterochromatin, placed adjacent to the rosy locus, produces a defect in transcription, leading to the formation of reduced quantities of rosy-locus-specific RNA transcripts.

Position-effect variegation is a consequence of a mosaicism of gene expression within a tissue. This mosaicism probably results from disruption of the initiation of transcription caused by the lack of accessibility of the necessary transcription factors. Although such mosaicism is known predominantly from *Drosophila,* it is now commonly regarded as a well-established general phenomenon, though this opinion rests more on faith than on fact. In *Drosophila,* the event determining the expression of variegation occurs early in development (Spofford, 1976). From this point on, the clonal proliferation of cells with different developmental programs leads to differential patterns of expression of variegation. Position-effect variegation is also known to occur in the case of X/autosome translocations in the mouse, though here, of course, it involves facultatively heterochromatinized chromatin. Evidence summarized by Catanach (1974) suggests that here, too, variegation is largely determined early in development and again leads to the production of clones that differ with regard to the activity of the rearranged autosomal genes. In the mouse system, however, the effect of the proximity of X heterochromatin on the adjoining autosomal region tends to diminish with time and/or the number of cell generations that have elapsed since the inactivation event, with loci furthest from the breakpoint being reactivated first. This contrasts with the situation in *Drosophila,* where reactivation is not known to occur.

3.1.3. Nucleotypic effects

Interspecific comparisons between related species that differ in genome size have led to the suggestion that the amount of DNA in a genome influences cell size, cell division time, and hence developmental time. There is no doubt that, in a number of plant systems, good correlations do exist between cell cycle time and nuclear DNA content. What is not clear, however, is the specific contribution that differences in heterochromatin content make to such nucleotypic effects. In one of the few studies where euchromatin and heterochromatin were separately analyzed for their effects on cell cycle duration, Nagl (1974) found that nuclear DNA content in the three plant genera *Anacyclus, Anthemis,* and *Artemisia* may increase through the addition of heterochromatin without a lengthening of the cell cycle. Indeed, the cell cycle time most commonly shortens (Table 1.19). Additionally, experimentally constructed strains of *Drosophila melanogaster* carrying substantial deletions of heterochromatin show no differences in developmental time (Miklos, 1982).

The influence of heterochromatin on nucleotype is also likely to be confounded by the fact that, in many plants and invertebrate animals, a majority of somatic cells are not diploid but endoploid and commonly exhibit a differential underreplication of heterochromatin. It is true that no such complication exists in vertebrate animals where a majority of somatic cells remain diploid and no differential replication of heterochromatin is known to occur. Even here, however, it is not at all clear what contribution quantitative differences in heterochromatin content in different species do make to nucleotypic effects. Certainly, within a species, cells with identical DNA and heterochromatin content may vary strikingly in both cell size and cell cycle time.

3.1.4. Effects on centromere behavior

In a series of papers, Vig and his associates (Fazal Farook and Vig, 1980; Vig, 1982; Figueroa and Vig, 1983; Vig and Zinkowski, 1985) have claimed that:

1. The chromosomes in a given genome separate at their centromeres in a genetically determined and nonrandom sequence.
2. Centromeric heterochromatin plays a key role in controlling this sequence. Specifically, the position of a given centromere in the sequence of separation is determined by an interaction between quantitative and qualitative aspects of centromeric heterochromatin.

Table 1.19. Relationship between heterochromatin content and mitotic cycle time in perennial (P) and annual (A) species in the Anthemidea, a group of composite plants

Species	Type	2C DNA (pg)	Nuclear vol. (μm^3)	Heterochromatin content (%)	Cell cycle (hours $\pm$ *SE*)
Anacyclus depressus	P	12.42	141.4	0.20 (1.62)	15.9 ± 2.36
Anacyclus clavatus	A	10.48	132.6	0.90 (8.55)	11.0 ± 0.98
Anacyclus radiatus	A	16.92	134.6	1.57 (6.84)	13.6 ± 1.40
Anthemis tinctoria	P	7.46	116.5	0.18 (2.43)	12.3 ± 1.25
Anthemis cota	A	15.78	111.5	1.43 (9.09)	6.5 ± 0.54
Anthemis austriaca	A	9.63	89.2	3.29 (34.11)	7.0 ± 0.61
Artemisia absinthium	P	7.28	75.3	3.31 (45.45)	9.5 ± 0.56
Artemisia annua	A	4.05	72.0	0.57 (14.13)	7.7 ± 0.35

Note: All species have $2n = 18$.
Source: Nagl, 1974.

These claims are based almost exclusively on the analysis of colcemid-treated cells. Despite this limitation, it has been concluded that this non-random separation of centromeres provides evidence for a function of the repetitive DNA centered around the centromere and also offers an explanation for nondisjunction leading to aneuploidy. Even given the reality of this effect in colcemid-treated cells (and the evidence offered by Vig is far from convincing), the question remains whether these observations have any significance for normal cell behavior. As long ago as 1962, Lima de Faria and Bose showed that colchicine treatment reversed the division sequence of mitotic chromosomes. In conventional anaphase behavior, the regions proximal to the centromeres separate first and the telomeres separate last. The actual order of chromosome separation during anaphase is then a function largely of chromosome size. Small chromosomes invariably separate before larger chromosomes since telomeres that are close to a centromere separate earlier than those farther away from a centromere. Indeed in the case of larger chromosomes, the telomeres frequently remain associated until mid to late anaphase. When colchicine is applied, this sequence is reversed. Now the telomeres separate first and the proximal regions last. Behavior at c-mitotic anaphase thus offers no rational basis for behavior at normal mitotic anaphase.

Given that centromeres separate first, not last, in a conventional division sequence, and that the order of separation of the chromosomes is a reflection of chromosome length, the key question is whether the amount of centric heterochromatin has any influence on this pattern of behavior. The chromosome complements of grasshoppers include a wide range of chromosome sizes and may also include substantial differences in the amount of centric heterochromatin. In *Frogattina australis,* for example, there are small blocks of centric heterochromatin on the eight largest members of the haploid set but large blocks of CMA-positive heterochromatin on the four smallest members of the complement (Figure 1.17). Yet these small chromosomes still separate first at both mitotic and meiotic anaphase.

A genuine effect of heterochromatin on centromere behavior has been demonstrated in *Sciara.* Here, sex is not determined at the time of fertilization. All eggs contain 3A + 1X and all sperm 3A + two genetically identical X chromosomes. During early embryogeny, one or both of the paternal X chromosomes are eliminated, at the seventh or eighth cleavage mitosis (see Figure 1.12). The chromosomes that are eliminated begin anaphase but fail to separate completely and are left lying in the middle of the spindle. This selective elimination depends upon the presence of a short heterochromatic region (H2) located in the proximal heterochromatin to the right of the centromere in the X chromosome (Figure 1.18). The H2 block also contains 50% of the rRNA cistrons. If this controlling

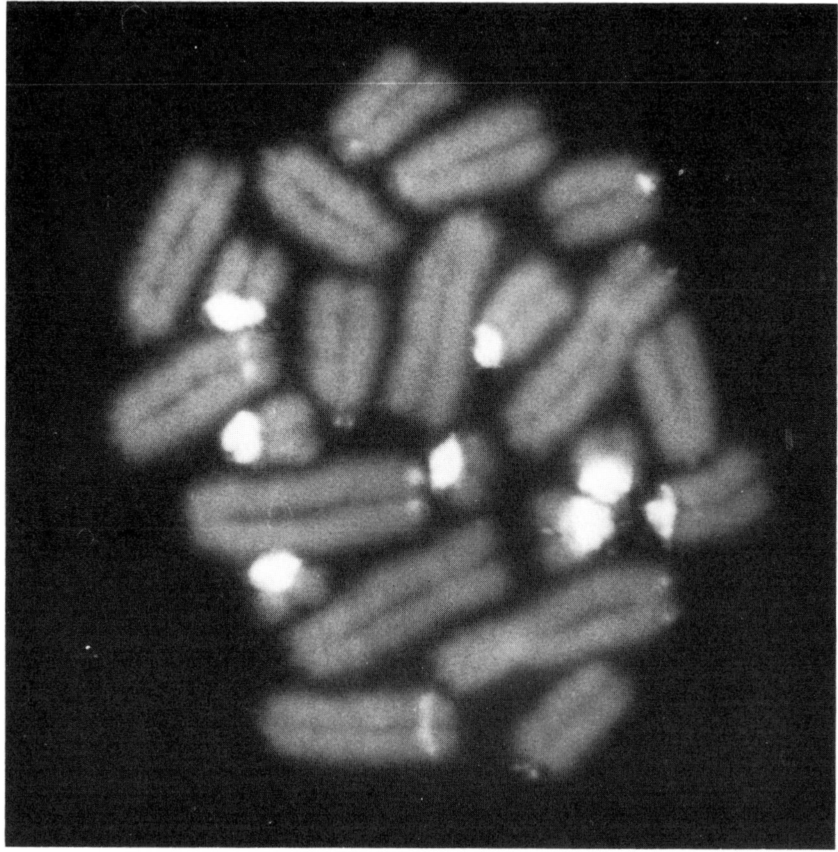

Figure 1.17. A male mitotic nucleus of the grasshopper *Froggattina australis* ($2n = 23$, XO) following treatment with the fluorescent dye chromomycin (CMA). Note the large, brightly fluorescent blocks of heterochromatin proximal to the centromeric regions of the four smallest chromosome pairs (8 to 11) and the lack of centric heterochromatin in the remaining members of the complement.

element is experimentally translocated either to the metacentric autosome IV or to the shorter rod II, it is able to induce their respective centromeres to undergo lagging and elimination (Crouse, 1979). However, it has no effect on the longer rod III whose proximal end carries its own substantial heterochromatic segment.

3.2. Germ-line effects

Because constitutive heterochromatin is invariably retained within the germ line, a number of workers have argued that one might expect to find

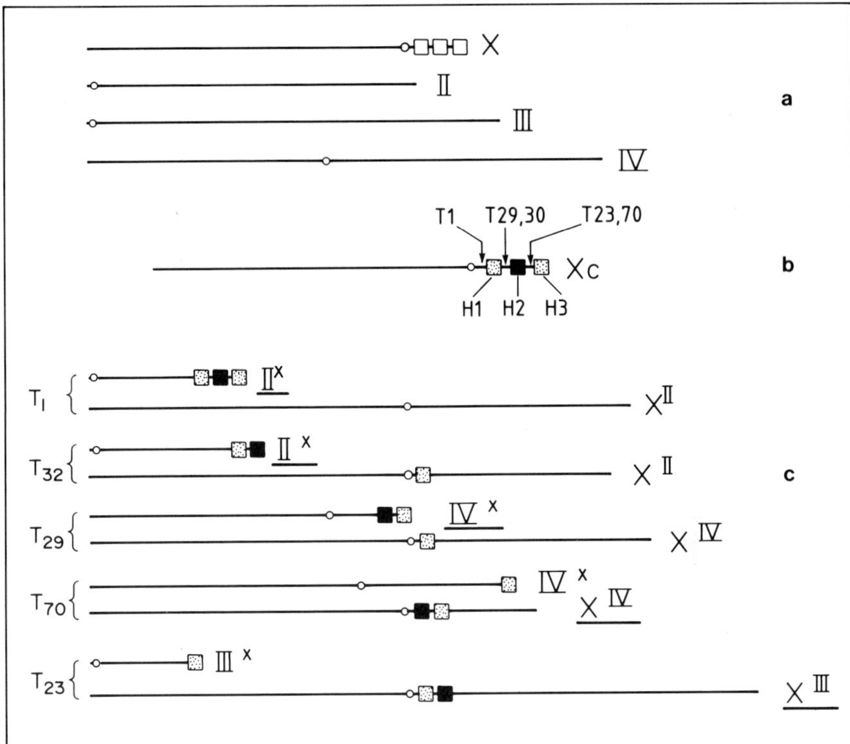

Figure 1.18. Control of centromere nondisjunction by the H2 block of heterochromatin, one of three such blocks (H1 to H3) present on the short arm of the X chromosome of *Sciara coprophila.* The breakpoints of five X autosome translocations (T1,X/II; T29,X/IV; T32,X/II; T70,X/IV; and T23,X/III) relative to these three blocks (b), lead to alterations (c) in the conventional mitotic set (a). The chromosomes that undergo nondisjunction in the standard and translocated stocks are underscored in each case. (After Crouse, 1979.)

heterochromatin effects specifically at meiosis. Indeed, a number of claims and counterclaims have been made in relation to the influence of heterochromatin on all three of the principal meiotic processes – that is, pairing, recombination, and segregation.

3.2.1. Effects on meiotic pairing

Two kinds of pairing, homologous and nonhomologous, are known at meiosis, and heterochromatin has been held to influence both of them.

3.2.1.1. Homologous pairing. The evidence to support the notion that prior association of homologous heterochromatic segments plays a role

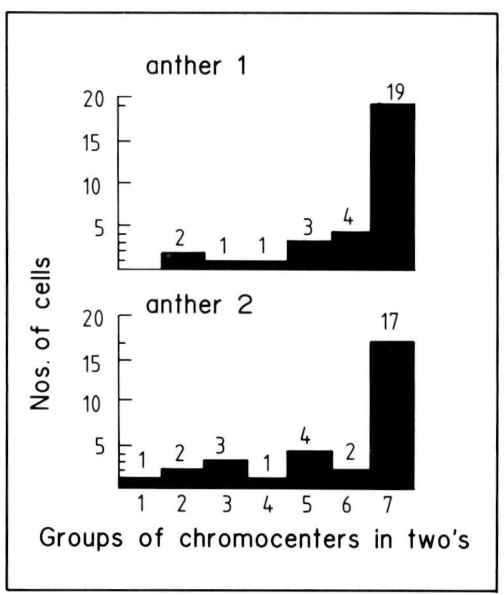

Figure 1.19. Frequencies of premeiotic nuclei with varying numbers of paired homologous chromocenters in 30 cells from each of two anthers of the plant *Salvia nemorosa.* (After Chauhan and Abel, 1968.)

in meiotic pairing is of variable quality. The argument is not a new one. As early as 1968, Chauhan and Abel presented observations which they believed lent credence to this conclusion. Their observations came from two plants, *Impatiens balsamina* and *Salvia nemorosa,* both with $2n$ = 14. In *Impatiens,* the interphase nuclei preceding leptotene showed from one to seven groups of chromocenters associated in two's. At early leptotene itself, however, only some of the deeply stained heterochromatic sections seemed to lie in pairs. In *Salvia,* about 50% of the premeiotic cells had chromocenters arranged in seven groups of two; here, because of size differences between pairs, which does not obtain in *Impatiens,* it was possible to confirm that homologous chromocenters were indeed paired (Figure 1.19). Although this evidence certainly supports the variable retention into leptotene of premeiotic associations between homologous blocks of heterochromatin, it is clear that these associations could not serve as a regular basis for subsequent meiotic pairing.

Macgregor and Kezer (1971) also stressed the possible role of centromeric heterochromatin in promoting the synapsis of centric regions at first meiotic prophase. The validity of this conclusion is, however, equally debatable since such pairing does not occur in many species where most commonly it is initiated at the telomeres. In mice, there is especially convincing evidence to indicate that homologous pairing is not determined by prior association of heterochromatic regions. C-banded preparations of *Mus dunni,* for example, demonstrate that the centric C-blocks are unpaired at leptotene. Moreover, as is commonly the case, synapsis starts from the telomeric ends and proceeds towards the centromeres (Pathak and Hsu, 1976). This applies also to the grasshopper *Stauroderus scalaris,* where there are especially pronounced blocks of centric heterochromatin on all the autosomes, and these regions are again the last to pair (John, 1976).

If heterochromatin is indeed crucial for the pairing of homologous chromosomes, then deletions of that heterochromatin might be expected to lead to serious pairing problems. It is possible to test this prediction by analyzing meiotic behavior in individuals of *Drosophila melanogaster* where the amount of heterochromatin can be modified experimentally. Yamamoto (1977) carried out such an analysis and showed that neither substantial deletions from, nor rearrangements of, specific blocks of heterochromatin caused any disruption of pairing in any of the autosomes or between the X chromosomes at female meiosis. In both of these situations, pairing was found to depend solely on euchromatic homology. It is true that at the onset of meiosis in female *Drosophila,* a compound chromocenter is present which involves the association of all the heterochromatic segments. This, however, plays no part in determining the subsequent behavior of homologues, though it has been considered to be an important factor in regulating the disjunctional properties of nonhomologues in the absence of homologous pairing (Novitski and Puro, 1978).

3.2.1.2. Nonhomologous pairing. Heterochromatic regions commonly associate nonspecifically at first meiotic prophase. This behavior has been particularly well documented in the cricket *Gryllus argentinus* (Drets and Stoll, 1974). Here, terminal C-bands of variable size are present in both arms of most chromosomes, together with smaller paracentromeric segments. All of these are heteropycnotic at first prophase of male meiosis, and many nonhomologous associations, involving two or more bivalents, are present at pachytene. Some such associations are still present at diplotene but now they are much fewer in number, and by first metaphase none are present. An equivalent behavior is known in many other species. Even so, it is worth noting that not all heterochromatic regions associate in all species; or, if such associations do develop, they are usually less frequent.

Apart from such transient nonhomologous meiotic association, which clearly is reminiscent of ectopic pairing, the pairing of heterochromatic regions may, on occasion, provide a basis for the segregation of homologues in the absence of chiasmata (Figure 1.20) as well as leading to the

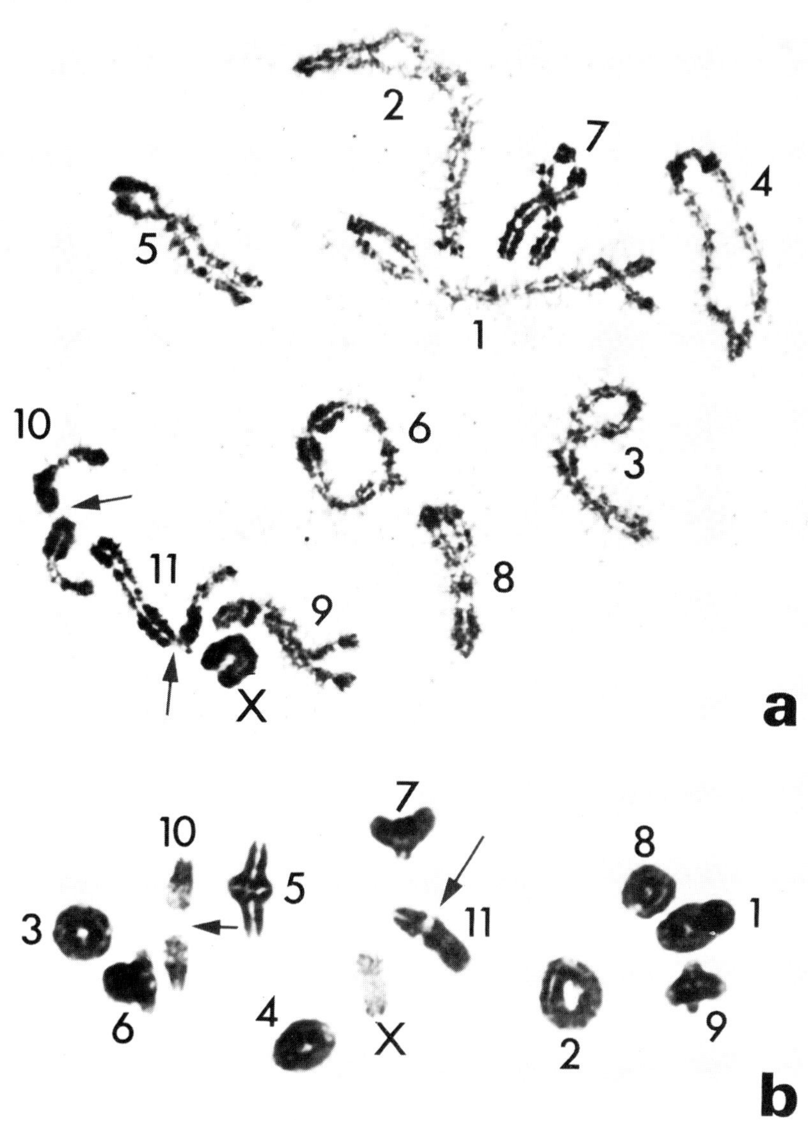

Figure 1.20. Male meiotic behavior at early diplotene (a) and first metaphase (b) of the two smallest autosome pairs (10 and 11) in the northern chromosome race of the grasshopper *Cryptobothrus chrysophorus* ($2n = 22 + XO$) in the presence of supernumerary heterochromatic segments. In the cells illustrated pair 10 is homozygous for distal heterochromatic segments whereas pair 11 is simultaneously heterozygous for a distal segment and homozygous for a proximal segment. Although partner homologues carrying these supernumerary segments do not pair

production of pseudomultiples between chiasmate bivalents (John and King, 1985a).

3.2.2. *Effects on meiotic recombination*

A number of different studies point convincingly to effects of heterochromatin on meiotic recombination. These effects are of three different kinds. First, chiasmata do not form between homologous segments of autosomal heterochromatin at meiosis (Table 1.20). For example, in the hedgehog *Erinaceus europaeus,* three pairs of autosomes carry large distal blocks of heterochromatin. At male meiosis, there is a complete pairing of all homologues including those with heterochromatic segments. The lack of chiasmata in these heterochromatic segments, however, leads to their early separation at diplotene (Natarajan and Gropp, 1971). Likewise, in *Ammospermophilus harrisi* and *A. interpres,* up to three bivalents per nucleus have obvious asynaptic regions which stain more darkly than the adjacent chromatin and are C-band positive (Mascarello, 1980). Some sex chromosomes are, however, exceptional in this respect. Thus, in several rodents, including European and Chinese hamsters (Vistorin, Gamperl, and Rosenkranz, 1977) and the Sitka deer mouse (Hale and Greenbaum, 1986), a single interstitial chiasma regularly forms between those regions of the X and Y that give positive C-banding. This implies that there may be some categories of heterochromatin that are able to undergo recombination and others that cannot. It is worth noting, however, that the sex chromosomes of the Chinese hamster are known to be deficient in repetitive DNA sequences (Arrighi, Hsu, Pathak, and Sawada, 1974).

Second, in cases of heterochromatic polymorphisms there is commonly a redistribution of chiasmata in bivalents carrying supernumerary heterochromatic segments (John and King, 1985b; Torre, Lopez Fernandez, Nichols, and Gosálvez, 1986) such that chiasmata either do not form, or else form less frequently, in the euchromatic regions adjacent to the heterochromatic segments, a behavior reminiscent of somatic position effects.

Third, statistically significant effects of heterochromatic supernumerary segments on mean cell chiasma frequency are now well known. The

Caption to Figure 1.20 (*cont.*)
in a conventional manner, the proximity of their heterochromatic blocks (distal in 10, proximal in 11, arrowheads) provides a mechanism (a) which ensures their subsequent segregation in the absence of chiasmata at first metaphase (b), where the two homologues of 10 lie apart but cooriented while the two homologues of 11 are associated proximally.

Table 1.20. Species demonstrating absence of autosomal meiotic recombination in defined heterochromatic regions

Species	Location of heterochromatin		
	Proximal	Interstitial	Distal
Plants			
1. *Plantago ovata*		+	
2. *Secale cereale*			+
3. *Anemone blanda*			+
Animals			
4. *Stauroderus scalaris*	+		
5. *Cryptobothrus chrysophorus* (northern race)			+
6. *Ammospermophilus harrisi*		+	
7. *Erinaceus europaeus*			+
8. *Drosophila melanogaster*	+		
9. *Atractomorpha similis*			+

Sources: (1) Hyde, 1953; (2) G. H. Jones, 1978; (3) Marks, 1974; (4) John, 1976; (5) John and King, 1977; (6) Marscarello, 1980; (7) Gropp, Citoler, and Geisler, 1969; (8) Baker, 1958; (9) John, unpublished.

most striking of these comes from the work on *Phryne cincta.* As has already been mentioned, in this species the X chromosome may be either euchromatic or β heterochromatic in appearance, depending on the conditions of rearing. The X also occurs in one of two structural forms containing, respectively, either one (X_s) or three (X_l) blocks of α heterochromatin. The X_s has a small α band near its proximal end (α_1), while the X_l has two, much larger, additional bands (α_2 and α_3) near its distal end (Figure 1.21). Consequently, at mitosis, X_l chromosomes are twice as long as X_s chromosomes. In polytene nuclei, however, little length difference is apparent between X_l and X_s despite the fact that they differ not only by their content of α heterochromatin but also by a series of complex inversions that lead to the formation of loops in the proximal and mid regions of X_xX_l polytene combinations. The point of significance is that the presence of the additional α blocks on one or both X chromosomes leads to a striking increase in the frequency of autosomal crossing over at female meiosis.

Crossing over in this case can be conveniently gauged cytologically by utilizing the pairs of paracentric inversions, present as polymorphisms in both of the large metacentric autosomes 2 and 3 (Figure 1.22), and by crossing females heterozygous for the coupled inversions to standard

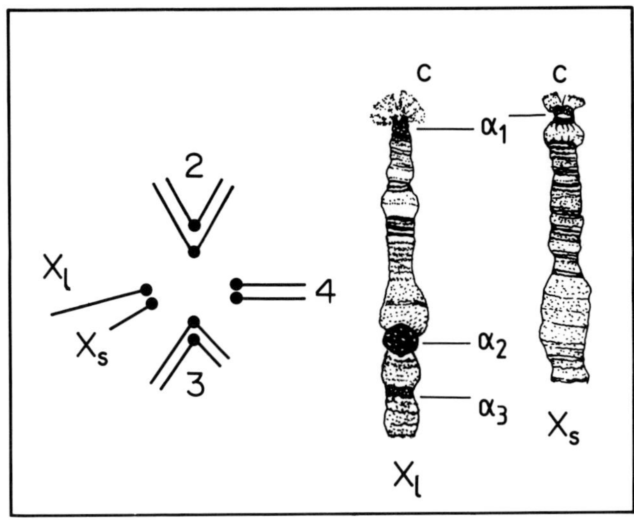

Figure 1.21. Organization of the short (X_s) and long (X_l) variants of the X chromosome of the gnat *Phryne cincta* ($2n = 8$) in mitotic (left) and polytene (right) nuclei. In the latter, the two homologues are distinguished by the presence of two additional blocks (α_2 and α_3) of α heterochromatin in the long variant. Note C = centromere. (After Wolf and Sokoloff, 1973.)

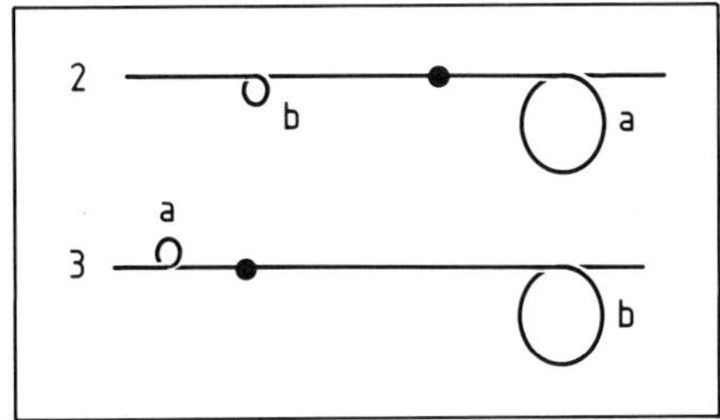

Figure 1.22. The location of the two inversions (a,b) present in each of autosomes 2 and 3 of *Phryne cincta* relative to the locations of the centromeres (shown solid).

males. Recombinant types, indicative of crossing over in the intercept between the two inversions, have only one of the maternal inversions present and can be easily identified by salivary gland analysis of progeny individuals. The results (Table 1.21) provide a striking demonstration of the effects of added α blocks.

In *Drosophila melanogaster,* both the amount of heterochromatin and its arrangement within the genome can be readily manipulated experimentally. By analyzing the effects of systematic deletions of the proximal

Table 1.21. Percent crossing over (% C.O.) between heterozygous (het.) linked inversions, a and b, in autosomes 2 and 3 of *Phryne cincta* in the presence (X_l) and absence (X_s) of extra α heterochromatin in the X chromosome

Constitution of F_1 female			% C.O. in a/b intercept	
X	2	3	2	3
1. X_sX_s	het. 2a, 2b	standard	1.0	—
X_sX_l	het. 2a, 2b	standard	9.9 ± 1.6	—
2: X_sX_s	standard	het. 3a, 3b	—	0.0
X_sX_l	standard	het. 3a, 3b	—	3.6
3. X_sX_s	het. 2a, 2b	het. 3a, 3b	2.5 ± 0.6	0.0
X_sX_l	het. 2a, 2b	het. 3a, 3b	42.0 ± 1.6	17.2 ± 1.9
4. X_sX_s	het. 2a, 2b	het. 3a, 3b	2.1	0.1
X_sX_l	het. 2a, 2b	het. 3a, 3b	41.9 ± 1.9	13.3 ± 1.3
5. X_sX_s	het. 2a, 2b	het. 3a, 3b	39.3 ± 2.0	13.01 ± 1.4
X_lX_l	het. 2a, 2b	het. 3a, 3b	38.5 ± 2.0	11.0 ± 1.3

Sources: (1–3) Wolf, 1973; (4, 5) Wolf and Wolf, 1969.

heterochromatin of the X chromosome, Yamamoto (1979) found that, not only was there a significant decrease in the intrachromosomal recombination in the proximal euchromatin of the X itself, but, additionally, there were accompanying increases in the frequency of crossing over in chromosome 3 and especially in the regions nearest to the centromere where crossing over is normally negligible (Figure 1.23). Specific portions of the Y heterochromatin and the basal X heterochromatin, when present as homologues for reversed acrocentric compound X chromosomes, also markedly and coordinately increase the frequency of crossing over between the elements of the compound chromosome (Sandler and O'Tousa, 1979).

3.2.3. Effects on chromosome segregation

In *Zea mays* ($2n = 20$), each of the ten chromosomes in the haploid set may be polymorphic for the presence of supernumerary heterochromatic segments which, classically, have been referred to as *knobs*. When chromosome 10, the smallest member of the complement, is heterozygous for the presence of one such knob at the distal end of its long arm (K10k10), it confers upon that arm the capacity to organize accessory spindle fibers in the vicinity of the knob. The result is the production of a neocentromere which functions in addition to the principal centromere at both first

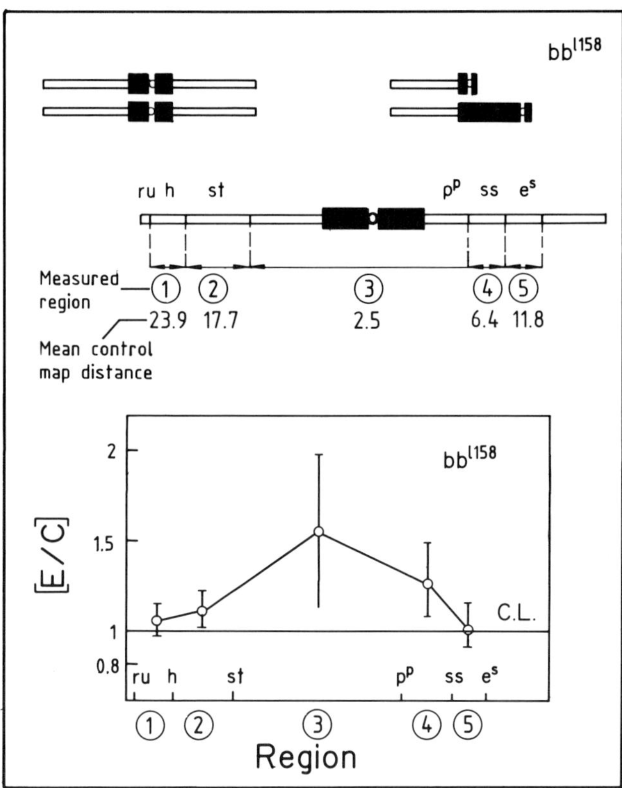

Figure 1.23. The effect of a heterozygous deletion (bb^{l158}) in the heterochromatin of the X chromosome of *Drosophila melanogaster* (upper) on recombination (lower) involving five regions of chromosome 3 (center). In the lower figure, [*E*/*C*] represents the ratio of map distance in the experimental relative to that of the control (C.L.), whereas the vertical lines denote 95% confidence limits. (After Yamamoto, 1979.)

and second anaphase of female meiosis. Chiasma formation between the centromere and the knob in such a heterozygote inevitably means that the two knobbed chromatids in a bivalent are drawn closer to the two first division spindle poles than are the other chromatids with the K10k10 bivalent. They retain this proximity throughout the second meiotic division; consequently the knobbed chromatids are incorporated preferentially into the two terminal products of the linear tetrad produced by meiosis (Figure 1.24). Since it is from one of these products that the functional egg nucleus is subsequently derived, and since a single chiasma commonly forms between the centromere and a knob, knobbed chro-

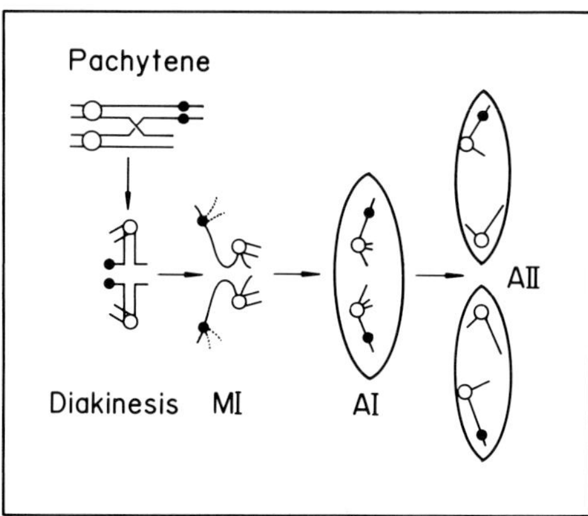

Figure 1.24. Preferential segregation of knobbed chromatids from a bivalent heterozygous for a heterochromatic knob at female meiosis in *Zea mays.*

matids are in fact transmitted by heterozygous females with a frequency of about 70%. Not only does the heterozygous knobbed 10 bring about a directed segregation of K10 itself, but additionally its presence causes all other chromosomes carrying heterozygous knobs to behave in a comparable fashion (Table 1.22), though they do not do so in the absence of K10.

A variety of effects on pairing, recombination, and segregation of standard (A) chromosomes are also known in cases involving supernumerary (B) chromosomes that have been described as heterochromatic simply because they display heteropycnosis at meiosis. No reference is made to these because, as pointed out by Jones and Rees (1982), the introduction of C-banding to define constitutive heterochromatin has made it clear that none of these B's is totally heterochromatic. Consequently, one cannot unequivocally ascribe these effects to heterochromatin per se.

3.3. Functions of heterochromatin

While heterochromatin evidently exerts effects on a number of cellular events both in the soma and in the germ line, it is important, as Maynard Smith (1982) has emphasized, to recognize that effects are not functions. Less cautious authors have, however, confused effects with functions. Brutlag (1980), for example, concludes that "heterochromatin shows many well defined *functions* in such diverse processes as chromosome pairing and segregation, position effect variegation, chromosome rearrangements, speciation and recombination" (italics mine). While some of these effects may indeed have functional connotations, it is clear that most do not.

Table 1.22. Preferential segregation of knobbed (K) chromatids in K/k heterozygotes at megasporogenesis of *Zea mays* in the presence of K10

Chromosome no.	Knobbed arm	Total no.		% K recovery
		Chrms. observed	K recovered	
1	S	136	79	58.1[a]
	L	108	68	63.0
2	S	59	35	59.3[a]
	L	93	55	59.1[a]
3	S	113	72	63.7
	L	169	116	68.6
4	L	106	87	82.1
5	L	77	56	72.7
6	L	212	138	65.1
8	L	90	57	63.3
9	L	207	148	71.5
10	L	142	106	74.6

[a]With the exception of the three indicated values, the recovery percentages are all significantly different from randomness.
Source: Longley, 1945.

Even worse, in a number of instances, functions have been ascribed to heterochromatin in the absence of meaningful demonstrable effects. The most blatant of such cases concerns arguments relating to the significance of so-called centric heterochromatin. Because of the common occurrence of constitutive heterochromatin proximal to the centromere, one of the earliest speculations was that such heterochromatin was concerned in some way with centric activity. This speculation appears to have had its origins in the observation that measurable differences exist in the capacity of different experimentally constructed X chromosomes from *Drosophila melanogaster* to break dicentric bridges, originating from double crossovers within heterozygous inversions, at oogenesis (Lindsley and Novitski, 1958). Specifically, centromeres derived from a normal X are weak, in the sense that such bridges persist late into telophase without breaking, whereas specially constructed attached X chromosomes, carrying a Y^L in one arm, are strong and lead to early breakage of dicentric bridges. As "weak" as centromeres may be in breaking dicentric bridges, they are evidently capable of functioning perfectly normally in terms of conventional anaphase behavior. Nevertheless, when this observation was coupled with the near ubiquity of centric heterochromatin, it was concluded that centromere strength might be determined by the amount of centric heterochromatin and that this was an important factor in controlling the kinetic activity of the centromere under normal circumstances. Embellishing the argument still further, Walker (1971) actually suggested that "satellite DNa confers a direct advantage on the chromsome which carries it, because such a chromosome survives the mechanical processes of, primarily, meiosis better than a sister chromosome not so well endowed."

The work of Yamamoto and Miklos (1977), who used a variety of deletions of X heterochromatin in *D. melanogaster,* indicated, to the contrary, that the amount of heterochromatin associated with the centromere of the X chromosome has no demonstrable effect on its kinetic activity and that chromosome segregation is not altered in any way from that of nondeleted controls. Similar conclusions can be drawn from cases involving polymorphisms for the amount of centric heterochromatin, as well as from species having different amounts of centric heterochromatin in different members of the same chromosome set.

Finally, several workers, including Hilliker, Appels, and Schalet (1980) and Manuelidis (1982) have speculated that heterochromatin may play an important role in determining the three-dimensional structure of the interphase nucleus and so many affect gene expression both directly and indirectly. An ordered arrangement of chromosomes in interphase nuclei has also been proposed as a means by which genetic activity is coordinated within a genome. The existence of compound chromocenters, of ectopic pairing, and of hot spots for exchange have all contributed to the opinion that an ordered chromosome arrangement is partially attributable to nonspecific associations of heterochromatin. Comings (1980), for example, concluded that heterochromatin is condensed at the inner surface of the nuclear membrane and at the periphery of the nucleolus whereas genetically active chromatin occupies the most central position. Bennett (1984), on the other hand, has come to the opposite conclusion. In the Indian muntjac, *Muntiacus muntjak,* the X and Y_2 chromosomes contain the largest C-banded regions. Here, the X heterochromatin is located centrally in both somatic metaphase configurations and in the interphase nucleus, whereas the Y_2 is located most frequently at the periphery of the metaphase plate (Verma, Jacob, and Babu, 1986). Thus, the distribution of different heterochromatic segments may vary even within the same species.

The question that obviously needs to be addressed is whether heterochromatin occurs at specific sites because it has a selectable function at that site or whether its location is simply a consequence of the historical events that led to its production at that site.

To analyze the role of heterochromatin in maintaining the organization of the interphase nucleus, it is necessary, as Lifschytz and Hareven (1982) have emphasized, to distinguish between the role of centromeres and telomeres independently of the heterochromatin associated with

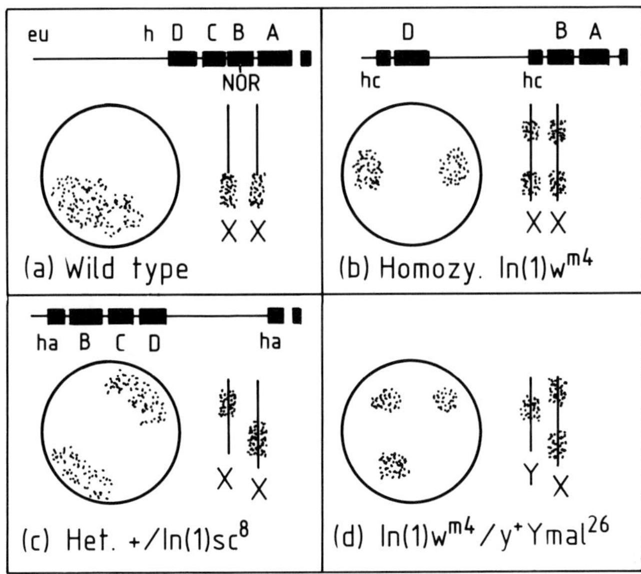

Figure 1.25. Positional localization in diploid interphase nuclei of specific heterochromatic regions of the X chromosome of *Drosophila melanogaster* marked by a labeled mr sequence probe specific for the proximal X heterochromatin. Whereas this probe recognizes and marks the nucleolus organizing region (NOR) of the X, it does not label the NOR of the Y. (After Lifschytz and Hareven, 1982.)

these chromosome organelles. This is now possible through the use of DNA probes that localize specifically to known heterochromatic regions, in conjunction with experimentally constructed rearrangements of these regions. Using this approach, Lifschytz and Hareven (1982) studied the effect of relocating heterochromatin on the ordered arrangement of interphase nuclei in the larval brain and the imaginal disk cells of *D. melanogaster,* neither of which normally contains chromocentral associations of X-, Y-, or autosomal-based heterochromatin.

XH10 is a heterochromatic clone composed of members of the is1 repeated gene family which serves as a specific marker for the NOR and the region spanning the NOR and the heterochromatin–euchromatin boundary of the X chromosome of *D. melanogaster* but is not found proximal to the NOR (Figure 1.25). With normal diploid female nuclei, the grains produced by this radiolabeled probe are concentrated at one site in close proximity to the nuclear membrane, giving either one, or sometimes two closely associated clusters of grains corresponding to the two X homologues. Inversion In(1)w^{m4} moves most of the heterochromatin from regions hC and hD to the distal euchromatic end of the X and so leads to the formation of two hybridization sites for XH10. One of these is equivalent to the NOR; the other, to the adjacent heterochromatin distal to the NOR. Hybridization now produces two separate, and most frequently opposite, clusters of silver grains in both female and male nuclei. Evidently the position which these segments of X heterochromatin occupy within the resting nucleus is determined simply by their topology in the mitotic chromosome and not by interaction with specific, prefixed sites on the nuclear membrane, as others have implied.

Inversion In(1)sc^8 moves both the NOR and the hChD segments to a distal site close to the euchromatic telomere of the X. In females heterozygous for this inversion, two opposite groups of grains are again most frequently found. One of these is equivalent to the normal X, the other to the inverted X, so demonstrating the complete independence of the same segment of heterochromatin when linked either to the centromere or to the euchromatic telomere.

The relation between homologous heterochromatic segments in heterologous chromosomes was studied in In(1)w^{m4}/y^+Ymal^{26} male nuclei. The y^+Ymal^{26} chromosome carries heterochromatic segments from the proximal region of the X as well as nucleoli of both X and Y origin. In nuclei from these males, hybridization with XH10 resulted in three groups of silver grains, two of which correspond to the inverted X and the third to the X heterochromatin which has been transferred to the Y. Thus, even when X-based heterochromatic segments are associated not with an X centromere or an X telomere but with a different chromosome, they nevertheless retain independence of distribution in the interphase nucleus.

These observations, drawn from segments that normally form part of the basal heterochromatin of the X chromosome, were confirmed and extended with the use of four other cloned probes. R8, for example, consists of clones with homology sites on both the Y and the major autosomes, whereas F3 shows homology to all heterochromatic regions. Yet neither of these multisite markers gave rise to chromocentral associations in diploid nuclei. Likewise, multisite sequence families on euchromatic arms, as defined by the mobile element cDm412, which is located at more than 30 sites, showed an even distribution over the entire nuclear area. Finally, the YDm12 clone exhibits a major homology site on the long arm of the Y together with several secondary sites within all the major heterochromatic blocks, but, additionally, is dispersed in many euchromatic sites, too. Here, labeling revealed a major cluster of grains equivalent to the Y^L site as well as a dispersed pattern reminiscent of cDm412.

From their observations, Lifschytz and Hareven (1982) concluded that, although under normal circumstances homologous segments at the base of the X do indeed align in close proximity, heterochromatin plays no role in this arrangement. Rather, the location of heterochromatic segments at interphase in these diploid nuclei is determined solely by their position on the mitotic chromosomes. Rearrangements of the X heterochromatin from a proximal to a distal position results in a completely different nuclear distribution, and, under these circumstances, there is no association between heterochromatic regions that are unquestionably homologous in sequence structure.

A polarized arrangement of chromosomes, so-called Rabl orientation, is, in fact, of common occurrence in intermitotic nuclei. It arises in telophase as a consequence of the orientation of all centromeres within a telophase group toward a specific spindle pole. During first meiotic prophase, this is commonly converted into a bouquet orientation in which the telomeres of all chromosomes are polarized toward a restricted area of the nuclear membrane. Neither form of polarization is, as some authors appear to believe, a consequence of the presence of heterochromatin in the regions involved, that is, the centromeres or the telomeres. Rabl orientation reflects the polarized nature of mitosis and exists regardless of whether or not heterochromatic regions occur at the centromeres. Bouquet orientation reflects a shift in the mutual position of telomeres and centromeres following the onset of meiosis (Virrki, 1974), and it, too, occurs regardless of the presence or absence of heterochromatin in the regions concerned.

4. Evolutionary perspectives

"For a Biologist the alternative to not thinking in evolutionary terms is not to think at all."

– Peter and Jean Medawar

It is commonplace to assume that valid insights into the evolutionary past may be revealed by the study of, or comparisons between, the DNA of modern forms. Attempting to assign evolutionary significance to observed similarities or differences in DNA sequences will, however, be valid only if the sequences concerned are either epigenetically or phylogenetically meaningful.

Peacock et al. (1978) have argued that "the observation of chromosome-specific distribution of sequences, of disjunct multisite occurrence of sequence species, and of sequence conservation argue strongly for functional roles of highly repeated sequence DNA in the molecular structure and evolution of heterochromatin." How valid then is this argument?

4.1. Heterochromatin and sex chromosome evolution

A majority of animals, and a minority of plants, with separate sexes have differentiated or heteromorphic chromosomes in one sex. Where the male carries a single heteromorphic pair they are referred to as X and Y chromosomes, whereas when the female carries a heteromorphic pair they are termed Z and W chromosomes.

It has long been a convention to assume that such cytologically differentiated sex chromosomes have arisen from a progenitor pair of homologues which initially were indistinguishable in mitotic morphology and showed complete pairing at meiosis. Lower vertebrates offer good confirmatory grounds for such an assumption since related species are known in fishes, amphibians, and reptiles both with and without sex chromosomes. It has also been generally agreed that the evolution of differentiated sex chromosomes was probably initiated by either partial or else complete prevention of recombination between the two homologues in question. This provides a basis for the differential accumulation of opposing sex determinants within the genome through the creation of genetically differential segments. This, in turn, has been assumed to lead to the loss of gene function of the loci within the differential segment, a process that has been referred to as *degeneration* and has been related to the heterochromatinization of the differential segment. The conventionally accepted hypothesis thus argues that, because of a lack of recombination between sex chromosomes which are initially identical in size and morphology, the Y or the W will, over time, become heterochromatic, lose most of its functions, and then decrease in size (Darlington, 1958).

4.1.1. Sex chromosome differentiation

In theory, at least three mechanisms are known which are capable of restricting recombination between homologous chromosomes: (1) genotypically determined chiasma localization; (2) structural rearrangements, and particularly pericentric inversions; and (3) differential heterochromatinization.

Our interest is specifically in the role of heterochromatinization in the process of sex chromosome differentiation. In some cases, this role is clear enough. Chironomids, culicids, and simulids do not, in general, exhibit conspicuously heteromorphic sex chromosomes, but a number of these organisms have been found to show subtle, but clear-cut, sex differences in polytene organization which involve heterochromatin differentiation. The following two examples illustrate the range of differences observed. Hägele (1985) demonstrated that, in a stock of *Chironomus thummi thummi,* region D_3e_1 of the F arm in chromosome 3 exhibits a

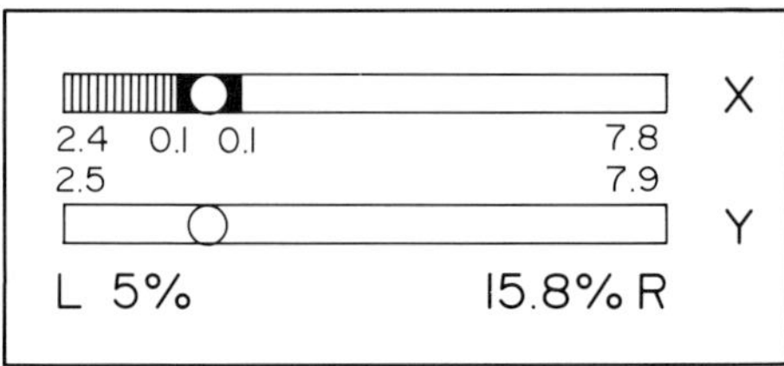

Figure 1.26. Organization of the X and Y chromosomes of the chironomid *Prodiamesa olivacea.* Although these chromosomes are indistinguishable at mitosis, they differ strikingly in their polytene behavior. The genetic Y is entirely α heterochromatic and fails to polytenize. The right arm of the X, too, is not represented in polytene nuclei and is also composed of heterochromatin. The left arm of the X, by contrast, is endoreplicated, though not banded, whereas the X centromere is surrounded by polytenizing material. Thus, of the total of 20.8% genomic DNA present in the X and Y chromosomes, only 2.6% is polytenized. (After Zacharias, 1981.)

male-specific difference in the size of a single polytene band which is selectively stained after Q banding. This difference is not detectable in mitotic chromosomes, and Hägele has concluded that this heterochromatic AT-rich band at D_3e_1 represents the site of the male sex determiner in this species.

In terms of relative DNA content, chromosome 3 of the chironomid *Prodiamesa olivacea* displays a rigorous sex linkage. In female mitosis and male meiosis, this chromosome comprises 21% of the total genome but is represented by only 6% in female and 3% in male polytene nuclei. In female polytene nuclei, the right arm of 3 is underreplicated and the minute element so formed, which does not join the chromocenter formed by the two other metacentrics, is composed only of the endoreduplicated 3L chromatids. In polynemic nurse cells, on the other hand, the right arm is visibly heterochromatic (Zacharias, 1979, 1981). In the male, one left arm of chromosome 3 is also excluded from polytenization; thus this represents a case of an incipient XX female, XY male sex heteromorphism (Figure 1.26). The X chromosome consists of a left arm, composed of 2.4% polytenized euchromatin and 0.1% polytenized β heterochromatin, and a right arm with another 0.1% β heterochromatin together with 7.8% nonpolytenizing heterochromatin. The genetic Y consists entirely of α heterochromatin and does not polytenize at all.

In other cases, however, the problem is confounded by the fact that two or more mechanisms may be involved in sex chromosome differentiation so that there is a need to distinguish between primary and secondary, reinforcing, mechanisms. This is well illustrated by newts of the genus *Triturus* (Amphibia: Urodela). In *T. alpestris alpestris,* for exam-

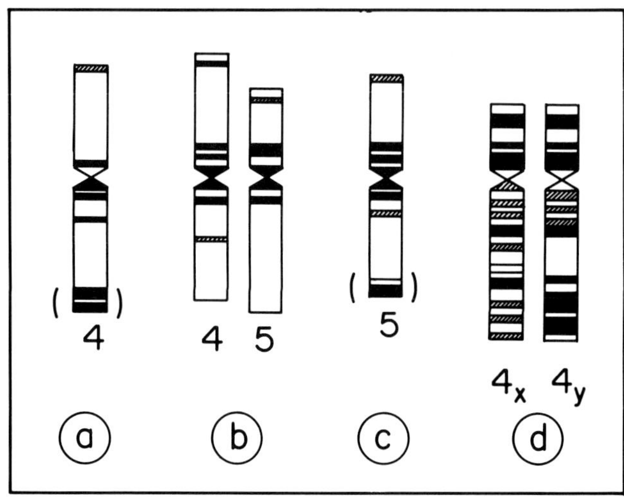

Figure 1.27. Homomorphic sex chromosomes in the Amphibia. In *Triturus alpestris alpestris* (a) and *Triturus vulgaris vulgaris* (c), the distal C-band-positive heterochromatic region which characterizes the male is enclosed by a bracket. In *Triturus helveticus helveticus* (b), note that the equivalent chromosomes (4 and 5) both lack cytologically distinguishable sex-specific regions. (After Schmid, Olert, and Klett, 1979.) A distinct form of differentiation is present in *Rana esculenta* (d), where homorphic sex chromosomes differ in respect of a late replicating region present in the long arm of the Y, but lacking in the X, as defined by the BrdU–Hoechst–Giemsa technique. These two chromosomes cannot be distinguished, however, by any other banding techniques. (After Schempp and Schmid, 1981.)

ple, males have one heteromorphic pair in which one homologue of chromosome 4 has a pair of distal C bands which are lacking in its partner and are missing from homologues in females (Schmid, Olert, and Klett, 1979). *Triturus vulgaris vulgaris* has an equivalent heteromorphism in chromosome 5 (Figure 1.27), whereas in *T. cristatus carnifex* and *T. marmoratus* both chromosomes 4 have C bands near the ends of their long arms, but in the male these are more numerous in one homologue than in the other (Sims, Macgregor, Pellatt, and Horner, 1984). At male meiosis, the heteromorphic long arms of bivalent 4 of *T. alpestris alpestris* are commonly (33% of all meiocytes) completely without chiasmata, whereas in *T. vulgaris vulgaris* chromosome 5 forms bivalents with chiasmata exclusively confined to their short arms (Schmid et al., 1979). However, *T. helveticus helveticus,* which does not have a heteromorphic sex pair, nevertheless also generally lacks chiasmata in the long arms of bivalent 5. Although Schmid et al. have concluded that C-band heteromorphism constitutes the first step in the differentiation of the sex chromosomes in these newts, it is clear that chromosome 5 homologues in the male of related species already show strong chiasma localization to the short arm even in the absence of any heterochromatic differentiation. It would be more logical in this case to conclude that the C-band differentiation has accumulated secondarily in regions where chiasma formation is already restricted on genotypic grounds.

A comparable differentiation of sex chromosomes has been reported in the largest pair in the cyprinodont fish *Poecilia sphenops* var. *melanistica.* Here females exhibit a W chromosome with a large segment of terminal heterochromatin that is missing from the otherwise indistinguishable Z (Haaf and Schmid, 1984). This segment, which is evident both at mitosis and at interphase, shows bright fluorescence with Q, DAPI, and H33258. What is not clear in this case is whether C-band differentiation was a primary or a secondary event, although it is known that no equivalent differentiation exists in other poeciliids.

An equivalent problem exists in the lizard genus *Cnemidophorus,* where only one species, *C. tigris,* has differentiated sex chromosomes. Here the X and Y differ in two respects, namely in centromere location and C-band characteristics (Bull 1978). The question that again needs to be resolved is whether the inversion of the centromere or the C-band heteromorphism came first, or whether the latter was coincidental with the former.

A distinct form of sex chromosome differentiation is present in anuran amphibians, in which Schempp and Schmid (1981) have used a modified BrdU–Hoeschst–Giemsa technique to identify somatic chromosome replication patterns. With this technique, the morphologically homomorphic pair 4 of *Rana esculenta* could be identified as a sex-specific system of the XX, XY type since males exhibited a distinctly asynchronous replicating chromosome pair whereas in females both homologues of pair 4 replicated synchronously (Figure 1.27d). This raises the interesting possibility that differential replication may represent a forerunner of heterochromatinization. The view that asynchronous replication may have been the cause, rather than the consequence, of heterochromatinization has also received support from Raman and Nanda (1986).

Within the Amphibia, some 25 species are now known to have heteromorphic sex chromosomes; 20 of these belong to the Urodela, whereas the remaining 5 belong to the Anura. Most of these show only minimal, often cryptic, differentiation. Highly differentiated sex chromosome pairs have, however, been described in three cases. In all five species of the neotenic salamander genus *Necturus,* the X and Y differ greatly in shape and size and in the amount of C-banding (Sessions and Wiley, 1985). The X is biarmed, whereas the Y is small, acrocentric, and almost entirely C-banded (Figure 1.28). Even so, at interphase it is dispersed and does not

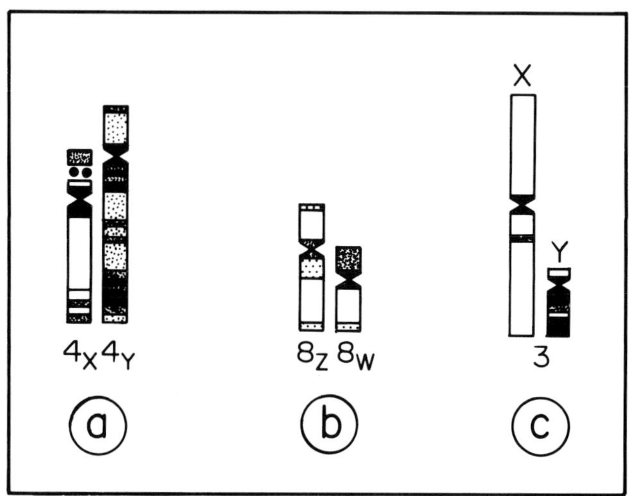

Figure 1.28. Markedly heteromorphic sex chromosomes in three amphibians as defined by C-banded mitotic preparations of (a) XY *Gastrotheca riobambae;* (b) ZW *Pyxicephalus adspersus;* and (c) XY *Necturus maculosus.* [(a) After Schmid, Haaf, Gerle, and Sims, 1983; (b) after Schmid, 1980b; (c) after Sessions, 1980.]

form a chromocenter. Well-differentiated sex chromosomes occur also in some plethodontid salamanders (Sessions, 1980), but these have a similar distribution of solely centric C-bands so that here it seems unlikely that heterochromatinization has played a primary role in sex chromosome evolution.

A second case of a highly differentiated sex chromosome pair has been uncovered in the marsupial frog, *Gastrotheca riobambae,* (Anura, Hylidae) by Schmid, Haaf, Geile, and Sims (1983). Here the Y is larger than the X, and, whereas the X has little heterochromatin, the Y is almost completely heterochromatic. The heterochromatin of the Y has a very heterogeneous composition and includes at least three cytochemically distinct types as defined by fluorochrome staining. Finally, the South African bullfrog *Pyxicephalus adspersus* has a ZW system in which the W is considerably smaller than the Z and about a half of it consists of constitutive heterochromatin. In a second species, *P. delandii,* the same chromosome pair, number 8, which is markedly heteromorphic in *P. adspersus* shows only slight heteromorphism between the two homologues. One is consistently a little more metacentric than the other, and there is more centric heterochromatin localized in the long arm of the more submetacentric than in its homologue (Schmid, 1980). In these three cases, again, we are faced with the problem of having to decide which of the differences between the heteromorphic chromosomes in question constituted the primary source of differentiation.

Despite problematical cases of the kind so far encountered, K. W. Jones and his co-workers, in a series of papers spanning the period 1976, to 1984, have consistently argued that heterochromatinization represents the first step in sex chromosome differentiation. This argument is based on an analysis of the sex chromosomes of snakes which, unquestionably, provides one of the most acceptable phylogenetic series in which to examine sex chromosome evolution. By comparing the skeletal characteristics of extant snakes with those of the oldest known fossils, herpetologists are agreed that the present-day pythons and boas (family Boidae) retain presumed "primitive" characters whereas the families Elapidae (cobras and mambas) and Viviperidae (vipers and pit snakes) are more specialized and hence derived. The Colubridae, which are mostly nonvenomous, occupy a morphologically intermediate position. Sex chromosome differentiation, which in all cases involves female heterogamety, parallels this morphological series. Thus, in boids only one case is known involving sex chromosome heteromorphism; colubrids have moderate to extensive heteromorphism whereas elapids and viviperids have highly differentiated Z and W chromosomes. Additionally, all snakes share a similar karyotype in which the sex chromosomes invariably correspond to the fourth pair of chromosomes in order of decreasing size so that all sex chromosome systems within snakes appear to be homologous.

In all snakes where the W is differentiated from the Z, either by allocycly in DNA replication or in morphology, the W forms a condensed body at interphase, though the frequency with which it does so varies from tissue to tissue (Ray Chaudhuri, Singh, and Sharma, 1971) and is most common in brain cells. In boids, where with but one exception no differentiated sex chromosomes exist, there is, as expected, usually no condensed sex chromatin body at interphase. However, in *Naja naja naja* there is a sex chromatin body despite the fact that there are no heteromorphic sex chromosomes and here one of the pair 4 homologues shows an asynchronous DNA replication pattern in females, a situation reminiscent of the one encountered earlier in the anuran amphibian *Rana esculenta.* Likewise in the boid *Eryx johni,* Ray Chaudhuri and Singh (1972) reported autoradiographic data which suggested that here, too, asynchrony in DNA replication serves to define presumptive sex chromosomes in the absence of any visible heterochromatinization.

Added to this, in the only known boid with differentiated sex chromosomes, namely *Acranthopis dumereli,* the sex pair is identical in length but differs in centromere position. G-banding indicates that the change in centromere location has resulted from a pericentric inversion (Mengden and Stock, 1980). Consequently, we can be confident that the primary event involved in the production of sex chromosome differentiation in this case cannot be heterochromatinization. An equivalent situation is known in at least two other reptiles, namely the lizard *Chemidophorus tigris* (Bull, 1978) and the gekko *Phylodactylus mamoratus* (King and Rofe, 1976).

In two major papers, Singh, Purdom, and Jones (1976, 1980a) claimed that in snakes, and by extension other vertebrates as well, heterochromatinization was the critical step leading to the initial differentiation of the sex chromosome pair, and that where other morphological differences existed between the sex chromosomes these would have arisen as secondary consequences of such heterochromatinization. This claim was based on the following facts.

In the colubrid *Ptyas mucosus,* where the Z and W are indistinguishable in length and in centromere location, and show neither asynchronous replication nor a sex chromatin body at interphase, the Z has only a small centric C-band whereas the whole of the W is C-band positive. Additionally, in all cases where the W forms a condensed sex chromatin body at interphase, it has been shown by in situ hybridization that the W is loaded with an interspersed repetitive DNA which can be recovered as a sex-limited satellite fraction from female snakes. The first such fraction to be isolated came from the banded krait, *Bungarus fasciatus,* and so was designated as Bkm, banded krait minor. This satellite was shown by in situ hybridization to be specifically located over the entire W of *B. fasciatus,* but no hybridization was observed in boids like *Python reticularis* and *Xenopeltus unicolor,* both of which lack differentiated sex chromosomes and do not form a sex chromatin body in interphase nuclei.

Since the contribution of the W chromosome to total genome size was considerably in excess of the amount of Bkm satellite in the genome, it followed that the Bkm sequences must be interspersed with other W associated sequences. Moreover, since boids that lack cytologically defined sex chromosomes also lack any sex bias in the amount of Bkm satellite, it was concluded that the evolution of differentiated sex chromosomes in snakes resulted from an increase in both the relative and the absolute amount of Bkm-satellite-related DNA, that genic inactivation of the W occurred before morphological divergence of the Z and W, and that this inactivation by heterochromatinization resulted, among other things, in asynchrony of replication pattern.

In boids, Bkm satellite sequences are scattered thoughout the entire genome in both sexes. *Elaphe radiata* and *Naja naja naja,* both of which lack heteromorphic sex chromosomes, nevertheless have a W chromosome that C-bands completely, and Bkm sequences are concentrated at an intercalary site on the W. This has led Singh et al. (1980a) to conclude that the W satellite originated from preexisting dispersed sequences present generally in the genome. Once having accumulated at a localized site within the W by tandem multiplication, satellite sequences have become secondarily distributed throughout this chromosome by internal chromosome rearrangements. Jones (1984) has therefore proposed that sex chromosome evolution was a consequence, rather than a cause, of chromosome inactivation as has been classically assumed. Thus, if heterochromatinization was indeed the initial event in sex chromosome divergence, then both the suppression of crossing over between the sex homologues and the loss of gene function in the W could have been achieved simultaneously rather than through a stepwise sequence of events.

Bkm-related sequences have also been shown to occur in birds, mammals, and *Drosophila* (Singh, Purdom, and Jones, 1980b). Moreover, in the mouse, radioactive probes indicated that these sequences were concentrated on the Y chromosome, whereas in *Drosophila* they were located at the base of the X euchromatin (Jones and Singh, 1982). This led to the claim that conservation of Bkm sequences had resulted from the restriction of sequence divergence because of functional constraints associated with sex determination and indicative of selection pressure (Peacock, Dennis, and Gerlach, 1982).

In 1982, Epplen, McCarrey, Sutou, and Ohno succeeded in cloning and sequencing a 2.5-kb fragment of a sex-specific satellite DNA from *Elaphe radiata* which was shown to have 26 and 12 copies, respectively, of the simple base quadruplets GATA and GACA as its only repetitious components. They also confirmed the sex-specific conservation of different subfragments of this cloned sequence in the chicken, the mouse, and the human and showed that, in the mouse, the elements that cross-hybridized with parts of the cloned snake DNA were concentrated in the pericentric region of the Y chromosome. Equivalent simple quadruplet repeats (Sqr's), belonging to the consensus family $\mathrm{GA}{}^{\mathrm{T}}_{\mathrm{C}}\mathrm{A}$, have, however, been found in all eukaryotes examined, though only in certain cases have they accumulated differentially in a heterogametic sex-specific manner (Schmidtke and Epplen, 1980).

Singh, Phillips, and Jones (1984) reported a sex-specific expression of Sqr analogous sequences in the mouse, but this was not confirmed by Schäfer et al. (1986a). Moreover, detailed comparisons between published sequences for DNA clones from snakes, mice, and *Drosophila* that exhibit hybridization to snake satellite DNA indicate that the organization of this simple repeat differs in each species so that, in fact, there has been no strict evolutionary conservation (Levinson, Marsh, Epplen, and Gutman, 1985). Moreover, in the mouse, Bkm sequences are also concentrated on autosome 17 as well as in the X chromosome, whereas in the human there are few Bkm sequences on the Y although there are concentrations on the X and on two autosomes, one of which (No. 6) is homologous to mouse autosome 17 (Kiel Metzger, Warren, Wilson, and Erickson, 1985). Finally, at least two mammals, *Bos taurus* and *Ovis ovis,* have no detectable homologies with the Bkm sequence (Miklos and Reed, personal communication).

The balance of evidence thus indicates that the sex-specific arrangement of Sqr's is fortuitous. The evolutionary conservation of significant stretches of Sqr's certainly sets them apart from other satellite DNAs, but their linkage with the heterogametic sex over long evolutionary distances, and in situations characterized by both female and male heterogamety, may simply reflect the preferential formation of long stretches of these sequences in regions with reduced crossing over (Schäfer, Ali, and Epplen, 1986); that is, the W and Y chromosomes appear to have acted simply as convenient sinks for the accumulation of Sqr's.

Evidence in support of this conclusion comes from observations on the specific DNA sequences present in the W chromosome of the domestic fowl *(Gallus gallus domesticus)* which forms a densely stained sex chromatin body at somatic interphase. In the White Leghorn chicken, Tone, Sakaki, Hashiguchi, and Mizuno (1984) isolated two female-specific repeating DNA units of 0.6 kb and 1.1 kb, by digesting genomic DNA with the restriction enzyme *XhoI.* Neither of these units is related to the Bkm sequence. They also showed that about 46% of the W chromosome is built up out of these two units, which are repeated approximately 14,000 and 6,000 times respectively. Sequences related to the 0.6-kb unit are also present in the male genome of the domestic fowl, but their repetition frequency is much lower and they do not appear to be localized at any one site since no significant hybridization occurs to the mitotic chromosomes of the male.

Both repeat units occur in the genomes of various species of domestic and jungle fowl, but they are not detectable in three other birds, belonging to different genera of the suborder Galli. Additionally, in an Egyptian line of domestic fowl, the Fayoni, the repetition frequency of the 0.6-kb sequence is much lower than that of the 1.1-kb sequence. Thus, in addition to sequences like the Bkm satellite, which is widely distributed in both vertebrates and invertebrates, the work of Tone et al. (1984) serves to identify a second category of sex-limited sequences which are restricted in distribution to a limited number of species. This category most certainly cannot have played any role in the initial divergence of the Z and W chromosomes of the fowl and, like Sqr's of other verebrates, is most sensibly interpreted as having accumulated preferentially by chance in a chromosome that has facilitated their expansion.

4.1.2. Sex chromosome functions

In mammals, the X and Y chromosomes of the heterogametic sex differ extensively in size and in genetic content. In humans, for example, the X is three times the length of the Y and encodes at least 100 genes that are not represented in the Y. Even so, the human Y has genetic activity of at least two types: First, it carries the gene or genes that switch development of the embryonic gonad towards a male mode. Additionally, either a distinct gene, or the testis-determining gene produces H-Y antigen. Both these functions are located on the short arm of the Y since iso-long Y chromosomes lead to a failure in testis formation.

In mammals generally, the Y is either the smallest or one of the smallest chromosomes in the complement. The extreme in size reduction is reached in certain marsupials. For example, in the marsupial mouse *Sminthopsis crassicaudata,* the Y is at, or only just above, the limit of resolution with the light microscope. Not too dissimilar situations obtain in *Didelphis albiventis,* a number of bats, the ocelot, and some marmoset monkeys. In the case of these minute Y chromosomes, therefore, one must assume that all nonessential material has been discarded. It comes as no surprise, therefore, to find that many Y chromosomes, and especially the long Y chromosomes, are predominantly heterochromatic. What is abundantly clear, however, is that the mammalian Y is not simply an array of inactivated genes homologous to those of the X, and thus the heterochromatinization of the Y cannot be explained merely as a matter of shutting down formerly active genes. Rather, major changes have occurred in the composition of the Y itself. The best evidence for this comes from primates. Here the Y chromosomes of different species, though clearly homologous, show greater variability in morphology, Q fluorescence, and C-banding pattern than any other set of homologous primate chromosomes (Miller, 1977). Moreover, recombinant DNA studies indicate that the evolutionary history of the human Y has been complex, involving considerable sequence rearrangement. Two prominent, male-specific, DNA fragments, 3.4 and 1.9 kb in size, can be recovered from human genomic DNA when digested with the *HaeIII* restriction endonuclease (Cooke, 1976; Kunkel, Smith, and Boyer, 1976). Y chromosomes of different length occur in different males, and this polymorphism most commonly involves the distal heterochromatic segments on the long arm of the Y. Using restriction analysis, McKay, Bobrow, and Cooke (1978) showed that the amount of 3.4-kb sequences present in the male genome was directly correlated with the length of the distal heterochromatic segment present on the Y (Table 1.23). The 3.4-kb fragment includes both Y chromosome specific (mr) and non–Y-specific (hr)

Table 1.23. Proportionality between length of the distal heterochromatic (het.) segment and the amount of 34-kb-type sequences present in three human males with Y chromosomes of varying length

Male no.	Mean length Y (arbitrary units)	Ratio het./nonhet.	Relative amount 3.4-kb sequence
1	3.5	1.0	1.0
2	4.0	1.7	2.0
3	7.5	5.6	5.5

Source: McKay, Bobrow, and Cooke, 1978.

sequences. Surprisingly, homologues of the Y-specific fragments are not present in the Y chromosomes of other primates (Kunkel and Smith, 1982). However, one subset of the non–Y-specific sequences forms stable duplexes with both human and ape autosomal DNA with no detectable mismatch. Three distinct DNA domains of the non–Y-specific sequences with concentrations on autosomes have been identified (Burk et al., 1985). These are located on three separate chromosomes, namely 1, 15, and 16. Each domain consists of a tandemly arranged cluster containing one set of repeated sequences that is homologous to those on the Y, and a second set that is not. Thus, the human Y has evolved, at least in part, by the incorporation or loss of specific fragments rather than by the gradual modification of the DNA present within the Y at its inception.

Likewise, the middle repetitive sequences present in the Y of *D. hydei* are not necessarily restricted to this chromosome. Of six recombinant clones derived from the Y, three (PY 1, 4, and 9) are Y specific and are members of the same family of repeats found associated with the loop forming "noose" fertility loci on Y^S. The other three (PY 3, 5, and 8) are also found preferentially on Y^S but have additional copies elsewhere in the genome, especially in the centric heterochromatin of the autosomes. These copies, however, are not replicated during polytenization. One of the clones (PY 3) is represented in the heterochromatin of the X and in a single polytene band on chromosome 4 (Vogt and Hennig, 1983).

In some organisms, function other than sex determination have been associated with the Y chromosome. Two such cases have been analyzed in some detail, in both of which the Y plays no part in sex determination. In the first of these, the plant *Rumex thyrsiflorus,* normal males are XYY in constitution whereas females are XX. Experimentally produced male plants, either monosomic (XY) or nullisomic (XO) for the Y chromosome, show normal viability and have a normal meiosis (Zuk 1970a, 1970b). Such plants are, however, sterile to varying degrees, which implies that the Y plays a role in controlling normal pollen development. The Y chromosomes of *R. thyrsiflorus* have been described as heterochromatic though they have not been C-banded. However, at somatic interphase of meristematic tissues, two large chromocenters are present in male plants (Zuk, 1969). At premeiosis these chromocenters become fuzzy and enter a diffuse stage which coincides with a period of intense RNA synthesis.

An analogous example of fertility control by the Y chromosome is known in *Drosophila*. Here, too, the Y is inactive in somatic development. Indeed, it can be totally dispensed with developmentally since XO males are phenotypically normal and fully viable in most species. The Y, however, is normally essential for the functional maturation of sperm, and XO flies are characteristically sterile. The reason for this is that the process of spermiogenesis is under the direction of RNA accumulated by Y-initiated transcription during the early stages of spermatocyte development. At this time, specific regions of the Y produce a series of loops which are active in RNA synthesis and which represent the morphological expression of active fertility loci (Hess and Meyer, 1968).

The situation has been most thoroughly analyzed in *D. hydei* (Hennig, 1985). Here the Y has no known phenotypic marker genes although it constitutes almost 10% of the genome. The fertility loci it contains are of two kinds. Five are loop-forming sites, four of which are located in the long arm of the Y (Figure 1.29a). These loci consist of mr sequences whose transcripts contribute significantly to the production of testis-specific RNA. All of them are contained within regions identified as weakly fluorescent after H33258 staining. Apart from the loop-forming sites, several other Y-linked loci are known to affect fertility, but these loci are functionally independent of the loop loci and reside between them (Hackstein et al., 1982).

The Y chromosome of *Drosophila melanogaster*, similarly, contains six male fertility factors, two on the short arm and four on the long; like those of *D. hydei*, they appear to be responsible for the formation of intranuclear loops (Gatti and Pimpinelli, 1983; Pimpinelli, Bonaccorsi, Gatti, and Sandler, 1986). When defined by a series of noncomplementing breakpoints and deficiencies, in conjunction with H33258-stained preparations of extended prophase neuroblast chromosomes, it is apparent that at least four of them occupy segments of considerable length (Figure 1.29b). For example, k_l-5 is some 4000 kb long, whereas taken together the six loci occupy about a third of the total length of the Y chromosome.

Hilliker et al. (1980) claim that the occurrence of functional regions, like the male fertility loci, within the Y chromosome make it clear that the concept of the absolute genetic inertness of heterochromatin is an oversimplification. Pimpinelli et al. (1986) express a similar view, point-

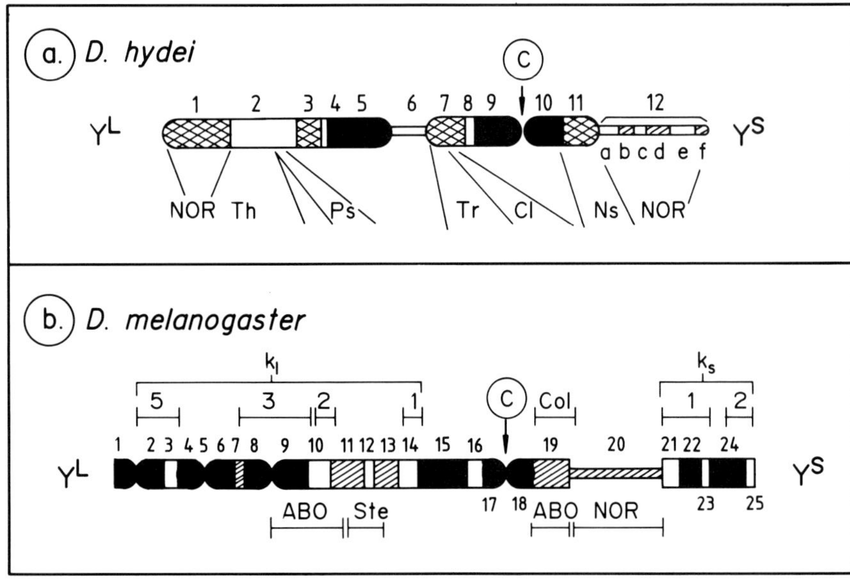

Figure 1.29. Hoechst 33258 banding patterns of the Y chromosomes of *Drosophila hydei* (a) and *Drosophila melanogaster* (b). Very bright fluorescent regions are shown solid, dull regions are hatched or crosshatched, whereas very dull regions are blank. Also indicated in (a) are the positions of the centromere (C), the nucleolus organizing region (NOR), and the five fertility loci, namely *Th* (threads), *Ps* (pseudonucleolus), *Tr* (tubular ribbons), *Cl* (clubs), and *Ns* (nooses). In (b), kl 1, 2, 3, 5, and k_s 1, 2 demarcate the fertility loci; the locations of one of the male pairing sites or collochores *(Col)*, the two ABO sites, and the Stellate locus *(Ste)* are also indicated. [(a) After Bonaccorsi, Pimpinelli, and Gatti; Hennig, 1985; (b) after Pimpinelli, Bonaccorsi, Gatti, and Sandler, 1986.]

ing out that even though the number of genetically detectable functions in the Y is small, they are nevertheless biologically very significant. This interpretation, however, very much depends on whether the functional segments in question are indeed part of the heterochromatin or whether they represent segments of DNA that have become intercalated within heterochromatin. Goldschmidt (1955), for example, took the view that "there are no heterochromatic loci in the strict sense of the term but only euchromatic loci surrounded by heterochromatin." As Hennig (1985) points out, the study of the Y of *D. hydei* in neuroblasts indicates that it is not entirely condensed in the interphase nucleus and that the chromosome regions carrying the fertility genes are less condensed at prophase than is the rest of the chromosome (Bonaccorsi, Pimpinelli, and Gatti, 1981), which suggests that they do not behave like heterochromatic regions.

In other species of *Drosophila,* the NOR is not present on the Y chromosome; thus its location within the Y of *D. melanogaster* should not be taken to indicate that it forms a necessary component of Y heterochromatin. Karpen and Laird (1986) have demonstrated that the heterochromatic location of ribosomal genes is most certainly not a requirement for their activity in *D. melanogaster.* Thus, single rRNA cistrons introduced into euchromatin by P-element-mediated germ-line transformation are able to transcribe and give rise to mini nucleoli in polytene cells. Additionally, at least five species of *Drosophila* are known in which XO males are both viable and fertile; these are *D. affinis, D. narragansett, D. annulimana, D. orbospiracula,* and *D. longala* (Voelker and Kojima, 1971). In *D. affinis,* males ordinarily possess a Y chromosome, but XO males, resulting from nondisjunction of the X in females of strains whose males themselves normally carry a Y, are known to be viable and fertile under laboratory conditions. Hess and Meyer (1968) had earlier reported that no male in the *D. obscura* group, to which *D. affinis* belongs, contained the complex of loops normally seen within the primary spermatocyte. Neither does the Y of *affinis* carry a bobbed locus.

In *D. melanogaster,* the Y also carries two pairing sites, or collochores, one in each of the two arms, through which the Y establishes a nonchiasmate relationship with the X chromosome, which has comparable pairing sites within its heterochromatin, at male meiosis. Although the composition of these pairing sites has never been defined, Cooper (1964), who first described them, stated quite categorically that they were to be regarded as "expressions of mappable linear differentiations within the heterochromatic regions, not as aspects of heterochromatin."

Finally, the heterochromatin of the Y chromosome of *D. melanogaster* includes elements capable of specific effects on euchromatic loci. The best known of these is ABO (Sandler, 1977). The autosomal ressive maternal effect mutation, abnormal oocyte *(abo),* is located in the euchromatin of 2L. When homozygous in the female, it leads to a marked reduction in the probability that an egg produced by such a female will develop into an adult. This effect can be compensated for if the fertilizing sperm carries either a normal ellele of *abo* or else a specific kind of element, called ABO, which is present in the genome in multiple copies, all within heterochromatic regions, including two locations in the Y (Figure 1.29b), one in the X, and one in 2R. In both the X and the Y, the ABO elements are located in Hoechst-dull regions which may again mean that they represent intercalary euchromatin.

Whereas the amelioration effect produced by a normal abo^{+} allele reduces postblastoderm mortality, the ABO elements themselves reduce only preblastoderm mortality. It will be recalled that in *D. melanogaster* there is a pronounced change in the conformation of interphase chromatin coincident with blastoderm formation. In the preblastoderm state, interphase heterochromatin does not stain deeply but does so in postblastoderm nuclei coincident with the onset of zygotic gene expression. This suggests that the preblastoderm rescue by ABO of the *abo*-induced egg defect may depend upon some activity of the ABO element at a time when conventional euchromatic loci are as yet inactive. Pimpinelli, Sullivan, Prout, and Sandler (1985) thus argue that one functional concomitant of the location of ABO within heterochromatin is the ability of the locus to act at a time when conventional euchromatic loci are unable to do so, though this need not mean, as they infer, that this observation demonstrates that heterochromatin is transcriptionally active at this time. They also report on preliminary data which suggest that *abo* may be one member of a cluster of linked genes – including holdup *(hup)*, wavoid 2 *(wd^2)*, and daughterless abnormal-oocyte–like *(dal)* – which interact with different, locus-specific heterochromatic factors; from this they conclude that the heterochromatin of *Drosophila* contains developmentally important genetic elements capable of acting, however, only at the very early stages of embryogenesis.

One consequence of the lack of a Y chromosome, or else of a large Y deficiency which deletes the fertility locus k_l-2, is the appearance of needle-shaped crystals in the primary spermatocyte (Hardy et al., 1984). Moreover, depending on whether the euchromatin of the X carries a *Ste^+* or an *Ste* allele of the Stellate locus (12C–13A), the needles may occur singly or as star-shaped aggregates, respectively. Livak (1984) isolated a DNA segment from the genome of *D. melanogaster,* which he designated as 2L.1, consisting of at least eight copies of a tandemly repeated 1250-bp sequence which he showed was present on both the X and the Y chromosomes. In the Y, the sequence mapped to a region between k_l-1 and k_l-2 where it was present in at least 80 copies. He also showed that there was a correspondence between the copy number of this sequence on the X chromosome and the presence of a particular allele of the Stellate locus, with low copy number corresponding to *Ste^+* and high copy number corresponding to *Ste.* The 2L.1 sequence was also present, though only in reduced copy number, in the Y chromosomes of *D. simulans* and *D. mauritiana,* but was absent from the X chromosome of these two species and was not represented at all in *D. erecta, D. teisseri,* and *D. yakuba.* This shows a pattern of distribution reminiscent of that observed for a number of mobile genetic elements in *Drosophila* and raises the interesting possibility that other so-called heterochromatic gene loci may have a similar origin so that their occurrence within constitutive heterochromatic should not be taken as evidence of heterochromatic genes. Regulatory interactions also exist between Y chromosome and autosomal loci in *D. hydei;* thus here, too, Y chromosome-associated sequences with additional autosomal copies may play a role in such interactions (Vogt and Hennig, 1983).

In summary, there are indeed both effects and functions associated with the Y chromosome of *Drosophila,* some, though not all, of which are equivalent to those of its presumed homologue, the X. The key question is whether these effects and functions are indeed properties of the Y heterochromatin itself or whether they reflect the presence of sequences that have become incorporated into this heterochromatin through the vagaries of chance. As we have seen earlier, the Sqr sequences present in the W and Y chromosomes of advanced vertebrates demonstrate how sequences capable of transcription may not only survive but also multiply within sex heterochromatin.

Unidentified factors that play a role in male fertility have been described in the Y chromosome of two other animals and so may be a more general phenomenon. In the tsetse fly *Glossina,* the heterochromatic Y is not involved in sex determination and its absence in males has no obvious morphological effect but again leads to sterility (Southern, 1980). Meiosis in XO males proceeds to completion, but the products are incapable of developing into functional sperm. The only time the Y is known to despiralize in this organism is for a period of nine to ten days during the G_2 period of premeiotic interphase.

In the KE inbred strain of mice, some 17% of the sperm are abnormal. Consequently, in successful matings, more than 20% of the ova remain unfertilized, many of them containing sperm within the previtelline space. By analyzing backcrosses between KE and normal KP strains, Krzanowska (1969) found a higher proportion of abnormal sperm in cases where the Y chromosome came from a KE male. By introducing the Y of a CBA strain, where only 6% of abnormal sperm are present, into a KE genetic background, she obtained a marked improvement in both sperm quality and fertility rate. The Y chromosome of the mouse is unusual for a mammalian sex chromosome because it does not C-band; nor does it contain the mouse satellite DNA present in the C-bands of all other members of the complement (Pardue and Gall, 1970). Precisely what it does contain has never been clarified, but Krzanowska's results suggest that it includes a factor, or factors, affecting sperm morphology and male fertility.

Although the Y chromosome is necessary for normal spermiogenesis in most species of *Drosophila,* this process is also known to be influenced by a large number of genes in both the X and the autosomes (Lindsley and Lifschytz, 1972; Kiefer, 1973). Some of these genes act directly, others appear to regulate the activity of the Y-linked loci, and mutants of these various genes produce disturbances similar to, though less frequent than, those caused by absence of the Y. However, in no case, including XO males, do all spermatids in any one individual show obvious defects.

In *D. melanogaster,* all translocations between the X and autosomes 2 or 3, other than those that interchange only the tips of the chromosomes or else involve a breakpoint in the proximal heterochromatin, are male sterile. About half of all X;Y and Y;A translocations are also male sterile but can be compensated for by adding a normal Y chromosome. In the case of T(X;2) and T(X;3) translocations, however, the resulting sterility is dominant and cannot be rescued by adding a segment of the X homologous to that removed by the translocation. This fact, coupled with the apparent nonspecific nature of the induced sterility, led Lifschytz and Lindsley (1972) to hypothesize that translocation sterility in *Drosophila* is related to a process of facultative heterochromatinization, and consequent inactivation, of the X prior to the onset of male meiosis, on the assumption that this inactivation is a prerequisite for normal spermiogenesis. Thus, if one of the functions of the X prior to its inactivation was to repress the activity of the Y, then X inactivation would lead automatically to Y activation. They also postulated that the basal heterochromatin of the X contained a controlling element, equivalent to the inactivation center of the mammal X, which was responsible for the heterochromatinization event since deficiencies that included both su(f) and bb also led to male sterility. Translocations that removed part of the X euchromatin from the control of this X-linked heterochromatic region would then be expected to lead either to a reactivation or else to an incorrectly timed inactivation of that X euchromatin.

There are five qualifications that need to be taken into account in assessing this intriguing hypothesis. First, although Lifschytz and Lindsley (1972) claim to have observed precocious condensation of the X in XO males of *D. melanogaster,* no photographic demonstration of this has ever been published. In *D. hydei,* Kremer, Hennig, and Dijkhof (1986) have shown that the euchromatin of the X is entirely decondensed in the primary spermatocyte stage, and even the heterochromatic arm becomes transiently decondensed during a considerable part of the first meiotic prophase. Added to this, autoradiographic studies involving [^{3}H]UdR incorporation in the primary spermatocytes of XO males of *D. hydei* indicates that the X is transcriptively active at this time (Hennig, 1967). Second, T(X;4) translocations do not produce male sterility. Third, if translocation sterility is indeed due to a nonspecific chromosomal effect, then one might predict, as Kiefer (1973) has pointed out, that different translocations would produce similar disturbances in spermiogenesis. According to Lifschytz and Lindsley, the phenotype of T(X;A) sterility invariably involves some degree of failure in sperm head elongation, whereas this is not common in mutational steriles. In the T(1;3)10;93B translocation studied by Kiefer (1973), however, the only defect was in the sperm tail. Fourth, if X inactivation were a prerequisite for Y activation, then Y loops would not be expected to form in Y autosome translocations; however, Kiefer reports that these loops were present in the T(1;3) translocation he studied. Finally, it is now clear that some deficiencies for the region including su(f) and bb are male fertile. From a study on male sterility, arising from interactions between a series of X chromosome deficiencies and duplications, Rahman and Lindsley (1981) concluded that, while there is no single site within the X heterochromatin that is required for male fertility, there are three regions at the base of the X which appear to be critical for the postulated inactivation of the Y in the primary spermatocyte. One of these is distal, and two are proximal, to the bobbed locus.

In mice and men, male carriers of interchanges between an autosome and either an X or a Y chromosome are also sterile (Eicher, 1970; Madan, 1983). In contrast to the situation in *Drosophila,* where the effect is invariably postmeiotic, in mammals it may occur earlier. In X;A heterozygotes, meiosis is arrested at pachytene, whereas in Y;A heterozygotes it occurs at the spermatid stage (Chandley, 1981). The precise cause of these disturbances is unknown. Ford (1970) suggested that, in the case of X;A interchanges, the physical disruption of the X might somehow interfere with the normal development of the sex vesicle at male meiosis and that this subsequently results in meiotic arrest, but the idea has never been seriously tested.

Purely autosomal interchanges may also lead to male sterility, at least in cases where one breakpoint is found near a centromeric or a telomeric end. Such asymmetrical interchanges lead to the preferential formation of chain multiples because of a failure of chiasma formation in the short arms of the pachytene pairing cross. Associated with this, Forejt and colleagues (Forejt and Gregorova, 1977; Forejt, Gregorova, and Goetz, 1981) found a high frequency of associations involving the unpaired arm of the translocation and the male sex bivalent both at pachytene and at diakinesis. They postulated that this association interferes with X inactivation and that sterility results from this (Forejt, 1984). Significantly there is no effect in the female, and there is certainly a good correlation between the severity of the effect on the sperm count and the percentage of chain configurations present at meiosis in carrier males (Chandley, 1981).

Human female carriers of X;A translocations in which the breakpoint lies between Xq 13 and Xq 26 are also infertile (Sarto, Therman, and Patau, 1973; Madan, 1983). Since fertile females with deletions of part or all of this critical region have been described, it is clear that it is not the break within this region that is important, and Madan (1983), favors a position-effect argument to explain the situation. Thus, while there are a variety of effects on fertility associated with rearrangements involving the X both in *Drosophila* and mammals, in no case has the situation yet been satisfactorily resolved.

4.2. Heterochromatin polymorphisms

Where large samples of individuals of the same species have been examined, it is not uncommon to find evidence for the occurrence of heterochromatic polymorphisms in which the amount of heterochromatin at a given chromosome location varies quantitatively in different individuals of the same population. Heterochromatin polymorphisms of this kind, which are ubiquitous in many eukaryotes, sometimes involve an apparently continuous variation in the amount of C-heterochromatin at specific sites (Verma and Dosik, 1980). At other times the variation is more discontinuous. For example, highly variable C-bands, showing both interindividual and interpopulational differences, occur on five different autosomes in European hedgehogs (Mandahl, 1978). Here, the variation involves both the presence and the absence of C-bands on a given pair of homologues as well as differences in the size of these C-bands, though only the second of these categories has ever been found within populations. The size and distribution of C-band segments are also singularly labile features of certain cetacean genomes where intraspecific heteromorphisms are both common and pronounced (Arnason, Lutley, and Sandholt, 1980). Perhaps the most spectacular polymorphism known to date is that present in the grasshopper *Atractomorpha similis;* in this species, all ten members of the haploid set can be affected and in a variety of ways (Figure 1.30) so that over 250 different cytomorphs can be distinguished within this one species (John and King, 1983).

At the other extreme are instances, like that in the genus *Triturus,* where the polymorphism is more confined, though no less impressive. Of the nine recognized species of newts belonging to this genus, two, *T. marmoratus* and *T. cristatus,* have an exceptionally pronounced and consistent heteromorphism in either the C-band pattern *(T. cristatus)* or both this pattern and the size *(T. marmoratus)* of the long arm of chromosome 1, the longest member of the complement (Figure 1.31). These regularly heteromorphic C-band regions also correspond to a region of heteromorphism for the distribution of distinctive lampbrush loops in oocytes, as well as to a region in which chiasmata never form. At zygotene of male meiosis, for example, the heteromorphic regions do not synapse and subsequently remain both asynaptic and achiasmate, apart from an occasional terminal association between the euchromatic tips of the long arms

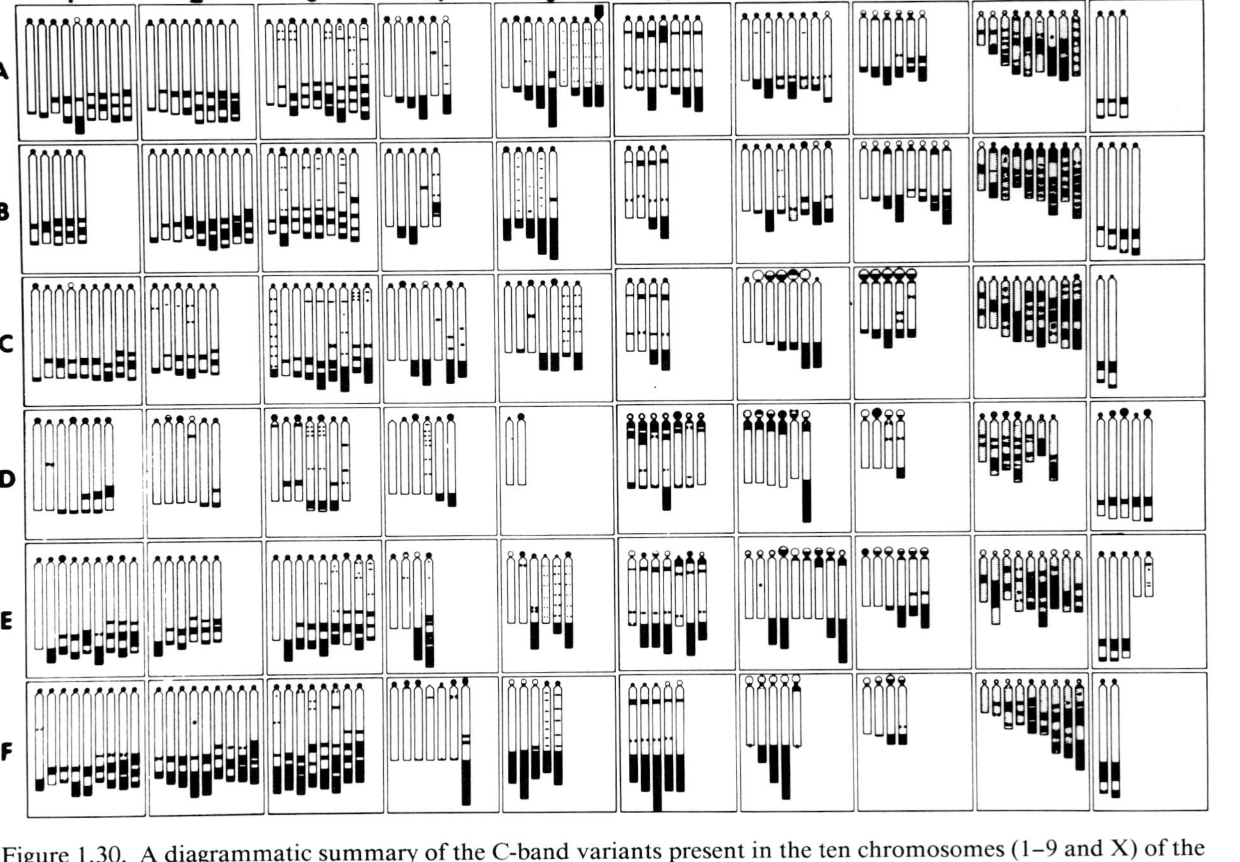

Figure 1.30. A diagrammatic summary of the C-band variants present in the ten chromosomes (1–9 and X) of the haploid set and the six cytotypes (A–F) present in Australian populations of the grasshopper *Atractomorpha similis*.

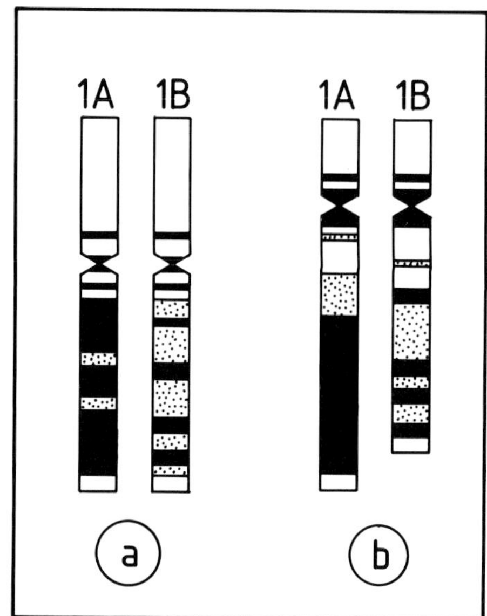

Figure 1.31. The C-banding patterns of the hetermorphic regions of pair 1 homologues (A and B) of (a) *Triturus cristatus carnifex* and (b) *Triturus marmoratus*. (After Sims, Macgregor, Pellatt, and Horner, 1984.)

(Sims, Macgregor, Pellatt, and Horner, 1984). Of particular interest, however, is the fact that individuals carrying identical morphs of chromosome 1 have been shown to arrest and die at the late tailbud stage of embryogenesis. That is, the segment evidently contains some type of recessive lethal system which maintains homologues in a state of permanent heteromorphism.

The existence of these obvious heterochromatic polymorphisms parallel observations in humans and other mammals, that regions containing large amounts of hr DNA, and which are C-band positive, tolerate loss or gain with only minimal phenotypic expression (Savage, Breckon, Goy, and Bigger, 1980). In humans, chromosomes 1, 9, 16, and the Y are particularly prone to polymorphisms for the size of the C-bands they contain. Carriers of such heterochromatic variants are generally free from phenotypic abnormalities. Additionally, the reproductive fitness of such carriers appears to be normal (Carothers et al., 1982) despite earlier claims to the contrary.

4.3. Species differences in heterochromatin content

Numerous cases now testify to the fact that a marked variability of constitutive heterochromatin, not only within but also between species, commonly occurs with no phenotypic effect. The mammals *Microtus agrestis* and *M. pennsylvanicus,* for example, are two species that show a 15% difference in constitutive heterochromatin content, yet live under similar ecological conditions and are so alike anatomically that taxonomists can distinguish them with certainty only in the presence of supporting cytological data (Schmid, 1967).

In the absence of overt effects on somatic development, and hence on phenotype, many have concluded, sometimes on the basis of circumstantial evidence (K. W. Jones, 1978; Peacock, Dennis, and Gerlach, 1982) but more often on the basis of no evidence at all (Fredga and Mandahl, 1972; Gatti, Pimpinelli, and Santini, 1976), that heterochromatin plays an important role in facilitating the evolution of new karyotypes and new species. This claim rests on arguments of two kinds:

1. *The presence of heterochromatin facilitates structural rearrangements which subsequently lead to reproductive isolation.*

In support of this claim, Yunis and Yasmineh (1971) pointed out that, in mammals, the subfamilies of rodents that possess large amounts of pericentric heterochromatin (Muridae, Cricetidae, and Microtunae) show a wide diversity of karyotypes and include a large number of species. By contrast, the Felidae (Carnivora) have only relatively small amounts of heterochromatin and include only a few species with very similar karyotypes. This proposal, based on broad general comparisons, fails, however, when comparisons are made between closely related species. Thus, in the rodent genus *Rattus, R. norvegicus* ($2n = 42$) has less than 5% of its genome as heterochromatin, and *R. sordidus* ($2n = 32$) and *R. villosissimus* ($2n = 50$) have even less. Yet these species differ by 11 major centric rearrangements, despite the fact that they are closely related and difficult to distinguish on morphological grounds (Miklos, Willcocks, and Baverstock, 1980). At the other extreme, in the Hawaiian *Drosophila,* where satellite sequences in some instances account for 70% of the genome, centric rearrangements are rare (Carson, Clayton, and Stalker, 1967) despite the fact that the centromere regions are loaded with repetitive DNA (Miklos and Gill, 1981).

There is certainly evidence to indicate that heterochromatic sites in some organisms may be associated with chromosome rearrangements. For example, in some Hawaiian drosophilids, pseudochromocenters are present in polytene nuclei in addition to the true chromocenter (Figure 1.32). Such pseudochromocenters result from the ectopic pairing of specific sites of IH on the X and on autosomes 2, 5, and 6. According to the species, pseudochromocenters form with frequencies of 50% *(D. hystricosa, D. mitchelli),* 75% *(D. furcifacies),* or even 100% *(D. biseriata).* The regions of centric heterochromatin involved in true chromocenter for-

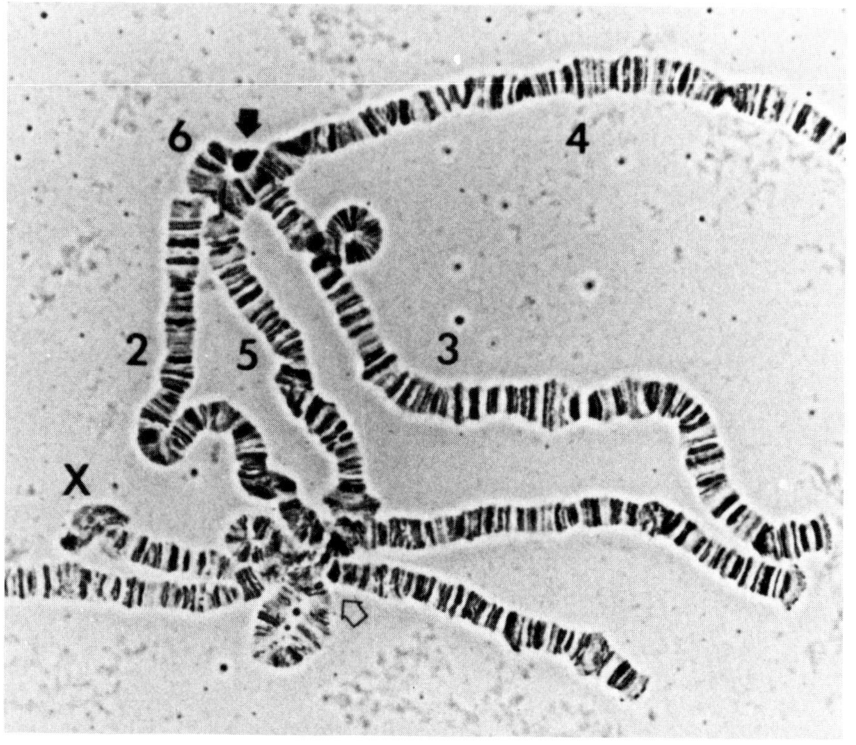

Figure 1.32. A polytene nucleus of *Drosophila biseriate* ($2n = 12$) carrying both a true chromocenter (solid arrow) and a pseudochromocenter (open arrow) formed between the X and autosomes 2 and 5. The centromere of the X has become detached from the true chromocenter in the preparation of the squash. (Photograph courtesy of Dr. J. S. Yoon.)

mation are also frequently associated with these pseudochromocenters (Yoon and Richardson, 1978). Of particular significance, all naturally occurring inversion breakpoints coincide with these sites; similarly, translocations, both reciprocal and nonreciprocal, also involve breakpoints corresponding to the IH sites involved in pseudochromocenter formation. Moreover, in hybrids between pairwise combinations of these four species, both the ectopic sites and the associations they give rise to are strongly conserved, even though the majority of chromosome arm regions are in fact asynapsed in polytene nuclei. Although in this case there can be no doubt that heterochromatin is indeed causally linked to chromosome rearrangement, there is no evidence that these rearrangements have played a meaningful role in either speciation or evolution in this group.

The repeated claim that heterochromatin has played a direct role in speciation by facilitating chromosome rearrangements also finds no support from the extensive data now available from pocket gophers (Table 1.24). Heterochromatin is a dominant component of the genome in the *Thomomys bottae* complex of pocket gophers, where it may account for up to 60% of the total genome (Patton and Sherwood, 1982). Yet, in every case where populations of *T. bottae* differing in heterochromatin content meet in nature, they hybridize freely and with extensive introgression. Moreover, whereas high levels of structural rearrangement have occurred in *T. talpoides,* a species in which the diploid number ranges from 40 to 60 but which has only very small amounts of centric heterochromatin, the reverse is true in *T. bottae.* Here diploid numbers are much more conservative, ranging from 74 to 78, and heterochromatin is much more extensive.

2. *Species specificity of heterochromatin confers an effective mechanism against hybridization and so constitutes an isolating agent in its own right.*

Thus, differences in either the content or the composition of heterochromatin may establish a fertility barrier that promotes evolutionary divergence and speciation (Corneo, 1976; Brutlag, 1980; Flavell, 1982; Peacock, Dennis, and Gerlach, 1982; Jarrell, 1984). Changes in the amount or the sequence composition of heterochromatin can, however, sensibly contribute to reproductive isolation only insofar as they alter the homology of chromosomes at meiosis in such a way that their pairing, recombination, or segregation is significantly disrupted. There is no hard data to support this, and the evidence from natural polymorphisms totally opposes it.

An alternative proposal, first mooted by K. W. Jones (1976), is that the changes that occur at speciation might favor changes in constitutive heterochromatin by encouraging saltatory events leading to the generation of new satellite sequences. This suggestion thus implies a reverse causation, with speciation favoring subsequent changes in heterochromatin composition or content, as opposed to the claim that changes in heterochromatin favor speciation. The numerous cases of heterochromatin polymorphism caution that speciation is certainly not a requirement for large-scale quantitative changes in heterochromatin content. Neither does speciation necessarily lead to such changes. In the genus *Ammospermophilus,* for example, whereas G-band comparisons indicate that karyotypes in different species have undergone extensive reconstruction in the course of evolutionary divergence, three chromosome pairs bearing substantial blocks of interstitial heterochromatin are common to all four

Table 1.24. Genome and heterochromatin variation in subspecies of the pocket gopher *Thomomys bottae* ($2n = 76$)

Subspecies	2C(pg)	Total no. arms	Total no. acrocentrics	Heterochromatin in metacentrics				Total no. metacentrics
				Total	Long arm	Short arm	Centric	
actuosus	8.42	116	32	20	10	4	10	44
fulvus	8.43–9.98	114–144	34–36	20	10–20	2–22	10	42–72
grahamensis	9.72	132	16	20	16	14	10	60
ruidosae	9.3							
bottae	9.35–11.43							
laticeps	10.03	148	0	20	22	24	10	76
planorum	10.74							
silvifugus	11.17							

Note: Acrocentric chromosomes have only centric heterochromatin.
Sources: Patton and Sherwood, 1982; Sherwood and Patton, 1982.

species and have not altered during speciation (Mascarello and Mazrimas, 1977).

Therefore, the claims that relate heterochromatin variation to speciation either are based on circumstantial evidence lacking in objectivity or else represent overgeneralizations from too limited a data base. What may appear as causative from a small sample has often turned out to be unsound when analyzed in a broader context.

4.4. The origin and evolution of heterochromatin

The questions of the origin and evolution of heterochromatin are inevitably bound up with the mode of origin of DNA sequence changes and their representation within the genome. This applies more particularly to the production of hr sequences.

Britten and Davidson (1976) suggested that repeated sequences were probably added to the genome as long tandem regions which subsequently underwent base substitutions and translocations, events that sometimes led to their loss from the hr class and their appearance in the single-copy fraction. Certainly not all of the single-copy fraction codes for proteins. In reality, however, we still know little about the precise events involved in the qualitative changes that go on within DNA. We are on stronger grounds when we turn to consider mechanisms leading to quantitative changes in the amount of heterochromatin within the genome. Here, at least five distinct processes have been identified; these are multiple replication, unequal exchange amplification, accumulation, and deletion.

4.4.1. Multiple replication

DNA sequences that, by whatever mechanism, are able to replicate differentially within the germ line will inevitably tend to accumulate in a genome unless they are detrimental to the organism. Consequently, it is not difficult to appreciate how multiple replication of a sequence, or sequences, could lead to the production of a supernumerary heterochromatic region on a single chromosome or even several different chromosomes. The production of enlarged polytene bands in *Chironomus thummi thummi* appears to represent a straightforward example of this principle.

The genome of *Ch. thummi thummi* has approximately 25% more DNA than its sister subspecies *Ch. thummi piger.* On the basis of its polytene banding pattern, it has been concluded that *Ch. th. piger* is phylogenetically older than *Ch. th. thummi;* thus there has been an increase in DNA content during the differentiation of these two subspecies (Keyl,

Table 1.25. Relative DNA content of 21 homologous polytene bands in the subspecies *Chironomus thummi thummi* and *Ch. th. piger*

Chromosome arm	Polytene band	No. of doublings in *th. thummi* 0	1 (×2)	2 (×4)	3 (×8)	4 (×16)
1R	c4.1			+		
	c3.6	+	+			
	c3.3		+			
1L	d3.8	+	+			
	e1.2	+	+			
	e1.4		+			
	e1.7		+			
2R	c2.12			+		
	c2.11		+			
	c2.4		+			
	c1.11		+	+		
	c1.8	+	+			
	c1.7	+	+			
	b5.28	+	+	+		
2L	c4.12		+			
	c4.7		+	+		
	c4.6			+		
	c4.3		+			
3	b1.13				+	
	b3.11		+	+	+	+
	b3.17		+			

Source: Keyl, 1965a.

1965a). From a comparison of the size of morphologically similar polytene bands in F_1 hybrids between the two subspecies, it is evident that the difference between the two genomes is the result of an increase in the size and DNA content of at least 30 individual polytene bands located proximal to the centromeric regions of all the major chromosomes. By spectrophotometry, Keyl (1965a) was able to show that, regardless of the absolute amount of DNA present, the enlarged bands of *Ch. th. thummi* in each case had a DNA level that represented a geometric multiple of that present in the homologous bands of *Ch. th. piger* (Table 1.25). The fact that the DNA content of individual bands followed a strict 1:2:4:8:16 ratio, with no intermediate values, points unmistakably to a mechanism based upon differential DNA replication.

Apart from centromeric C-bands, which are present in both subspecies, the polytene bands of *Ch. th. thummi* with an enlarged DNA content also give positive C-banding, whereas the homologous bands of *Ch. th. piger* do not (Schmidt, Vistorin, and Keyl, 1980). That is, the multiplication of band size is correlated with a visible heterochromatinization of the band in the polytene nucleus (Figure 1.33). Both genomes contain multiple, tandemly repeated, copies of a 120-bp sequence, though at very different repetition frequencies. This repeat is some 5.5 times more frequent in *Ch. th. thummi,* and in situ hybridization indicates that the extra sequences are in the C-bands. Apart from the difference in the degree of repetition, there is also a sequence divergence of about 35% in the 120-bp unit between the two subspecies (Schaeffer and Schmidt, 1981).

A somewhat similar phenomenon has been described in the plant genus *Nicotiana.* The tetraploid *Nicotiana tabacum* combines two different diploid genomes, referred to as S and T, which are represented in present-day diploids by *N. sylvetris* (S genome) and by *N. otophora* and *N. tomentosiformis* (T genomes). Both *sylvestris* and *tabacum* have only small amounts of heterochromatin. By contrast, *N. otophora* has large segments of heterochromatin at two terminal and three subterminal sites on the five largest chromosome arms in the haploid set. *Nicotiana tomentosiformis* has either one or two large terminal heterochromatic segments. When any of these large blocks from either species is introduced by hybridization into a *tabacum* genome, they increase in size in a small number of hybrid cells. The extent of this increase varies considerably in different cells of any one hybrid individual, but maximally they produce giant megachromosomes up to 15 times normal length (Gerstel and Burns, 1966; Burns and Gerstel, 1973). These megachromosomes are mechanically inefficient and are not transmitted by mitosis. They arise de novo in different cells as a consequence of multiple replication in the interphase immediately preceding their appearance.

4.4.2. Unequal exchange

Smith (1976) was the first to propose that a pattern of tandem repeats, equivalent to that observed in heterochromatin, is the natural state for DNA whose sequence is not maintained by selection. He also argued that one mechanism by which tandemly repeated DNA might arise and evolve was through random unequal exchange, within the germ line, between the daughter molecules produced by DNA replication. At the time he made this suggestion, there was no direct evidence for spontaneous unequal sister chromatid exchange within the germ line, and his arguments were based on a computer simulation of the consequences of such presumed exchange. This not only confirmed his predictions but

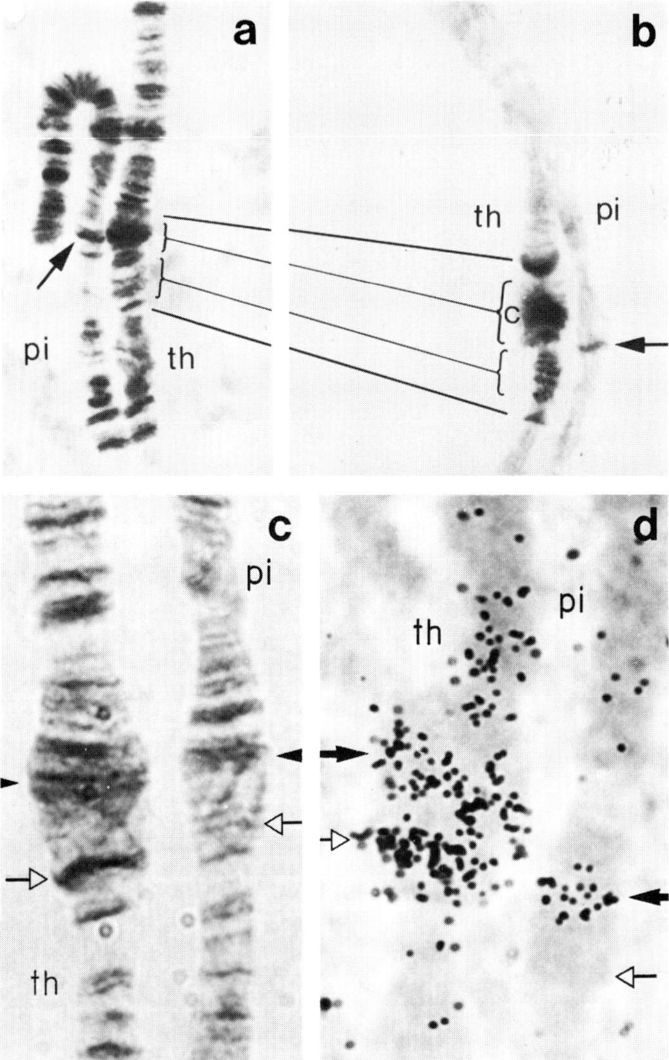

Figure 1.33. A comparison of orcein-stained (a) and C-banded (b) homologous polytene regions of chromosome 3 in the F_1 hybrid between *Chironomus thummi thummi* (th) and *Ch. th. piger* (pi) to illustrate the dark-staining enlarged bands of *thummi thummi.* Homologous bands in the two subspecies are connected by lines while the centromere region is marked by an arrow. Also shown is the centromere region (solid arrow) of chromosome 1 of the same F_1 hybrid following successive orcein staining (c) and in situ hybridization with satellite II DNA from *Glyptotendipes barbipes* (d). Note that the pi chromosome is labeled exclusively at the centromere but not at band 1d3 (open arrow) which is enlarged and heavily labeled. [Photographs (a) and (b) courtesy of Dr. K. Hägele. Photographs (c) and (d) courtesy of Dr. E. R. Schmidt.]

Table 1.26. Ratio of Y chromosome length to the length of the F chromosome group in 10 cells from a 10-year-old boy with borderline mental retardation, and 10 cells from his father and his brother

Cell type	Y/F length ratio	Standard deviation
Fibroblasts from proband[a]		
Short Y	0.72	0.07
Long Y	1.03	0.07
Lymphocytes from proband[a]		
Short Y	0.73	0.06
Long Y	1.06	0.09
Lymphocytes from father	0.83	0.07
Lymphocytes from brother	0.88	0.09

[a]In both the fibroblasts and the lymphocytes from the proband, some 25% of the cells had a long Y and the remainder a short Y.
Source: Wahlström, Kyllerman, Hansson, and Taranger, 1985.

additionally accounted for the patterns of periodicity commonly evinced by hr sequences.

Since crossing over in autosomal heterochromatin at meiosis does not occur, this makes unequal exchange at meiosis an unlikely source of variability for constitutive heterochromatic regions, as Kurnit (1979) pointed out. Consequently, if unequal exchange is indeed a source of quantitative variation in heterochromatin content, then it must occur in mitotic germ-line lineages. Although conclusive evidence for the occurrence of unequal exchange leading to heterochromatin variation is still lacking, a number of studies on humans indicate that it is the most likely explanation for the heterochromatic variants observed in chromosomes 1, 9, 16, and the Y.

There is wide interindividual variation in the length of the human Y chromosome. Wahlström, Kyllerman, Hansson, and Taranger (1985) reported the existence of two separate Y chromosome populations, with distinct cytomorphological characteristics, in lymphoblasts and fibroblasts from a single individual. In both of these populations, the Y chromosomes differed in length from those of the father and the brother of this mosaic individual (Table 1.26). The most probable explanation of this situation was held to be unequal somatic exchange, since somatic mosaicism of this kind is difficult to explain other than by unequal exchange. A second, more direct, line of evidence, and one that is rarely possible in any other organism, involves the study of family pedigrees with direct comparisons between parents and offspring in regard to the amount of heterochromatin present in specific chromosomes. The first such a study was undertaken by Craig Holmes, Moore, and Shaw (1975) who examined five pedigrees, comprising 10 segregating sibships, involving a total of 33 offspring. They identified seven new C-band variants, not represented in either parent, with an additional seven showing a C-band pattern that had visibly altered from that of one of the parents. More recently, Simi and Tursi (1982) studied the inheritance of C-band pattern in 30 normal families each consisting of both parents and one child, and, in the case of one family, a pair of dizygotic female twins. In three of these families, four new variants were observed in the offspring.

In both these studies, gains and losses of heterochromatin were identified, involving chromosomes 1 and 16 in the case of Craig Holmes et al. (1975), and chromosomes 1 and 9 in the case of Simi and Tursi (1982). Both sets of authors were of the opinion that the variants arose by some form of unequal exchange. Significantly, in Bloom's syndrome some 7% of the observed mitotic exchanges that are associated with this defect are known to involve the centric heterochromatin of chromosomes 1, 9, and 16 (Kuhn, 1976).

4.4.3. Amplification

The preservation of both fragment length and the locations of restriction sites within the hr DNA out of which a heterochromatic region is constructed indicates that the repeat arrays of tandem sequences must have arisen through a series of amplification events (Dennis, Dunsmuir, and Peacock, 1980; Arnason, Höglund, and Widegren, 1984; John, Appels, and Contreras, 1986).

Amplification is a blanket term used to describe large-scale increases in the copy number of DNA sequences. While multiple replication is certainly implicated in amplification, some form of exchange, most likely unequal sister chromatid exchange, is also usually associated with such replication. Thus either rereplication stimulates unequal exchange or else unequal exchange leads to rereplication (Hamlin, Milbrandt, Heintz, and Azizkhan, 1984).

European populations of the house mouse, *Mus musculus,* are known to be polymorphic for an interstitially located and extra segment in chromosome 1, the largest member of the complement. The size of this segment varies between 6% and 30% of the length of the chromosome. It is C-band positive, though somewhat less intensely stained than the centric C-bands which characterize this species. It gives bright fluorescence with

MM and weak fluorescence with H33258, the reverse of that shown by the centric heterochromatin. Finally, in situ hybridization indicates that this supernumerary segment does not contain detectable quantities of mouse satellite DNA. Traut, Winking, and Adolph (1984) therefore regard it as a homogeneously staining region (HSR), that is, a sequence of integrated amplified DNA equivalent to the HSRs found in cell lines under strong artificial selection, as well as in tumor cells.

In some Hawaiian *Drosophila,* not only are the breakpoints for chromosome inversions known to be clustered at intercalary heterochromatic loci, but additionally there are characteristic cytomorphological changes, most commonly the deposition of additional heterochromatin, associated with these breakpoints (Yoon and Richardson, 1978). Moreover, in three different species of Hawaiian *Drosophila, D. formella, D. disjuncta,* and *D. recticilia,* Bamai (1975a, 1975b, 1977) observed an exclusive and consistent association between a heterozygous inversion, with a breakpoint located in the vicinity of the proximal centric heterochromatin, and the occurrence of an amplification of the proximal heterochromatin which transforms the chromosome in question from an acrocentric to a metacentric state (Figure 1.34). In several species of chironomids, too, an increase in the DNA content of an individual polytene band occurs in association with specific inversions. In each case, the increase is restricted to bands located at an inversion breakpoint (Keyl, 1965b).

Y and W chromosomes are characteristically rich in heterochromatin. This is not normally the case for X or Z chromosomes. Most placental mammals, for example, have a functional euchromatic X of similar size and comprising about 5% of the total haploid female genome. A few species, mainly of rodents and artiodactyls (Wurster, Benirschke, and Noelke, 1968; Fredga, 1970; Wurster, Snapper, and Benirschke, 1971), have larger X chromosomes. Ohno (1969) originally suggested that these enlarged X chromosomes had been produced by a literal doubling, trebling, and quadrupling of the basic 5% X. This, however, proved too simplistic an explanation since the amount of extra material was not simply a geometric multiple of the 5% X and at least some of these enlarged X's have been shown to have arisen through X autosome translocations (Wahrman, Richler, Neufeld, and Friedmann, 1983). More commonly, the increase in size of the enlarged X's is due to the presence of additional constitutive heterochromatin, as evidenced by C-banding. What this case highlights is the need to explain how the segments in question became heterochromatic since neither the original X nor, in the case of the X autosome translocation products, the translocated autosome arm, were heterochromatic at the outset. If, for a moment, we leave aside the translocation X's, there is evidence from two rodents for amplification leading to enlarged X chromosomes. In the Australian hopping mice, three spe-

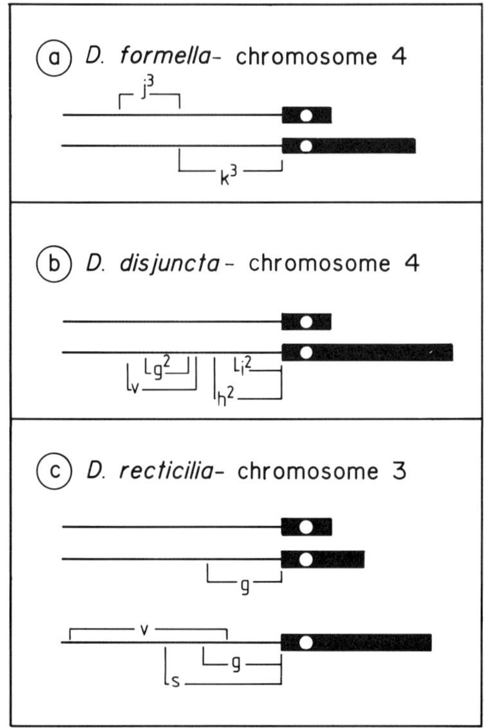

Figure 1.34. The correlation between inversions having a breakpoint at the euchromatin–heterochromatin junction and the enlargement of the short-arm heterochromatin in three species of Hawaiian *Drosophila*. In *D. formella* (a), there is a consistent relationship between the presence of k^3 and the heterochromatic enlargement of the short arm of chromosome 4 in both classes of homozygotes, $4j^3/4j^3$ and $4k^3/4k^3$, as well as in the $4j^3/4k^3$ heterozygote. In *D. disjuncta* (b), only the $4^+/4vg^2h^2i^2$ heterozygote is known, but an additional inversion (k), with breakpoints confined to euchromatin, has no influence on the size of the short arm in either the heterozygous ($4^+/4k$) or the homozygous (4k/4k) states. Finally, in *D. recticilia* (c), only 3g/3g homozygotes and 3g/3gs heterozygotes have been identified, so that the basic form of chromosome 3 shown for this species is hypothetical. (After Bamai, 1975a, 1975b, and 1977.)

cies of *Notomys* show polymorphisms for the size of the X chromosome. C-banding indicates that this variation is due entirely to added heterochromatin formed by the amplification of the short arm of the X (Table 1.27). The Y chromosome, which is a totally heterochromatic element in this genus, also showed variation in size, due to additional heterochromatin (Baverstock, Watts, and Hogarth, 1977). Indeed a number of other mammals with enlarged X's due to added heterochromatin also have

Table 1.27. X chromosome polymorphism in three species of Australian hopping mice, genus *Notomys*

		% Length relative to (X + A) complement	
Species	X type	Total X	Euchromatic X
N. alexis	Small subacrocentric	6.7 ± 0.2	5.3 ± 0.1
	Medium submetacentric	8.0 ± 0.3	5.1 ± 0.2
	Large metacentric	11.1 ± 0.2	6.2 ± 0.2
N. cervinus	Small submetacentric	8.4 ± 0.1	5.0 ± 0.1
	Medium metacentric	9.5 ± 0.2	4.9 ± 0.1
	Large metacentric	11.1 ± 0.1	5.0 ± 0.2
N. fuscus	Small subacrocentric	8.7 ± 0.2	6.5 ± 0.1
	Medium submetacentric	10.3 ± 0.2	6.1 ± 0.1

Source: Baverstock, Watts, and Hogarth, 1977.

enlarged Y chromosomes (Fredga, 1970). Sharma and Raman (1973) had earlier reported a case similar to that of *Notomys* in *Bandicota bengalensis bengalensis.*

4.4.4. Accumulation

Drosophila miranda has a multiple, X_1X_2Y male, $X_1X_1X_2X_2$ female, sex chromosome system in which the metacentric Y chromosome has an arm of largely autosomal origin (Figure 1.35). It is thus a neo-Y chromosome and, because of this, is polytenized in salivary gland nuclei, unlike the completely heterochromatic Y chromosomes of other drosophilids. Its detailed polytene banding pattern differs considerably, however, from that of its unfused homologue, the X_2 chromosome. The polytene neo-Y is markedly bent and twisted as a result of a series of weak points which appear underreplicated and show ectopic pairing. Additionally, in several regions the band interband organization is transformed into a diffuse state (Steinemann, 1982). Some 6% of the genome of *D. miranda* is composed of satellite DNA. When cRNA fractions synthesized from unfractionated genomic DNA were hybridized in situ to mitotic preparations, label was restricted to the centromeric regions of the X_1 and the autosomes in females whereas the males showed basically the same pattern with the exception of the neo-Y which was prominently labeled throughout. Since the cRNA was made from the unfractionated genome, it is clear that the hybridizing sequences were repetitive.

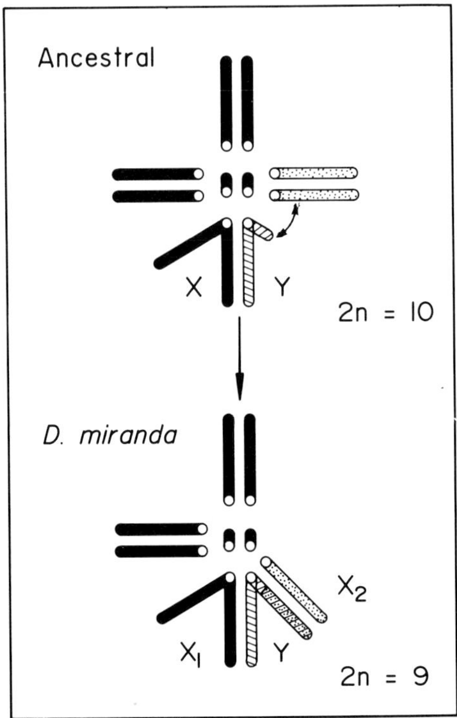

Figure 1.35. The invasion of repetitive DNA sequences (hatched areas) of the Y chromosome into the translocated autosomal arm (stippled areas) involved in the production of the X_1X_2Y multiple sex chromosome system of *Drosophila miranda*. (After Steinemann, 1982.)

In situ hybridization to polytene nuclei, using the same cRNA probe, gave strong labeling over the β heterochromatin of the chromocenter in both males and females; additionally, in males, the entire length of the Y was again labeled whereas the X_2 was unlabeled. Since the X_2 does not label in either mitotic or polytene preparations, whereas the entire Y chromosome is labeled in both, it follows that the autosomal material translocated onto the Y now contains repetitive sequences that are not represented in its free homologue. The implication of these results is that repetitive sequences of the progenitor Y have somehow invaded, and have accumulated in, the autosomal material transferred to it so that these sequences are now present throughout the entire neo-Y.

An equivalent process of accumulation, involving the invasion of repetitive sequences present in the centromeric region of the X into autosomal material translocated onto it, would explain the behavior of X;A translocation products in mammals. It could also account for the cases of euchromatin transformation that have been identified where C-band material has been added to chromosomes but with no accompanying size increase and an evident reduction in the amount of euchromatin (King, 1980).

4.4.5. Deletion

It is sometimes difficult to decide whether alterations in genome size resulting from differences in the amount of heterochromatin have been caused by a gain or loss mechanism. As we have already seen, at least two mechanisms are capable of leading to the loss of heterochromatin; these are unequal exchange and elimination by direct excision. The latter, so far as is known, is confined to presomatic lineages; to be of evolutionary significance, however, the loss would have to occur from the germ line.

Sentis, Santos, and Fernandez Piqueras (1986) have described what appears to be partial germ-line loss of heterochromatin within an individual of the tettigoniid *Baetica ustulata.* This species is known to be polymorphic for C-band heterochromatin in three of its medium-sized chromosomes: M3, 4, and 5. In a single male, where all cells of the gastric caeca (soma) were uniformly heterozygous for C-bands on all three medium pairs, both testes proved to be mosaic with a proportion of spermatogonia and spermatocytes carrying one of five novel combinations of C-band variants in addition to the pattern observed in the gastric caeca, which was assumed to represent the zygotic type. Some of these variants were illustrated as having lost heterochromatin (Figure 1.36), whereas others showed an evident gain. In the case of the M3 and M4 variants, the loss was explained in terms of mitotic exchange involving the euchromatic segments between individual C-bands. This, however, would have to be coupled with the differential survival of the products of exchange. Moreover, as the authors themselves point out, the M5 variant involving total loss of the terminal supernumerary segment could not be explained in the same way and its mode of production is unresolved since it is not apparent how an unequal exchange could have operated in this case.

Despite this lack of understanding of the mechanics of loss, comparisons between closely related species provide further circumstantial evidence that heterochromatin loss during evolution has occurred. For example, three sibling species of *Drosophila* show evident changes in the DNA content of both autosomes and sex chromosomes (Table 1.28). On the assumption that *D. eohydei* is more comparable with the ancestral state, Zacharias, Hennig, and Leoncini (1982) argue for an overall loss of DNA sequences during the evolution of this species cluster. Since the euchromatic regions of all three species are virtually indistinguishable in

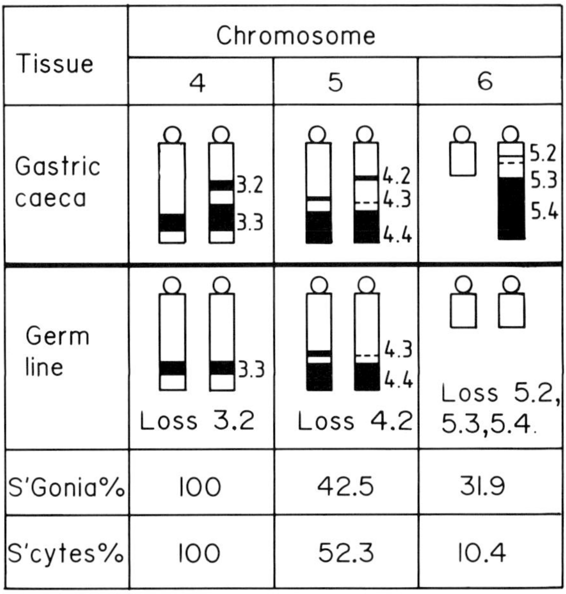

Tissue	Chromosome		
	4	5	6
Gastric caeca	3.2 3.3	4.2 4.3 4.4	5.2 5.3 5.4
Germ line	3.3 Loss 3.2	4.3 4.4 Loss 4.2	Loss 5.2, 5.3,5.4.
S'Gonia%	100	42.5	31.9
S'cytes%	100	52.3	10.4

Figure 1.36. Frequency of spontaneous loss of C-bands in three heteromorphic chromosome pairs (4, 5, and 6) within the germ line of an individual of the tettigoniid *Baetica ustulata.* (After Sentis, Santos, and Fernandez Piqueras, 1986.)

terms of polytene morphology, the loss must have involved the removal of heterochromatin. Certainly the changes in the DNA content of the X correlate with the variability in the length of the heterochromatic arm which is known to be enriched in hr sequences. Thus whereas in *D. hydei* the metacentric X has one euchromatic and one heterochromatic arm, the X of D. neohydei lacks almost all of the heterochromatic arm. There are also differences in the amount of heterochromatin associated with the autosomal centromeres as well as in the number of IH regions within the autosomes (Hennig, 1985).

The Y chromosome, on the other hand, has evidently grown in size and its growth is not related to the transfer of hr sequences from elsewhere in the genome since the Y in this species group contains little or no hr DNA. This growth is assumed to have resulted from the duplication of sequences within the Y itself. A comparison of the Y chromosomes of *D. hydei* and *D. neohydei* indicates, however, that there has also been an evident internal reorganization of its components (see fig. 16 of Hennig, 1985). Thus, in these two species the Y loops are entirely different, both in their location and their morphology. Despite the 20% difference in genome size between *D. hydei* and *D. neohydei,* fertile hybrids can be obtained between them. Consequently, by backcrossing, it is pos-

Table 1.28. Comparative genome sizes in three sibling species of the *Drosophila hydei* cluster

		Total genome		Percent		
Species	Sex	%	2C (pg)	X	Y	Autosomes
D. eohydei	♂	102.3	0.49	11.3	5.9	85.1
	♀	107.8	0.52			
D. hydei	♂	100.0	0.48	17.9	9.14	72.9
	♀	108.4	0.52			
D. neohydei	♂	83.9	0.40	9.1	8.7	66.2
	♀	84.3	0.40			

Source: Zacharias, Hennig, and Leoncini, 1982.

sible to introduce a *neohydei* Y into a background of *hydei* autosomes. Under these circumstances, the structure of the Y loops approaches those of *hydei,* indicating that loop morphology is not determined solely by the genetic structure of the loop locus itself but rather is also influenced by the products of autosomal genes.

The *Drosophila virilis* species group is subdivided into two clusters, the *virilis* and *montana* phylads, with the latter assumed to have been derived from the former. Since members of the *virilis* phylad are generally characterized by larger amounts of C-heterochromatin (Table 1.29), it has been argued that, in this case too, heterochromatin has been lost during evolution (Mahan and Beck, 1986). Here the loss has predominantly involved the centric heterochromatin of the autosomes since in all species of both phylads the Y is a large and totally heterochromatic element.

These various mechanisms of gain and loss suggest three possibilities for the evolution of heterochromatin within a genome:

1. Heterochromatin forms an additive component of genome evolution in the sense that its addition leads to an increased genome size. This is particularly well illustrated in placental mammals (Hsu and Arrighi, 1971). Here genome size is remarkably constant overall and usually shows only minor variations. In the few groups with significant deviations, the differences are most commonly due to increases in heterochromatin content (Table 1.30).

Table 1.29. Percent heterochromatin (Het.) content relative to genome length in C-banded neuroblast chromosomes from two related phylads of the *Drosophila virilis* group

virilis phylad			*montana* phylad		
	% Het.			% Het.	
Species	♀	♂	Species	♀	♂
D. virilis	48.3	49.2	*D. littoralis*	37.5	23.5
D. americana texana	40.5	36.1	*D. borealis*	36.9	31.7
D. a. americana	38.5	38.2	*D. ezoana*	33.0	26.5
D. novamexicana	34.4	28.4	*D. flavomontana*	23.3	40.4
			D. lacicola	12.9	28.3
Means	39.9	36.1		35.1	26.5

Source: Mahan and Beck, 1986.

Table 1.30. Variation in genome size and heterochromatin content in placental mammals

Species	1C (pg)	% C-band material
Peromyscus crinitus (deer mouse)	3.5	6
Muntiacus muntjak vaginalis (barking deer)	2.5	15
Homo sapiens (human)	3.5	17
Microtus agrestis (vole)	3.2	26
Mus musculus (mouse)	3.5	27
Mesocricetus auratus (golden hamster)	3.5	34
Peromyscus eremicus (deer mouse)	4.7	36
Dipodomys ordii monoensis (kangaroo rat)	5.5	58

Source: John and Miklos, 1987.

2. Heterochromatin accumulates at the expense of euchromatin, as appears to have happened in several sex chromosome systems, including the W chromosome of snakes and the neo-Y of *Drosophila miranda,* by a permanent heterochromatinization of euchromatin as a result of its invasion by repetitive sequences. A comparable process may have also occurred in cases where C-bands appear to have originated and expanded at the expense of a reduced euchromatin content.

3. Heterochromatin is lost from the genome. Examples include the reduction in size of W and Y chromosomes in many species as well as the more general loss evidenced by comparisons of related species in cases where the direction of evolutionary change can be inferred with some confidence. There is a claim on record for a wholesale transformation of heterochromatin into euchromatin, that is, a deheterochromatinization process with no change in genome size (Tanaka, 1969). This is a claim that has not, and cannot, be taken seriously because of a lack of meaningful evidence to support it.

4.4.6. The distribution of heterochromatin within the genome

Superimposed on these quantitative changes are two very general rules governing heterochromatin distribution which are not explicable in terms of the mechanisms identified for generating quantitative change. The first of these was noted by Heitz (1933, 1935), who pointed out that, both in drosophilids and in a number of plant species, heterochromatin tends to accumulate at similar sites on nonhomologous members of the same chromosome set. Such *equilokal heterochromatie* is supported most obviously by the predominance of heterochromatin at centromeres and/or telomeres in many species. A second feature is that when both proximal and distal C-bands occur within the same species, they regularly differ in composition (John, Appels, and Contreras, 1986). This is evident both from the differential fluorescence they display, on the one hand (John, King, Schweizer, and Mendelak, 1985), and more especially from the restriction of particular repeat sequences to specific regions of the genome. That is, not only are C-bands equilocally distributed but also their sequence composition reflects their location.

It is characteristic of cetaceans, for example, that C-bands are mainly, though not exclusively, present in terminal or interstitial positions and only a small number are centromerically located. The difference in C-heterochromatin content between toothed and baleen whales is then accounted for by the presence of much more terminal heterochromatin in the latter (Table 1.31). Two types of satellite DNA, with buoyant densities in neutral cesium chloride of $\rho = 1.702/1.703$ and $\rho = 1.710/1.711$, respectively, have been isolated from three of the balaenopteroid species (minke, sei, and fin whales). Although there was no evident sequence homology between these two types, homo- and heterologous DNA/cRNA hybrids within each group gave virtually identical melting-curve profiles, indicating the conservation of at least a major part of the sequences in the three species.

In situ hybridization revealed that the two satellite types have different and well-defined sites of localization. In all three species, the 1.702/1.703 satellites were located in centromeric or paracentromeric C-bands in a

Table 1.31. Heterochromatin (Het.) content of whales (Cetacea)

Mysticeti (baleen whales)				Odontoceti (toothed whales)			
	Percent				Percent		
Species	Het.	X	Y	Species	Het.	X	Y
Balaenoptera acutirostrata (minke)	25	4.7	1.9	*Stenella dubia* (spotted dolphin)	12	5	1.0
Balaenoptera borealis (sei)	25	4.7	1.0	*Tursiops gilli* (bottle-nosed dolphin)	15	4.9	m[a]
Balaenoptera physalis (fin)	30	8.5	1.0	*Globicephala macrorhyncha* (short-finned pilot)	10	54.5	m
Mesoplodon europaeus (Gervais's beaked)	17	12.7	?	*Phocoena phocoena* (harbor porpoise)	10	5.0	m

Note: The increase of heterochromatin content in baleen whales is due to the presence of additional terminal C-bands. With the exception of *M. europaeus* ($2n = 42$), all other species have $2n = 44$.
[a] m = minute.
Sources: Arnason, 1974; Arnason, Benirschke, Nichols, and Mead, 1977.

few chromosome pairs, whereas the 1.710/1.711 satellites were present at all terminal and some interstitial sites. (Arnason, Purdom, and Jones, 1978). Similarly, in *Atractomorpha similis,* a 537-bp *TaqI* restriction fragment, composed of a trimer of tandemly arranged 179-bp units and cloned from the satellite-1 DNA of this species, hybridizes in situ to all the distal C-bands, whether terminal or subterminal, but to none of the proximal C-bands, either centromeric or paracentromeric. Neither is there any hybridization to the heterochromatic short arms which are sometimes present on the otherwise telocentric chromosomes (John, Appels, and Contreras, 1986).

It follows from these regularities of, and restrictions to, heterochromatin distribution that there must be mechanisms capable of generating a similar, if not identical, array of hr sequences at equivalent sites on nonhomologous chromosomes in a given set and that these mechanisms play an important role in the origin and evolution of heterochromatin. In theory, there are three different ways of achieving such differential but equilocal patterns:

1. By the simultaneous or progressive multiple replication or amplification of similar sequences scattered at particular sites throughout the genome. This is precisely what appears to have happened in the case of *Chironomus thummi thummi.* That independent rereplication, involving different blocks of heterochromatin at similar sites, is indeed plausible is also supported by the production of megachromosomes in *Nicotiana.*
2. By the amplification of hr sequences at a restricted site followed by the spread of such amplified sequences, to other equivalent though nonhomologous sites, while chromosomes are in close association. This implies that the distribution of heterochromatic sequences reflects the way in which chromosomes are organized, and so are able to interact physically, within the nucleus. Schweizer and Loidl (in press) draw attention to the fact that relic or Rabl polarization of chromosomes in the intermitotic nucleus and the development of bouquet polarity in the meiotic nucleus both provide an opportunity for close physical proximity of specific chromosome regions, as too of course does the association of NOR regions through nucleolar fusion. Such associations offer a heightened opportunity for either indirect sequence transfer by sequence mobility or direct exchange of chromosome material between equilocal sites, an exchange which incidentally might be expected to be occasionally coupled with chromosome rearrangement, as is most likely in the case of centric fusions. Schweizer and Loidl (in press) also suggest that, in the case of nonhomologues which differ in arm length, the neighborhood rela-

tionship of telomeres and interstitial regions provide for the transfer of telomeric sequences to interstitial sites during Rabl polarization. This model thus implies that telomeres and centromeres function not only as sites where sequence amplification is initiated, but additionally as sites from which sequences are dispersed elsewhere in the genome either through the proximity afforded by polarization or else through the direct physical contact it creates between nonhomologues.

3. By the preferential accumulation of repeated sequences in the neighborhood of regions that show restricted recombination (Charlesworth, Langley, and Stephan, 1986). The essence of this proposal is twofold: First, that centromeres and telomeres represent two major sites that have been specifically selected for reduced rates of crossing over; and second, that the rate of loss of hr DNA increases with the rate of unequal exchange between such sequences. Macgregor, Horner, Owen, and Parker (1973) had earlier made a similar comment, namely that satellite sequences tend to accumulate in chromosome regions where the chances of crossing over, or structural rearrangement, are relatively slight. As we noted in the case of sex chromosome differentiation, the problem here is essentially one of priority; that is, do hr sequences tend to accumulate in regions where crossing over is restricted, or does the accumulation of hr sequences lead to a restriction in crossing over? Moreover, despite the claim of Charlesworth et al. (1986), in many organisms telomeric regions are most certainly not regions of low crossing over (Rasmussen and Holm, 1980). On the contrary, distal chiasma localization is exceptionally common in eukaryotes. In his detailed analysis of chiasma distribution in truxaline grasshoppers, for example, Southern (1967) showed that when only one chiasma forms, it is invariably located at the distal end of all chromosome arms. Indeed, even in *Drosophila* while there is certainly reduced recombination in the vicinity of the centromere, recombination increases as the regions monitored approach the ends of the autosomal arms. Only in the X chromosome is recombination lowest at the telomere (Lindsley and Sandler, 1977). Furthermore, recall that the polymorphisms in chromosomes 1, 9, and 16 of humans are most plausibly explained as a result of unequal exchange in regions of specifically centric heterochromatin, and thus, despite the fact these regions show no crossing over, they appear commonly to experience unequal exchange.

What remains unexplained by any of the above proposals is why some species have been able to increase their repetitive DNA content, and hence their heterochromatin content, whereas others have not, even though they are closely related. Within the grasshopper genus *Atractomorpha,* for example, only *similis* shows the spectacular development of distal C-bands. In its two sister species, *australis* and *hypoestes,* heterochromatin is confined to the centromere regions. Equally unexplained is why in some cases specific chromosomes have been excluded from the equilocality that affects other members of the same complement.

5. Conclusions

"The chromosomes of higher cells have long held a central role in the imagination of our better biologists. We know them, however, only at a semimolecular level."

–James D. Watson

There is no reason to doubt that the developmental and evolutionary significance of the facultatively heterochromatinized state lies in its capacity to shut off normal gene function. We still do not know how this is achieved at the molecular level. Nor do we know whether all forms of facultative heterochromatinization involve the same molecular mechanism. DNA from somatic cells containing a functional hypoxanthine–guanine phosphoribosyltransferase *(Hprt)* gene on an active X chromosome can transform hypoxanthine–guanine phosphoribosyltransferase deficient ($HPRT^-$) cells, whereas DNA from the inactive S cannot (Venolia and Gartler, 1983). By contrast, mature sperm from a number of placental and marsupial mammals functions in HPRT transformation (Venolia et al., 1984). This implies that the X chromosome of the male sperm is not inactivated in the same manner as the X of female somatic cells, so that different forms of X inactivation may exist even within the same species.

With constitutive heterochromatin we are dealing with a very different problem. It was Muller and Gershenson (1935) who first suggested that the inert regions of *Drosophila* "consist essentially of nongenic material" rather than inactivated or "degenerated" chromatin, a claim that has certainly proved prophetic. Chromosome segments composed of clusters of hr, or sometimes mr, sequences clearly cannot be involved in conventional coding functions. Why, therefore, such sequences should accumulate in the form of highly condensed heterochromatin remains something of an enigma since such condensation would appear to be irrelevant to the lack of any functional capability. Gatti, Smith, and Baker (1983) have identified a temperature-sensitive X-linked mutant in *Drosophila* whose wild-type allele appears necessary for the mitotic condensation of heterochromatin but not of euchromatin. This indicates that condensation is not simply an inherent property of the repetitive sequences present within heterochromatic regions. Moreover, there is no evidence that sequences naturally inserted into heterochromatin, as opposed to those inserted experimentally, lose their functional capability.

The critical question that needs to be addressed is whether constitutive heterochromatin is, or is not, in any sense a functional component of the genome. Many biologists have been, and some still are, reluctant to even consider the possibility that heterochromatin per se is "inert." Initially, it was argued, on commonsense grounds, that anything that is completely without function would inevitably be discarded by selection, though innumerable examples had long existed from comparative anatomy to indicate that this was certainly not inevitable. Nevertheless, on the basis of the inevitability argument, it was deemed not credible that heterochromatin, and the sequences it contained, should have been conserved in phylogeny unless such conservation implied functional significance. This belief carries with it the necessary rider that all the DNA within a genome has some function and is a product of natural selection. It is the view held by those who see the genome as an integrated evolutionary entity rather than as a somewhat serendipidous mixture of DNA sequences, determined not only by components that maintain basic developmental mechanisms but also by historical events and accidents of the past (Gehring, 1985).

For Mascarello and Mazrimas (1977), to take one example, the near ubiquity of centromeric heterochromatin left little doubt that "it serves some function or, at the very least, reflects the activity or status of some important component of the genome." More recently, Jones and Flavell (1982) argued that repetitive sequences would be readily tolerated only at those sites in the genome where they served a useful function, a theme that has been repeated by many others. Conservation of structure, however, is not compelling evidence for function (Miklos, 1985). Neither is transcription. Thus, the satellite sequences that make up the permanently heteromorphic region of chromosome 1 in *Triturus cristatus carnifex* show extensive readthrough, or overrun, transcription in lampbrush oocytes (Varley et al., 1980) although the transcripts produced have no functional significance. Indeed, transcription by readthrough from upstream structural gene promoters appears to be a common property of amphibian satellite DNAs (Wu, Murphy, and Gall, 1986).

The alternative point of view is that the eukaryote genome, or at least restricted portions of it, has the ability to accommodate developmental or evolutionary accidents which lead, among other things, to the formation of heterochromatic regions that have no functional role. In this connection, it is worth recalling that some organisms, including the mammals *Microtus pennsylvanicus* and *Ellobius lutescens,* function perfectly adequately with only very small amounts of heterochromatin (Yunis and Yasmineh, 1971). Indeed, in several species of felids, C-bands are absent altogether (Pathak and Wurster Hill, 1977), whereas in the opossum they are confined to the sex chromosomes (the autosomes have no heterochromatin; see Sinha, Kakati, and Pathak, 1972).

It is now clear that evolutionary change can be directed or aimless, gradual or abrupt, selective, neutral, or driven. The fallacy has been to focus specifically on one of these to the exclusion of all others. Even Darwin himself did not believe that all products of evolution were adaptive. He most certainly did attribute overwhelming importance to natural selection as the mechanism responsible for evolutionary change of an adaptive kind but did not exclude or reject the influence of other factors. Most of the variation in the hr and mr sequences found within and between species is not easily interpreted in terms of fitness or adaptive change. From the evidence we have reviewed, it is highly improbable that hr DNAs have a selectable function at centromeres or telomeres despite repeated claims that they do. Rather, their common occurrence at these regions suggests either that they possess characteristics favorable to the accumulation of hr sequences or that the mechanisms leading to the amplification of hr sequences operate preferentially at these regions. Neither is there convincing evidence that the presence of hr DNA has served as a means of promoting rapid karyotype change, or that gross changes to chromosome morphology, resulting from the accumulation of heterochromatin, have served as effective reproductive isolating mechanisms and so have stimulated speciation.

Although hr sequences make no direct contribution to the phenotype, it has been suggested that eventual phenotypic effects result from the mere presence, or position, of such sequences within the genome. The former forms the basis of the nucleotypic argument, whereas the latter underlies the claim that position effects can have evolutionary as well as developmental significance (Wilson, White, Carlson, and Cherry, 1977). Neither case is convincing. Some kinds of heterochromatin have long been associated with subtle quantitative phenotypic effects. Mather (1941, 1944) was the first to show that, in *Drosophila melanogaster,* Y chromosomes of different origin exerted a small but significant influence on the combined numbers of chaetae on the ventral surfaces of the fourth and fifth abdominal segments. He concluded that the Y heterochromatin, and probably all heterochromatin, contained linked combinations of distinctive polygenes with individually small, but additive, effects on quantitative parameters. Barigozzi (1951) subsequently showed that Y chromosomes of different origin also influenced the frequency of unicellular hairs on a given portion of the wing, as well as the size of the corneae of the compound eyes, of *D. melanogaster.* Both these characters are formed from a fixed number of cells. Each wing hair corresponds to a single cell whereas the corneae are constructed from two cells each. Barigozzi concluded that the increased mean number of wing hairs, and the reduced size of the corneae, in the male compared to the female was the result of Y-linked polygenes which affected cell size. There is nothing in the sequence structure of Y heterochromatin that provides a sensible basis for these hypothetical polygenes, and the precise causation of these effects has yet to be clarified.

There remains the possibility that heterochromatin serves a specifically germ-line function and so, understandably, is without significance for somatic development and hence viability. This is seemingly supported by two facts: first, its invariable retention within the germ line in those cases where it is eliminated from the soma; second, the several effects that heterochromatin may exert on recombination within the germ line. Bostock and Sumner (1978) have argued that if satellite DNA, and by implication heterochromatin, has no function of any kind, then means to eliminate it from the germ line would have evolved. The fact that this has not occurred suggests to them that heterochromatin has a function related to meiosis.

In *Drosophila melanogaster,* rather atypically as it turns out in retrospect, the heterochromatic blocks of each pair of homologues have their own specific arrangement of satellite sequences, as revealed by the in situ hybridization of uncloned probes. This led to the proposition that these distinctive sequence patterns provided a recognition system which regulated the pairing between homologues (Brutlag, Appels, Dennis, and Peacock, 1977; Peacock et al., 1978). Although the complex 359-bp satellite of this species was initially mapped to all chromosomes, the subsequent use of cloned probes has shown that it is located exclusively in the heterochromatin of the X (Hilliker and Appels, 1982) and so cannot possibly play a role in the pairing between the X and the Y. Added to this, the as yet uncloned AATAT and AAGAG sequences are located close to the centromere of the X so that the regions of this chromosome that include the male pairing sites, as defined by Cooper (1964), are unquestionably devoid of both these satellites. As far as the autosomes are concerned, deficiencies and rearrangements of the heterochromatic regions have no effect on pairing, which is governed by euchromatic, not heterochromatic, homology (Yamamoto and Miklos, 1978; Hawley, 1980; Szauter, 1984). Likewise, in *D. hydei,* one or both ends of the Y chromosome are usually associated with the end of the heterochromatic arm of the X at first metaphase of male meiosis. Here, as in *D. melanogaster,* there is no evidence for any DNA homology in the regions of association between the sex chromosomes (Kremer et al., 1986).

If, therefore, heterochromatin has a functional role in meiosis, this role is most certainly not involved in the pairing of homologues. There remain the well-demonstrated effects of heterochromatin, both positive and negative, on chiasma formation and crossing over. In *Ascaris lumbricoides,* there is evidence that the hr DNA, which is confined to the germ line, is not transcribed at meiosis (Muller et al., 1982). If, therefore, this satellite affects processes specific to the germ line, then it must do so by means other than transcription. Consequently, until we have a much clearer idea of how heterochromatic effects on recombination are mediated in molecular terms, and whether these effects do indeed have evolutionary consequences, it would be prudent to exercise some caution before attributing a functional role to the retention of heterochromatin within the germ line. Indeed, there is an alternative explanation that urgently needs to be tested.

P elements are a family of mobile genetic sequences, present in the genome of *D. melanogaster,* whose transposition is known to be tissue specific. It occurs at a high frequency in the germ line but does not occur at all in somatic tissues (Engels, 1983), despite the fact that P elements show somatic transcription (Laski, Rio, and Rubin, 1986). The reason for this is that the P-element transcript has a germ-line-specific splice that is required for transposase activity. Whether the splicing machinery is specially modified in the germ line, or whether special conditions exist within the germ line to facilitate splicing, is not known, but, whatever the precise explanation, it indicates that conditions within the germ line constrain the mRNA splicing of the P-element transcript. Could it be that the germ line also lacks an enzyme, or enzymes, necessary for the deletion of heterochromatin so that its retention here is an indication not of function, but rather of the special molecular conditions that obtain within germ-line tissues?

There is then a very real possibility that heterochromatin per se has no function in either development or evolution. This would imply that it is, in effect, nothing more than an evolutionary relic, a by-product of molecular events within the genome involving DNA sequences whose replication and multiplication are not constrained by normal cell control mechanisms. Cyclical rounds of amplification, divergence, and deletion might then account for many of the qualitative and quantitative changes experienced by heterochromatin over time (Singer, 1982). Further change might also result from the variable occurrence within heterochromatin of mobile genetic elements which are recombinogenic (Miklos, 1985) and about which, as yet, we still have very little detailed information. This stems largely from the failure to couple in situ studies on the polytene chromocenter with comparable studies on mitotic chromosomes in *Drosophila.* Consequently, authors persistently fail to discriminate between labeling in α versus β regions of the chromocenter.

Although heterochromatin is no longer simply what "an investigator chooses to make it or imagine it" (Cooper, 1956), the fact remains that there are very obvious differences between the regions we can now define in rather more precise terms as heterochromatin. The issue to be resolved is whether the basis of this heterogeneity rests on the structure of heterochromatin per se or whether it is a consequence of the sequestering within that heterochromatin of sequences of diverse origin which, through the wanderings of genomic components, have found a temporary, or even permanent, resting place within a region that is not constrained by the inevitable restrictions imposed upon euchromatin by development and differentiation. This implies that different heterochromatic regions embody distinct genetic systems, not so much by virtue of their own inherent molecular structure but through sequences that have been introduced into the heterochromatic environment and continue to function in that environment. This could well account for the very real effects and functions which, mistakenly in the past, were attributed to heterochromatin itself.

6. References

Ananiev, E. V., Gvozdev, V. A., Ilyin, Y. U., Tchurikov, N. A., and Georgiev, G. P. (1978). Reiterated genes with varying location in intercalary heterochromatin regions of *Drosophila melanogaster*. *Chromosoma, 70,* 1–18.

Ananiev, E. V., Barsky, V. E., Ilyin, Y. V., and Churikov, N. A. (1981). Localization of nucleoli in *Drosophila melanogaster* polytene chromosomes. *Chromosoma, 81,* 619–628.

Appels, R. (1982). The molecular cytology of wheat rye hybrids. *Int. Rev. Cytol., 80,* 93–132.

Arnason, U. (1974). Comparative chromosome studies in Cetacea. *Hereditas, 77,* 1–36.

Arnason, U., Benirschke, K., Nichols, W. W., and Mead, J. G. (1977). Banded karyotypes of three whales: *Mesoplodon europaeus, M. carlhubbsi* and *Balaenoptera acutirostrata. Hereditas, 87,* 187–200.

Arnason, U., Purdom, I. F., and Jones, K. W. (1978). Conservation and chromosomal locatization of DNA satellites in balaenopterid whales. *Chromosoma, 66,* 141–159.

Arnason, U., Lutley, R., and Sandholt, B. (1980). Banding studies on six killer whales: an account of C band polymorphism and G band patterns. *Cytogenet. Cell Genet., 28,* 71–78.

Arnason, U., Höglund, M., and Widegren, B. (1984). Conservation of highly repetitious DNA in cetaceans. *Chromosoma, 89,* 238–242.

Arrighi, F. E., Hsu, T. C., Pathak, S., and Sawada, H. (1974). The sex chromosomes of the Chinese hamster: constitutive heterochromatin deficient in repetitive DNA sequences. *Cytogenet. Cell Genet., 13,* 268–274.

Back, F. (1976). The variable condition of euchromatin and heterochromatin. *Int. Rev. Cytol., 45,* 25–64.

Baker, J. R., and Callan, H. G. (1950). Heterochromatin. *Nature, 166,* 227.

Baker, W. K. (1958). Crossing over in heterochromatin. *Am. Natur., 92,* 59–60.

Baker, W. K. (1971). Evidence for position effect suppression of the ribosomal

RNA cistrons in *Drosophila melanogaster. Proc. Natl. Acad. Sci. USA, 68,* 2472–2476.

Bamai, V. (1975a). The relationship between metaphase heterochromatin and polytene inversions in *Drosophila. Experientia, 31,* 779–780.

Bamai, V. (1975b). Heterochromatin and multiple inversions in a *Drosophila* chromosome. *Can. J. Genet. Cytol., 17,* 15–20.

Bamai, V. (1977). Chromosomal polymorphisms of constitutive heterochromatin and inversions in *Drosophila. Genetics, 85,* 85–93.

Barigozzi, C. (1951). The influence of the Y chromosome in quantitative characters of *Drosophila melanogaster. Heredity, 5,* 415–432.

Barlow, P. W. (1978). The interrelationship of the cycles of chromosome condensation and reduplication in cell growth processes. *Nucleus, 21,* 1–11.

Barnes, S. R., James, A. A., and Jamieson, G. (1985). The organisation, nucleotide sequence and chromosome distribution of a satellite DNA from *Allium cepa. Chromosoma, 92,* 185–192.

Barr, H. J., and Ellison, J. R. (1972). Ectopic pairing of chromosome regions containing chemically similar DNA. *Chromosoma, 39,* 53–61.

Baverstock, P. R., Watts, C. H. S., and Hogarth, J. T. (1977). Polymorphism of the X chromosome, Y chromosome and autosomes in the Australian Hopping mice, *Notomys alexis, N. cervinus* and *N. fuscus* (Rodentia, Muridae). *Chromosoma, 61,* 243–256.

Baverstock, P. R., Adams, M., Polkinghorne, R. W., and Gelder, M. (1982). A sex linked enzyme in birds: Z chromosome conservation but no dosage compensation. *Nature, 296,* 763–766.

Bedo, D. G. (1975). C banding in polytene chromosomes of *Simulium ornatipes* and *S. melatum* (Diptera:Simulidae). *Chromosoma, 51,* 291–300.

Bedo, D. G. (1980). C, Q and H banding in the analysis of Y chromosome rearrangements in *Lucilia cuprina* (Wiedermann) (Diptera:Calliphoridae). *Chromosoma, 77,* 299–308.

Bedo, D. G. (1982). Differential sex chromosome replication and dosage compensation in polytene trichogen cells of *Lucilia cuprina* (Diptera: Calliphoridae). *Chromosoma, 77,* 299–308.

Bedo, D. G. (1986). Polytene and mitotic chromosome analysis in *Ceratitis capitata* (Diptera; Tephritidae). *Can. J. Genet. Cytol., 28,* 180–188.

Beermann, S. (1977). The diminution of heterochromatic chromosome segments in *Cyclops* (Crustacea, Copepoda). *Chromosoma, 60,* 297–344.

Beerman, W. (1965). Riesenchromosomen. *Plasmatologia, Bd. VI/D.* Vienna: Springer-Verlag.

Belyaeva, E. S. P., Ananiev, E. V., and Gvozdev, V. A. (1984). Distribution of mobile dispersed genes (*mdg* 1 and *mdg* 3) in the chromosomes of *Drosophila melanogaster. Chromosoma, 90,* 16–19.

Berendes, H. D., and Keyl, H. G. (1967). Distribution of DNA in heterochromatin and euchromatin of polytene nuclei of *Drosophila hydei. Genetics, 57,* 1–13.

Berlowitz, L. (1965). Analysis of histone in situ in developmentally inactivated chromatin. *Proc. Natl. Acad. Sci. USA, 54,* 476–480.

Berlowitz, L., Loewus, M. W., and Pallotta, D. (1968). Heterochromatin and genetic activity in mealy bugs. I. Compensation for inactive chromatin by increase in cell number. *Genetics, 60,* 93–99.

Bennett, M. D. (1984). Nuclear architecture and its manipulation. In *Proc. 16th Stadler Symp.,* ed. J. P. Gustafson, pp. 469–502. New York: Plenum Press.

Biessmann, H., Kruger, P., Schröpfer, C., and Spindler, E. (1981). Molecular cloning and preliminary characterization of a *Drosophila melanogaster* gene from a region adjacent to the centromeric β heterochromatin. *Chromosoma, 82,* 493–503.
Bolshakov, V. N., Zharkikh, A. A., and Zhimulev, I. F. (1985). Intercalary heterochromatin in *Drosophila.* II. Heterochromatic features in relation to local DNA content along the polytene chromosomes of *Drosophila melanogaster. Chromosoma, 92,* 200–208.
Bonaccorsi, S., Pimpinelli, S., and Gatti, M. (1981). Cytological dissection of sex chromosome heterochromatin of *Drosophila hydei. Chromosoma, 84,* 391–403.
Bostock, C. J., and Christie, S. (1976). Analysis of the frequency of sister chromatid exchange in different regions of chromosomes of the kangaroo rat *(Dipodomys ordii). Chromosoma, 56,* 275–287.
Bostock, C. J., and Clarke, E. M. (1980). Satellite DNA in large marker chromosomes of methotrexate resistant mouse cells. *Cell, 19,* 709–715.
Bostock, C. J., and Sumner, A. T. (1978). *The Eukaryotic Chromosome.* Amsterdam: North Holland.
Bridges, C. B. (1935). Salivary chromosome maps. *J. Hered., 26,* 60–64.
Brinkley, B. R., Valdivia, M. M., Tousson, A., and Brenner, S. L. (1984). Compound kinetochores of the Indian muntjac: evolution by linear fusion of unit kinetochores. *Chromosoma, 91,* 1–11.
Britten, R., and Davidson, E. H. (1976). DNA sequence arrangement and preliminary evidence on its evolution. *Fed. Proc., 35,* 2151–2157.
Brown, S. (1983). Heterochromatin in the human genome: a quantitative study. Honor's thesis, Dept. Biology, Tulane University.
Brown, S. W. (1966). Heterochromatin. *Science, 151,* 417–425.
Brown, S. W. (1969). Developmental control of heterochromatinization in coccids. *Genetics* [*Suppl.*] *61,* 191–198.
Brutlag, D. L. (1980). Molecular arrangement and evolution of heterochromatic DNA. *Annu. Rev. Genet., 14,* 121–144.
Brutlag, D., Appels, R., Dennis, E. S., and Peacock, W. J. (1977). Highly repeated DNA in *Drosophila melanogaster. J. Mol. Biol., 112,* 31–47.
Bull, J. (1978). Sex chromosome differentiation: an intermediate stage in a lizard. *Can. J. Genet. Cytol., 20,* 205–209.
Burk, R. D., Szabo, P., O'Brien, S., Nash, W. G., Yu, L., and Smith, K. D. (1985). Organization and chromosomal specificity of autosomal homologs of human Y chromosome repeated DNA. *Chromosoma, 92,* 225–233.
Burns, J. A., and Gerstel, D. U. (1973). The formation of megachromosomes from heterochromatic blocks of *Nicotiana tomentosiformis. Genetics, 75,* 497–502.
Carothers, A. D., Buckton, K. E., Collyer, S., deMay, R., Frackiewicz, A., Piper, J., and Smith, L. (1982). The effect of variant chromosomes on reproductive fitness in man. *Clin. Genet., 21,* 280–289.
Carrano, A. V., and Wolff, S. (1975). Distribution of sister chromatid exchanges in the euchromatin and heterochromatin of the Indian muntjac. *Chromosoma, 53,* 361–369.
Carson, H. L., Clayton, F. E., and Stalker, H. D. (1967). Karyotype stability and speciation in Hawaiian *Drosophila. Proc. Natl. Acad. Sci. USA, 87,* 1280–1285.
Cattanach, B. M. (1974). Position effect variegation in the mouse. *Genet. Res., 23,* 291–306.
Chandley, A. C. (1981). The origin of chromosomal aberrations in man and their potential for survival and reproduction in the adult human population. *Ann. Genet. (Paris), 24,* 5–11.

Chandra, H. S. (1971). Inactivation of whole chromosomes in mammals and coccids: some comparisons. *Genet. Res., 18,* 265–276.
Chandra, H. S. (1985). Is human X chromosome inactivation a sex determining device? *Proc. Natl. Acad. Sci. USA, 82,* 6947–6949.
Chandra, H. S. (1986). X chromosomes and dosage compensation. *Nature, 319,* 18.
Charlesworth, B., Langley, C. H., and Stephan, W. (1986). The evolution of restricted recombination and the accumulation of repeated DNA sequences. *Genetics, 112,* 947–962.
Chauhan, K. P. S., and Abel, W. O. (1968). Evidence for the association of homologous chromosomes during premeiotic stages in *Impatiens* and *Salvia. Chromosoma, 25,* 297–302.
Church, K. (1979). The grasshopper X chromosome. II. Negative heteropycnosis, transcription activities and compartmentalization during spermatogonial stages. *Chromosoma, 71,* 359–370.
Citoler, P., Gropp, A., and Natarajan, A. T. (1972). Timing of DNA replication of autosomal heterochromatin in the hedgehog. *Cytogenetics, 11,* 53–62.
Cohen, E. H., and Kaplan, G. C. (1982). Analysis of DNAs from two species of the *virilis* group of *Drosophila* and implications for satellite DNA evolution. *Chromosoma, 87,* 519–534.
Comings, D. E. (1972). The structure and function of chromatin. *Adv. Hum. Genet., 3,* 237–431.
Comings, D. E. (1980). Arrangement of chromatin in the nucleus. *Hum. Genet., 53,* 131–143.
Comings, D. E., and Okada, T. A. (1976). Fine structure of the heterochromatin of the kangaroo rat *Dipodomys ordii* and examination of the possible role of actin and myosin in heterochromatin condensation. *J. Cell Sci., 21,* 465–477.
Cooke, H. J. (1976). Repeated sequence specific to human males. *Nature, 262,* 182–186.
Cooper, D. W., Johnston, P. G., Sharman, G. B., and Vandeberg, J. L. (1977). The control of gene activity on eutherian and metatherian X chromosomes: a comparison. In *Reproduction and Evolution,* 4th Symp. Comp. Biol. Rep., ed. J. H. Calaby and C. H. T. Tyndale Biscoe, pp. 81–877. Canberra: Aust. Acad. Sci.
Cooper, J. E. K. (1977a). Chromosomes and DNA of *Microtus.* II. Confirmation of deletion of constitutive heterochromatin in *M. agrestis* cells in vitro. *Can. J. Genet. Cytol., 19,* 537–541.
Cooper, J. E. K. (1977b). Chromosomes and DNA of *Microtus.* III. Heterochromatin rearrangement in *M. agrestis* bone marrow clones. *Chromosoma, 62,* 269–278.
Cooper, K. W. (1956). Phenotypic effects of Y chromosome hyperploidy in *Drosophila melanogaster* and their relation to variegation. *Genetics, 41,* 242–264.
Cooper, K. W. (1964). Meiotic conjunctive elements not involving chiasmata. *Proc. Natl. Acad. Sci. USA, 52,* 1248–1255.
Cordeiro, M., Wheeler, L., Lee, C. S., Kastritsis, C. D., and Richardson, R. H. (1975). Heterochromatic chromosomes and satellite DNAs of *Drosophila nasutoides. Chromosoma, 51,* 65–73.
Cordeiro Stone, M., and Lee, C. S. (1976). Studies on the satellite DNA of *Drosophila nasutoides:* their buoyant densities, melting temperatures, reassociation rates and localizations in polytene chromosomes. *J. Mol. Biol., 104,* 1–24.

Corneo, G. (1976). Do satellite DNAs function as sterility barriers in eukaryotes. *Evol. Theory, 1,* 261–265.
Cowell, J. K., and Hartmann Goldstein, I. J. (1980). Modification of the DNA content in translocated regions of *Drosophila* polytene chromosomes. *Chromosoma, 81,* 55–64.
Craig, J. W., and Tolley, E. (1986). Steroid sulphatase and the conservation of mammalian X chromosomes. *Trends Genet. 2,* 201–204.
Craig Holmes, A. P., Moore, F. B., and Shaw, M. W. (1975). Polymorphisms of human C band heterochromatin II. Family studies with suggestive evidence for somatic crossing over. *Am. J. Hum. Genet., 27,* 178–189.
Crouse, H. V. (1979). X heterochromatin subdivision and cytogenetic analysis in *Sciara coprophila* (Diptera, Sciaridae). II. The controlling element. *Chromosoma, 74,* 219–239.
Crouse, H. V., Brown, A., and Mumford, B. C. (1971). L chromosome inheritance and the problem of chromosome imprinting in sciara (Sciaridae, Diptera). *Chromosoma, 34,* 324–339.
Darlington, C. D. (1958). *Evolution of Genetic Systems,* 2nd ed. Edinburgh: Oliver and Boyd.
Davis, A. H., Kidd, G. H., and Carter, C. E. (1979). Chromosome diminution in *Ascaris suum:* two fold increase of nucleosomal histone to DNA ratios during development. *Biochim. Biophys. Acta, 565,* 315–325.
Deharveng, L., and Lee, B. H. (1984). Polytene chromosomal variability of *Bilobella aurantica* (Collembola) from Sainte Baume population (France). *Caryologia, 37,* 51–67.
Demerec, M. (1940). Genetic behavior of euchromatic segments inserted into heterochromatin. *Genetics, 25,* 618–627.
Dennis, E. S., Dunsmuir, P., and Peacock, W. J. (1980). Segmental amplification in a satellite DNA: restriction enzyme analysis of the major satellite of *Macropus rufogriseus. Chromosoma, 79,* 179–198.
Deumling, B. (1981). Sequence arrangement of a highly methylated satellite DNA of a plant, *Scilla:* a tandemly repeated inverted repeat. *Proc. Natl. Acad. Sci. USA, 78,* 338–342.
Dev, V. G., Miller, D. A., Charen, J., and Miller, O. J. (1974). Translocation of centromeric heterochromatin in the T(10;13)199H stock of *Mus musculus* and localization of chromosome break points. *Cytogenet. Cell Genet., 13,* 256–267.
Dorn, R., Heymann, S., Lindigkeit, R., and Reuter, G. (1986). Suppressor mutation of position effect variegation in *Drosophila melanogaster* affecting chromatin properties. *Chromosoma, 93,* 398–403.
Drets, M. E., and Stoll, M. (1974). C banding and nonhomologous associations in *Gryllus argentinus. Chromosoma, 48,* 367–390.
Dyer, A. F. (1964). Heterochromatin in American and Japanese species of *Trillium.* II. The behavior of H segments. *Cytologia, 29,* 171–190.
Eastman, E. M., Goodman, R. M., Erlanger, B. F., and Miller, O. J. (1980). The organization of DNA in the mitotic and polytene chromosomes of *Sciara coprophila. Chromosoma, 79,* 293–314.
Eicher, E. M. (1970). X autosome translocations in the mouse: total inactivation versus partial inactivation of the X chromosome. *Adv. Genet., 15,* 175–259.
Endow, S. A., and Gall, J. G. (1975). Differential replication of satellite DNA in polyploid tissues of *Drosophila virilis. Chromosoma, 50,* 175–192.
Engels, W. R. (1983). The P family of transposable elements in Drosophila. *Annu. Rev. Genet., 17,* 315–344.

Epplen, J. T., McCarrey, J. R., Sutou, S., and Ohno, S. (1982). Base sequence of a cloned snake hr chromosome fragment and identification of a male specific putative mRNA in the mouse. *Proc. Natl. Acad. Sci. USA, 79,* 3798–3802.

Fazal Farook, S. A., and Vig, B. K. (1980). Sequence of centromere separation: analysis of mitotic chromosomes of *Crepis capillaris* and *Haplopappus gracilis. Biol. Ziol., 99,* 675–682.

Figueroa, M. L., and Vig, B. K. (1983). Sequence of centromere separation: lack of colcemid effect on the Chinese hamster genome. *Cytogenet. Cell Genet., 36,* 627–632.

Flavell, R. B. (1982). Sequence amplification, deletion and rearrangement: major sources of variation during species divergence. In *Genome Evolution,* ed. G. A. Dover and R. B. Flavell, pp. 301–323. London: Academic Press.

Ford, C. E. (1970). The cytogenetics of the male germ cells and the testis in mammals. In *The Human Testis,* ed. E. Rosemberg and C. A. Paulsen, pp. 139–147. New York: Plenum Press.

Forejt, J. (1984). X inactivation and its role in male sterility. *Chromosomes Today, 8,* 117–127.

Forejt, J., and Gregorova, S. (1977). Meiotic studies of translocations causing male sterility in the mouse. I. Autosomal reciprocal translocations. *Cytogenet. Cell Genet., 19,* 159–179.

Forejt, J., Gregorova, S., and Goetz, P. (1981). XY pair associates with the synaptonemal complex of autosomal male sterile translocations in pachytene spermatocytes of the mouse *(Mus musculus). Chromosoma, 82,* 41–53.

Fraccaro, M., Hansson, K., Hutten, M., Lindsten, J., and Tieplo, L. (1969). Heterochromatin in preimplantation mouse embryos. *Exp. Cell Res., 55,* 427–430.

Fredga, K. (1970). Unusual sex chromosome inheritance in mammals. *Philos. Trans. R. Soc. Lond.* [*B*], *259,* 15–36.

Fredga, K., and Mandahl, N. (1972). Autosomal heterochromatin in some carnivores. In *Chromosome Identification,* Nobel Symp. No. 23, ed. T. Caspersoon and L. Zech, pp. 104–117. London: Academic Press.

Frenster, J. H. (1974). Ultrastructure and function of heterochromatin and euchromatin. In *The Cell Nucleus,* ed. H. Busch, pp. 565–580. New York: Academic Press.

Gall, J. G. (1973). Repetitive DNA in *Drosophila.* In *Molecular Cytogenetics,* ed. B. A. Hamkalo and J. Papaconstantenou, pp. 59–74. New York: Plenum Press.

Gall, J. G., Cohen, E. H., and Polan, M. L. (1971). Repetitive DNA sequences in *Drosophila. Chromosoma, 33,* 319–344.

Gartler, S. M., and Riggs, A. D. (1983). Mammalian X chromosome inactivation. *Annu. Rev. Genet., 17,* 155–190.

Gatti, M., and Pimpinelli, S. (1983). Cytological and genetic analysis of the Y chromosome of *Drosophila melanogaster.* I. Organization of the fertility factors. *Chromosoma, 88,* 349–373.

Gatti, M., Pimpinelli, S., and Santini, G. (1976). Characterization of *Drosophila* heterochromatin I. Staining and decondensation with Hoechst 33258 and quinacrine. *Chromosoma, 57,* 351–375.

Gatti, M., Smith, D. A., and Baker, B. S. (1983). A gene controlling condensation of heterochromatin in *Drosophila melanogaster. Science, 221,* 83–85.

Gehring, W. J. (1985). Homeotic genes, the homeo box and the genetic control of development. *Cold Spring Harbor Symp. Quant. Biol., 50,* 243–251.

Gerstel, D. U., and Burns, J. A. (1966). Chromosomes of unusual length in hybrids between two species of *Nicotiana*. *Chromosomes Today, 1,* 41–56.
Goday, C., and Pimpinelli, S. (1986). Cytological analysis of chromosomes in the two species *Parascaris univalens* and *D. equorum*. *Chromosoma, 94,* 1–10.
Goday, C., Ciofi Luzzatto, A., and Pimpinelli, S. (1985). Centromere ultrastructure in germ line chromosomes of *Parascaris*. *Chromosoma, 91,* 121–125.
Goldschmidt, R. B. (1955). *Theoretical Genetics*. Berkeley: University of California Press.
Gropp, A., and Citoler, P. (1969). Pattern of autosomal heterochromatin. In *Comparative Mammalian Cytogenetics,* ed. K. Benirschke, pp. 267–276. New York: Springer-Verlag.
Gropp, A., and Natarajan, A. T. (1972). Karyotype and heterochromatin pattern of the Algerian hedgehog. *Cytogenetics, 11,* 259–269.
Gropp, A., Citoler, P., and Geisler, M. (1969). Karyotypvariation und Heterochromatinmuster bei Igelen (Erinaceus und Hemiechinus). *Chromosoma, 27,* 288–307.
Gvozdev, V. A., Ananiev, E. V., Kotelyanskaya, A. E., and Zhimulev, I. F. (1980). Transcription of intercalary heterochromatin regions in *Drosophila melanogaster* cell culture. *Chromosoma, 80,* 177–190.
Haaf, T., and Schmid, M. (1984). An early stage of ZW/ZZ sex chromosome differentiation in *Poecilia sphenops* var. *melanistica* (Poecilidae: Cyprinodontiformes). *Chromosoma, 89,* 37–41.
Hackstein, J. H. P., Leoncini, O., Beck, H., Peelen, G., and Hennig, W. (1982). Genetic fine structure of the Y chromosome of *Drosophila hydei*. *Genetics, 101,* 257–277.
Hägele, K. (1977). Differential staining of polytene chromosome bands in *Chironomus* by Giemsa banding methods. *Chromosoma, 59,* 207–215.
Hägele, K. (1985). Identification of a polytene chromosome band containing a male sex determiner of *Chironomus thummi*. *Chromosoma, 91,* 167–171.
Hale, D. W., and Greenbaum, I. F. (1986). The behavior and morphology of the X and Y chromosomes during prophase I in the Sitka deer mouse *(Peromyscus sitkensis)*. *Chromosoma, 94,* 235–242.
Hamlin, J. L., Milbrandt, J. D., Heintz, N. H., and Azizkhan, J. C. (1984). DNA sequence amplification in mammalian cells. *Int. Rev. Cytol., 90,* 31–82.
Hardy, R. W., Lindsley, D., Livak, K. J., Lewis, B., Siverstein, A. L., Joslyn, G. L., Edwards, J., and Bonaccorsi, S. (1984). Cytogenetic analysis of a segment of the chromosome of *Drosophila melanogaster*. *Genetics, 107,* 591–610.
Hartmann Goldstein, I. J. (1967). On the relationship between heterochromatinization and variegation in *Drosophila* with special reference to temperature sensitive periods. *Genet. Res., 10,* 143–159.
Hartmann Goldstein, I. J., and Goldstein, D. J. (1979). Effect of temperature on morphology and DNA content of polytene chromosomes in Drosophila. *Chromosoma, 71,* 333–346.
Hatch, F. T., and Mazrimas, J. A. (1977). Satellite DNA and cytogenetic evolution. In *Molecular Human Cytogenetics,* ICN-UCLA Symp. 7, ed. R. S. Sparkes, D. E. Comings, and C. F. Fox, pp. 395–414. New York: Academic Press.
Hawley, R. S. (1980). Chromosomal sites necessary for normal levels of meiotic recombination in *Drosophila melanogaster*. I. Evidence for and mapping of the sites. *Genetics, 94,* 625–646.
Hayman, D. L., and Martin, P. G. (1974). *Mammalia I: Monotremata and Mar-*

supialia. Animal Cytogenetics, ed. B. John, Vol. 4: *Chordata,* Part 4. Stuttgart: Gebrüder Borntraeger.

Hayman, D. L., Martin, P. G., and Waller, P. F. (1969). Parallel mosaicism of supernumerary chromosomes and sex chromosomes in *Echymipera kalabu* (Marsupialia). *Chromosoma, 27,* 371–380.

Hazelrigg, T., Levis, R., and Rubin, G. M. (1984). Transformation of white locus DNA in *Drosophila:* dosage compensation, zeste interaction and position effects. *Cell, 36,* 469–481.

Heitz, E. (1928). Das Heterochromatin der Moose. I. *Jahrb. Wissensch. Bot., 69,* 762–818.

Heitz, E. (1933). Die somatische Heteropknose bei *Drosophila melanogaster* und ihre genetische Bedeutung. *Z. Zellforsch., 20,* 237–287.

Heitz, E. (1934). Über α and β Heterochromatinization sowie Konstahz und Bau der Chromomeren bei Drosophila. *Biol. Zentralbl., 54,* 588–609.

Heitz, E. (1935). Die Herkunft der Chromozentren. *Planta, 18,* 571–635.

Henikoff, S. (1981). Position effect variegation and chromosome structure of a heat shock puff in *Drosophila. Chromosoma, 83,* 381–393.

Hennig, W. (1967). Untersuchungen zur Struktur und Funktion des Lampenbursten Y Chromosoms in der Spermatogenese von Drosophila. *Chromosoma, 22,* 294–357.

Hennig, W. (1985). Y chromosome function and spermatogenesis in *Drosophila hydei. Adv. Genet., 23,* 179–234.

Hess, O., and Meyer, G. F. (1968). Genetic activities of the Y chromosome in *Drosophila* during spermatogenesis. *Adv. Genet., 14,* 171–223.

Hessler, A. Y. (1958). V type position effects at the light locus in *Drosophila melanogaster. Genetics, 43,* 395–403.

Hilliker, A. J., and Appels, R. (1982). Pleiotropic effects associated with the deletion of heterochromatin surrounding rDNA on the Y chromosome of *Drosophila. Chromosoma, 86,* 469–490.

Hilliker, A. J., Appels, R., and Schalet, A. (1980). The genetic analysis of *Drosophila melanogaster* heterochromatin. *Cell, 21,* 607–619.

Hsu, T. C., Schmidt, W., and Stubblefield, E. (1964). DNA replication sequences in higher animals. In *The Role of Chromosomes in Development,* ed. M. Locke, pp. 83–112. New York: Academic Press.

Hsu, T. C., and Arrighi, F. E. (1971). Distribution of constitutive heterochromatin in mammalian chromosomes. *Chromosoma, 34,* 243–253.

Huettner, A. F. (1933). Continuity of the centrioles in *Drosophila melanogaster. Z. Zellforsch. Mikros. Anat., 19,* 119–134.

Hyde, B. B. (1953). Differentiated chromosomes of *Plantago ovata. Am. J. Bot., 40,* 8809–8815.

Illmensee, K. (1972). Developmental potencies of nuclei from cleavage, preblastoderm and syncytial blastoderm transplanted into unfertilized eggs of *Drosophila melanogaster. Wilhelm Roux Archiv., 170,* 267–298.

Jablonka, E., Goitein, R., Marcus, M., and Cedar, H. (1985). DNA hypomethylation causes an increase in DNase I sensitivity and an advance in the time of replication of the entire inactive X chromosome. *Chromosoma, 93,* 152–156.

James, T. C., Elgin, S. C. R., and Eissenberg, J. C. (1986). Cloning of a heterochromatin-specific chromosomal protein gene in *Drosophila melanogaster. The EMBO Crete Workshop* (Abstract).

Jarrell, G. H. (1984). Adaptive radiation of whales: is constitutive heterochromatin an intrinsic isolating mechanism? *Evol. Theory, 7,* 53–62.

John, B. (1976). Myths and mechanisms of meiosis. *Chromosoma, 54,* 295–325.
John, B., and King, M. (1977). Heterochromatin variation in *Cryptobothrus chrysophorus.* I. Chromosome differentiation in natural populations. *Chromosoma, 64,* 219–239.
John, B., and King, M. (1983). Population cytogenetics of *Atractomorpha similis.* I. C band variation. *Chromosoma, 88,* 57–68.
John, B., and King, M. (1985a). Pseudoterminalization, terminalization and the origin of terminal associations. *Chromosoma, 93,* 89–99.
John, B., and King, M. (1985b). The interrelationship between heterochromatin distribution and chiasma distribution. *Genetics, 66,* 183–194.
John, B., and Miklos, G. L. G. (1979). Functional aspects of heterochromatin and satellite DNA. *Int. Rev. Cytol., 58,* 1–114.
John, B., and Miklos, G. L. G. (1987). *The Eukaryote Genome in Development and Evolution.* London: Allen and Unwin.
John, B., King, M., Schweizer, D., and Mendelak, M. (1985). Equilocality of heterochromatin distribution and heterochromatin heterogeneity. *Chromosoma, 91,* 185–200.
John, B., Appels, R., and Contreras, N. (1986). Population cytogenetics of *Atractomorpha simils.* II. Molecular characterisation of the distal C band polymorphisms. *Chromosoma, 94,* 45–58.
Johnston, P. G., and Robinson, E. S. (1985). Glucose 6 phosphate dehydrogenase expression in heterozygous kangaroo embryos and extraembryonic membranes. *Genet. Res., 45,* 205–208.
Jones, G. H. (1978). Giemsa C banding of rye meiotic chromosomes and the nature of 'terminal' chiasmata. *Chromosoma, 66,* 45–57.
Jones, J. D. G., and Flavell, R. B. (1982). The mapping of highly repeated DNA families and their relationship to C bands in chromosomes of *Secale cereale. Chromosoma, 86,* 595–612.
Jones, K. W. (1976). Repetitive DNA sequences in animals, particularly primates. *Chromosomes Today, 5,* 305–313.
Jones, K. W. (1978). Speculations on the functions of satellite DNA in evolution. *Z. Morphol. Anthropol., 69,* 143–171.
Jones, K. (1984). The evolution of sex chromosomes and their consequences for the evolutionary process. *Chromosomes Today, 8,* 241–255.
Jones, K. W., and Singh, L. (1982). Conserved sex associated repeated DNA sequences in vertebrates. In *Genome Evolution,* ed. G. A. Dover and R. B. Flavell, pp. 135–154. London: Academic Press.
Jones, R. N., and Rees, H. (1982). *B Chromosomes.* London: Academic Press.
Karpen, G., and Laird, C. (1986). Autonomous function of a single rRNA cistron. *27th American Drosophila Conference, Asilomar, California* (Abstract).
Kato, H. (1979). Preferential occurrence of sister chromatid exchanges at heterochromatin euchromatin junctions in the wallaby and hamster chromosomes. *Chromosoma, 74,* 307–316.
Kaufmann, B. P. (1939). Distribution of induced breaks along the X chromosome of *Drosophila melanogaster. Proc. Natl. Acad. Sci. USA, 25,* 571–577.
Kerem, B., Goitein, R., Richler, C., Marcus, M., and Cedar, H. (1983). In situ nick translation distinguishes between active and inactive X chromosomes. *Nature, 5921,* 89–90.
Kerem, B., Goitein, R., Diamond, G., Cedar, H., and Marcus, M. N. (1984). Mapping of DNase I sensitive regions on mitotic chromosomes. *Cell, 38,* 493–499.
Keyl, H.-G. (1965a). Duplikationen von Untereinheiten der Chromosomalen

DNS während der Evolution von *Chironomus thummi*. *Chromosoma, 17,* 139–180.

Keyl, H.-G. (1965b). A demonstrable local and geometric increase in the chromosomal DNA of *Chironomus*. *Experientia, 21,* 191–193.

Kiefer, B. I. (1973). Genetics of sperm development in *Drosophila*. In *Genetic Mechanisms of Development,* ed. F. H. Ruddle, pp. 47–102. New York: Academic Press.

Kiel Metzger, K., Warren, G., Wilson, G. N., and Erickson, R. P. (1985). Evidence that the human Y chromosome does not contain clustered DNA sequences (Bkm) associated with heterogametic sex determination in other vertebrates. *N. Engl. J. Med., 313,* 242–245.

King, M. (1980). C banding studies on Australian hylid frogs: secondary constriction structure and the concept of euchromatin transformation. *Chromosoma, 80,* 191–217.

King, M., and John, B. (1980). Regularities and restrictions governing C band variation in acridoid grasshoppers. *Chromosoma, 76,* 123–150.

King, M., and Rofe, R. (1976). Karyotypic variation in the Australian *Gekko Phyllodactylus marmoratus* (Gray) (Gekkonidae: Reptilia). *Chromosoma, 54,* 75–87.

Kit, S. (1961). Equilibrium sedimentation in density gradients of DNA preparations from animal tissues. *J. Mol. Biol., 3,* 711–716.

Kornher, J. S., and Kauffman, S. A. (1986). Variegated expression of the *Sgs 4* locus in *Drosophila melanogaster*. *Chromosoma, 94,* 205–216.

Kremer, H., Hennig, W., and Dijkhof, R. (1986). Chromatin organization in the male germ line of *D. hydei*. *Chromosoma, 94,* 147–161.

Krzanowska, H. (1969). Factors responsible for spermatozoan abnormality located on the Y chromosome in mice. *Genet. Res., 13,* 17–24.

Kuhn, E. M. (1976). Localization by Q banding of mitotic chiasmata in cases of Bloom's syndrome. *Chromosoma, 57,* 1–11.

Kunkel, L. M., and Smith, K. D. (1982). Evolution of human Y chromosome DNA. *Chromosoma, 86,* 209–228.

Kunkel, L. M., Smith, K. D., and Boyer, S. H. (1976). Human Y chromosome specific reiterated DNA. *Science, 191,* 1189–1190.

Kurnit, D. M. (1979). Satellite DNA and heterochromatin variants: the case for unequal mitotic crossing over. *Hum. Genet., 47,* 169–186.

Laird, C. (1973). DNA of *Drosophila* chromosomes. *Annu. Rev. Genet., 7,* 177–204.

Lakhotia, S. C. (1984). Replication in *Drosophila* chromosomes. XII. Reconfirmation of underreplication of heterochromatin in polytene nuclei by cytofluorometry. *Chromosoma, 89,* 63–67.

Lakhotia, S. C., and Jacob, J. (1974). EM autoradiographic studies on polytene nuclei of *Drosophila melanogaster*. *Exp. Cell Res., 86,* 253–263.

Laski, F. A., Rio, D. C., and Rubin, G. M. (1986). The tissue specificity of *Drosophila* P element transposition is regulated at the level of mRNA splicing. *Cell, 44,* 7–19.

Lee, J. C., and Yunis, J. J. (1971). Cytological variations in the constitutive heterochromatin of *Microtus agrestis*. *Chromosoma, 35,* 117–124.

Lefevre, G. (1976). A photographic representation and interpretation of the polytene chromosomes of *Drosophila melanogaster* salivary glands. In *The Genetics and Biology of Drosophila, Vol. 1a,* ed. M. Ashburner and E. Novitski, pp. 31–66. London: Academic Press.

Levinson, G., Marsh, J. L., Epplen, J. T., and Gutman, G. A. (1985). Cross hybridizing snake satellite *Drosophila* and mouse DNA sequences may have arisen independently. *Mol. Biol. Evol., 2,* 494–504.
Lifschytz, E., and Hareven, D. (1982). Heterochromatin markers: arrangement of obligatory heterochromatin, histone genes and multisite gene families in the interphase nucleus of *Drosophila melanogaster. Chromosoma, 86,* 443–455.
Lifschytz, E., and Lindsley, D. L. (1972). The role of X chromosome inactivation during spermatogenesis. *Proc. Natl. Acad. Sci. USA, 69,* 182–186.
Lima de Faria, A., and Bose, A. (1962). The role of telomeres at anaphase. *Chromosoma, 13,* 315–327.
Lima de Faria, A., and Jaworska, H. (1968). Late DNA synthesis in heterochromatin. *Nature, 217,* 138–142.
Linder, P., McCaw, K. K., and Hecht, F. (1975). Parthenogenetic origin of benign ovarian teratomas. *N. Engl. J. Med., 292,* 63–66.
Lindsley, D., and Novitski, E. (1958). Localization of the genetic factors responsible for the kinetic activity of X chromosomes of *Drosophila melanogaster. Genetics, 43,* 790–798.
Lindsley, D. L., and Lifschytz, E. (1972). The genetic control of spermatogenesis in *Drosophila.* In *Proc. Int. Symp. on the Genetics of the Spermatozoan,* ed. R. A. Beatty and S. Gluecksohn Waelesch, pp. 203–220. Obtained from the Department of Genetics, University of Edinburgh, U. K.
Lindsley, D. L., and Sandler, L. (1977). The genetic analysis of meiosis in female *Drosophila melanogaster. Philos. Trans. R. Soc. Lond.* [*B*], *277,* 295–312.
Livak, K. J. (1984). Organization and mapping of a sequence in the *Drosophila melanogaster* X and Y chromosomes that is transcribed during spermatogenesis. *Genetics, 107,* 611–634.
Longley, A. E. (1945). Abnormal segregation during megasporogenesis in maize. *Genetics, 30,* 100–113.
Macgregor, H. C., and Kezer, J. (1971). The chromosomal localization of a heavy satellite DNA in the testis of *Plethodon cinereus cinereus. Chromosoma, 33,* 167–182.
Macgregor, H. C., Horner, H., Owen, C. A., and Parker, I. (1973). Observations on centromeric heterochromatin and satellite DNA in salamanders of the genus *Plethodon. Chromosoma, 43,* 329–348.
Madan, K. (1983). Balanced structural changes involving the human X: effect on sexual phenotype. *Hum. Genet., 63,* 216–221.
Mahan, J. T., and Beck, M. L. (1986). Heterochromatin in mitotic chromosomes of the *virilis* group of *Drosophila. Genetics, 68,* 113–118.
Mandahl, N. (1978). Variation in C stained chromosome regions in European hedgehogs (Insectivora: Mammalia). *Hereditas, 89,* 107–128.
Manuelidis, L. (1982). Repeated DNA sequences and nuclear structures. In *Genome Evolution,* ed. G. A. Dover and R. B. Flavell, pp. 263–283. London: Academic Press.
Marks, G. E. (1974). Giemsa banding of meiotic chromosomes in *Anemone blanda* L. *Chromosoma, 49,* 113–119.
Mascarello, J. T. (1980). Meiotic chromosomes of antelope squirrels. *Cytogenet. Cell Genet., 26,* 104–107.
Mascarello, J. T., and Mazrimas, J. A. (1977). Chromosomes of antelope squirrels (genus *Ammospermophilus*): A systematic banding analysis of four species with unusual constitutive heterochromatin. *Chromosoma, 64,* 207–217.
Masubuchi, M. (1976). Differential replication of the chromosomes of *Pellavicinia longispina,* liverwort. *Cytologia, 41,* 523–541.

Mather, K. (1941). Variation and selection of polygenic characters. *J. Genet., 41,* 159–193.
Mather, K. (1944). The genetical activity of heterochromatin. *Proc. R. Soc. Lond.* [*B*] *132,* 308–332.
Mayfield, J. E., and Ellison, J. R. (1975). The organization of interphase chromatin in Drosophilidae. The self adhesion of chromatin containing the same DNA sequences. *Chromosoma, 52,* 37–48.
Maynard Smith, J. (1982). Overview: unsolved evolutionary problems. In *Genome Evolution,* ed. G. A. Dover and R. B. Flavell, pp. 375–382. London: Academic Press.
McKay, R. D. G., Bobrow, M., and Cooke, J. H. (1978). The identification of a repeated DNA sequence involved in the karyotype polymorphism of the human Y chromosome. *Cytogenet. Cell Genet., 21,* 19–32.
McLaren, A. (1983). Does the chromosomal sex of a mouse germ cell affect its development? In *Current Problems in Germ Cell Differentiation,* 7th Symp. Br. Soc. Dev. Biol., ed. A. McLaren and C. C. Wylie, pp. 225–240. Cambridge University Press.
Mengden, G. A., and Stock, A. D. (1980). Chromosomal evolution of G and C chromosome banding of some colubrid and bird genera. *Chromosoma, 79,* 53–64.
Miklos, G. L. G. (1982). Sequencing and manipulating highly repeated DNA. In *Genome Evolution,* ed. G. A. Dover and R. B. Flavell, pp. 41–68. London: Academic Press.
Miklos, G. L. G. (1985). Localized highly repetitive DNA sequences in vertebrate and invertebrate genomes. In *Molecular Evolutionary Genetics,* ed. R. J. MacIntyre, pp. 241–321. New York: Plenum Press.
Miklos, G. L. G., and Gill, A. C. (1981). The DNA sequences of cloned complex satellite DNAs from Hawaiian *Drosophila* and their bearing on satellite DNA sequence conservation. *Chromosoma, 82,* 409–427.
Miklos, G. L. G., and John, B. (1979). Heterochromatin and satellite DNA in man: properties and prospects. *Am. J. Hum. Genet., 31,* 264–280.
Miklos, G. L. G., Willcocks, D. A., and Baverstock, P. R. (1980). Restriction endonuclease and molecular analysis of three rat genomes with special reference to chromosome rearrangements and speciation problems. *Chromosoma, 76,* 339–363.
Miklos, G. L. G., Healy, M. J., Pain, P., Howells, A. J., and Russell, R. J. (1984). Molecular and genetic studies in the euchromatin heterochromatin transition region of the X of *Drosophila melanogaster.* II. A cloned entry point near the uncoordinated *(unc)* locus. *Chromosoma, 89,* 218–227.
Miller, D. A. (1977). Evolution of Primate chromosomes. *Science, 198,* 1116–1124.
Moore, G. D., Sinclair, D. A., and Grigliatti, T. A. (1983). Histone gene multiplicity and position effect variegation in *Drosophila melanogaster. Genetics, 105,* 327–344.
Moore, K. L., and Barr, M. L. (1954). Nuclear morphology according to sex in human tissues. *Acta Anat., 21,* 197–208.
Moritz, K. B., and Roth, G. E. (1976). Complexity of germline and somatic DNA in *Ascaris. Nature, 259,* 55–57.
Mottus, R., Reeves, R., and Grigliatti, T. A. (1980). Butyrate suppression of position effect variegation in *Drosophila melanogaster. Mol. Gen. Genet., 178,* 465–469.
Muller, F., Walker, P., Aeby, P., Neuhaus, H., Felder, H., Black, E., and Tobler,

H. (1982). Nucleotide sequence of satellite DNA contained in the eliminated genome of *Ascaris lumbricoides. Nucleic Acid Res., 10,* 7493–7510.

Muller, H. J., and Gershenson, S. M. (1935). Inert regions of chromosomes as the temporary products of individual genes. *Proc. Natl. Acad. Sci. USA, 21,* 69–75.

Murray, J. D., and McKay, G. M. (1979). Y chromosome mosaicism in pouch young of the marsupial greater glider (Marsupialia: Petauridae). *Chromosoma, 72,* 329–334.

Nagl, W. (1974). Mitotic cycle time in perennial and annual plants with various amounts of DNA and heterochromatin. *Dev. Biol., 39,* 342–346.

Nagl, W. (1977). Nuclear structures during cell cycles. In *Mechanisms and Control of Cell Division,* ed. T. L. Rost and E. M. Gilford, pp. 147–193. Stroudsburg, Penna.: Dowden, Hutchinson and Ross.

Nagl, W. (1979). Condensed interphase chromatin in plant and animal cell nuclei: fundamental differences. *Plant Syst. Evol.* [*Suppl.*], *2,* 247–260.

Nagl, W. (1985). Chromatin organization and the control of gene activity. *Int. Rev. Cytol., 94,* 21–56.

Nagl, W., and Fusenig, H. P. (1979). Types of chromatin organization in plant nuclei. *Pl. Syst. Evol.* [*Suppl.*], *2,* 221–233.

Natarajan, A. T., and Gropp, A. (1971). The meiotic behaviour of autosomal heterochromatic segments in hedgehogs. *Chromosoma, 35,* 143–152.

Newton, M. E. (1985). Heterochromatin diversity in two species of *Pellia* (Hepaticae) as revealed by C, Q, N and Hoechst 33258 banding. *Chromosoma, 92,* 378–386.

Nix, C. E. (1973). Suppression of transcript of ribosomal RNA astrons of *Drosophila melanogaster* in a structurally rearranged chromosome. *Biochem. Genet., 10,* 1–12.

Novitski, E., and Puro, J. (1978). A critique of theories of meiosis in the female of *Drosophila melanogaster. Hereditas, 89,* 51–67.

Nur, U. (1966). Non replication of heterochromatic chromosomes in a mealy bug, *Planococcus citri* (Coccoidea: Homoptera). *Chromosoma, 19,* 439–448.

Nur, U. (1967). Reversal of heterochromatinization and the activity of the paternal chromosome set in the male mealy bug. *Genetics, 56,* 375–389.

Nur, U. (1980). Evolution of unusual chromosome systems in scale insects (Coccoidea: Homoptera). In *Insect Cytogenetics,* Symp. Royal Entomol. Soc., Lond., Vol. 10, ed. R. L. Blackman, G. M. Hewitt, and M. Ashburner, pp. 97–117. Oxford: Blackwell Scientific.

Ohno, S. (1969). Evolution of sex chromosomes in mammals. *Annu. Rev. Genet., 3,* 495–524.

Ohno, S., Stenius, C., and Christian, L. (1964). The XO as the normal female of the creeping role *(Microtus oregoni). Chromosomes Today, 1,* 182–187.

Pardue, M. L., and Gall, J. G. (1970). Chromosomal location of mouse satellite DNA. *Science, 168,* 1356–1358.

Pathak, S., and Hsu, T. C. (1976). Chromosomes and DNA of *Mus.* The behavior of constitutive heterochromatin in spermatogenesis of *M. dunni. Chromosoma, 57,* 227–234.

Pathak, S., and Wurster Hill, D. H. (1977). Distribution of constitutive heterochromatin in Carnivores. *Cytogenet. Cell Genet., 18,* 245–254.

Patton, J. L., and Sherwood, S. W. (1982). Genome evolution in pocket gophers (genus *Thomomys*) I. Heterochromatin variation and speciation potential. *Chromosoma, 85,* 149–162.

Pavan, C., and Breuer, M. E. (1952). Polytene chromosomes in different tissues of *Rhynchosciara. J. Hered., 43,* 151–157.

Peacock, W. J., Lohe, A. R., Gerlach, W. L., Dunsmuir, P., Dennis, E. S., and Appels, R. (1978). Fine structure and evolution of DNA in heterochromatin. *Cold Spring Harbor Symp. Quant. Biol., 42,* 1121–1135.

Peacock, W. J., Dennis, E. S., and Gerlach, W. L. (1982). DNA sequence changes and speciation. In *Mechanisms of Speciation,* ed. C. Barigozzi, pp. 123–142. New York: Alan R. Liss.

Pimpinelli, S., Sullivan, W., Prout, M., and Sandler, L. (1985). On biological functions mapping to the heterochromatin of Drosophila melanogaster. *Genetics, 109,* 701–724.

Pimpinelli, S., Bonaccorsi, S., Gatti, M., and Sandler, L. (1986). The peculiar genetic organization of *Drosophila* heterochromatin. *Trends Genet., 2,* 17–20.

Plaut, W. (1963). On the replicative organization of DNA in the polytene chromosome of *Drosophila melanogaster. J. Mol. Biol., 7,* 632–635.

Rahman, R., and Lindsley, D. L. (1981). Male sterilizing interactions between duplications and deficiencies for proximal X chromosome material in *Drosophila melanogaster. Genetics, 99,* 49–64.

Raman, R., and Nanda, I. (1986). Mammalian sex chromosomes I. Cytological changes in the chiasmatic sex chromosomes of the male musk shrew *Sincus murinus. Chromosoma, 93,* 367–374.

Rasmussen, S. W., and Holm, P. B. (1980). Mechanics of meiosis. *Hereditas, 93,* 187–216.

Ray Chaudhuri, S. P., Singh, L., and Sharma, T. (1971). Evolution of sex chromosomes and formation of W chromatin in snakes. *Chromosoma, 33,* 239–251.

Ray Chaudhuri, S. P., and Singh, L. (1972). DNA replication pattern in sex chromosomes of snakes. *Nucleus, 15,* 200–210.

Redfern, C. P. F. (1981). Satellite DNA of *Anopheles stephensi Liston* (Diptera: Culcidae). Chromosomal location and underreplication in polytene nuclei. *Chromosoma, 82,* 561–581.

Renkawitz, R. (1978a). Characterization of two moderately repetitive DNA components localized within the β heterochromatin of *D.hydei. Chromosoma, 66,* 225–236.

Renkawitz, R. (1978b). Two highly repetitive DNA satellites of *D. hydei* localized within the α heterochromatin of specific chromosomes. *Chromosoma, 66,* 237–248.

Renkawitz Pohl, R., and Bialojan, S. (1984). A DNA sequence of *Drosophila melanogaster* with a differential telomeric distribution. *Chromosoma, 89,* 206–211.

Reuter, G., Werner, W., and Hoffmann, H. J. (1982). Mutants affecting position effect heterochromatinization in *Drosophila melanogaster. Chromosoma, 85,* 539–551.

Reuter, G., Wolff, I., and Friede, B. (1985). Functional properties of the heterochromatic segments inducing position effect variegation in *Drosophila melanogaster. Chromosoma, 93,* 132–139.

Ris, H., and Korenberg, R. D. (1979). Chromosome structure and levels of chromosome organization. In *The Structure and Replication of Genetic Material,* ed. D. M. Prescott and L. Goldstein, pp. 268–361. New York: Academic Press.

Roth, G. E., and Moritz, K. B. (1981). Restriction enzyme analysis of the germ line limited DNA of *Ascaris suum. Chromosoma, 83,* 169–190.

Rothfels, K. H., and Freeman, D. M. (1977). The salivary gland chromosomes of seven species of *Prosimulium* (Diptera, simulidae) in the mixtum (III L-1) group. *Can. J. Zool., 55,* 482–507.

Rubin, G. M. (1978). Isolation of a telomeric DNA sequence from *Drosophila melanogaster. Cold Spring Harbor Symp. Quant. Biol., 42,* 1041–1046.

Rushlow, C. A., Bender, W., and Chovnick, A. (1984). Studies on the mechanism of heterochromatin position effect at the rosy locus of *Drosophila melanogaster. Genetics 108,* 603–615.

Sabour, M. (1972). RNA synthesis and heterochromatinization in early development of a mealy bug. *Genetics, 70,* 291–298.

Samols, D., and Swift, H. (1979). Genomic organization in the flesh fly *Sarcophaga bullata. Chromosoma, 75,* 129–143.

Sandler, L. (1977). Evidence for a set of closely linked autosomal genes that interact with sex chromosome heterochromatin in *Drosophila melanogaster. Genetics, 86,* 567–582.

Sandler, L., and O'Tousa, J. (1979). Heterochromatic effects on the behavior of reversed acrocentric compound X chromosomes in *Drosophila melanogaster. Genetics, 91,* 537–551.

Sarto, G. E., Therman, E., and Patau, K. (1973). X inactivation in man: a woman with t(Xq^-; $12q^+$). *Am. J. Hum. Genet., 25,* 262–270.

Savage, J. R. K., Breckon, G., Goy, P., and Bigger, T. R. L. (1980). Tolerance of autosomal imbalance by the Syrian hamster *(Mesocricetus auratus). Cytogenet. Cell Genet., 27,* 88–97.

Schaeffer, J., and Schmidt, E. R. (1981). Different repetition frequencies of a 120 base pair DNA element and its arrangement in *Chironomus thummi thummi* and *Ch. thummi piger. Chromosoma, 84,* 61–66.

Schäfer, D. A., and Priest, J. H. (1984). Reversal of DNA methylation with 5 AzaC alters chromosome replication patterns in human lymphocyte and fibroblast cultures. *Am. J. Hum. Genet., 36,* 534–545.

Schäfer, R., Böttz, E., Becker, A., Bartelo, F., and Epplen, J. T. (1986a). The expression of the evolutionarily conserved GATA/GACA repeats in mouse tissues. *Chromosoma, 93,* 496–501.

Schäfer, R., Ali, S., and Epplen, J. T. (1986b). The organization of the evolutionarily conserved GATA/GACA repeats in the mouse genome. *Chromosoma, 93,* 502–510.

Schalet, A., and Lefevre, G. (1973). The localization of 'ordinary' sex linked genes in section 20 of the polytene X chromosome of *Drosophila melanogaster. Chromosoma, 44,* 183–202.

Schempp, W., and Schmid, M. (1981). Chromosome banding in Amphibia. VI. BrdU replication patterns in *Anura* and demonstration of XX/XY sex chromosomes in *Rana esculenta. Chromosoma, 83,* 697–710.

Sherwood, S. W., and Patton, J. L. (1982). Genome evolution in pocket gophers (genus *Thomomys*) II. Variation in cellular DNA content. *Chromosoma, 85,* 163–179.

Schmid, M. (1980a). Chromosome banding in Amphibia. IV. Differentiation of GC and AT rich chromosome regions in *Anura. Chromosoma, 77,* 83–103.

Schmid, M. (1980b). Chromosome banding in Amphibia V. Highly differentiated ZW/ZZ sex chromosomes and exceptional genome size in *Pyxicephalus adspersus* (Anura, Ranidae). *Chromosoma, 80,* 69–96.

Schmid, M., Olert, J., and Klett, C. (1979). Chromosome banding in Amphibia III. Sex chromosomes in *Triturus. Chromosoma, 71,* 29–55.

Schmid, M., Haaf, T., Geile, B., and Sims, S. (1983). Chromosome banding in Amphibia. VIII. An unusual XY/XX sex chromosome system in *Gastrotheca riobambae* (Anura, Hylidae). *Chromosoma, 88,* 69–82.

Schmid, W. (1967). Heterochromatin in mammals. *Arch. Julius Klaus Stiftung Vererb., 42,* 1–60.

Schmidt, E. R. (1980). Two AT rich satellite DNAs in the chironomid *Glyptotendipes barbipes* (Staeger). Isolation and localization in polytene chromosomes of *G. barbipes* and *Chironomus thummi. Chromosoma, 79,* 315–328.

Schmidt, E. R., and Keyl, H. G. (1981). In situ binding of AT rich repetitive DNA to the centromeric heterochromatin in polytene chromosomes of chironomids. *Chromosoma, 82,* 197–204.

Schmidt, E. R., Vistorin, G., and Keyl, H-G. (1980). An AT rich DNA component in the genomes of *Ch. thummi thummi* and *Ch. thummi piger. Chromosoma, 76,* 33–45.

Schmidtke, J., and Epplen, J. T. (1980). Sequence organization of animal nuclear DNA. *Hum. Genet., 55,* 1–18.

Schnedl, W., and Czaker, R. (1974). Centromeric heterochromatin and comparison of G banding in cattle, goat and sheep chromosomes (Bovidae). *Cytogenet Cell Genet., 13,* 246–255.

Schneider, E., Keukamp, U., and Pera, F. (1973). Loss of heteropycnosis of the constitutive heterochromatin in specifically activated cells of the thyroid gland of *Microtus agrestis. Chromosoma, 41,* 167–173.

Schwarzacher, T., Mayr, B., and Schweizer, D. (1984). Heterochromatin and nucleolus organiser region behavior at male pachytene of *Sus scrofa domestica. Chromosoma, 91,* 12–19.

Schweizer, D. (1980). Fluorescent chromosome banding in plants: applications, mechanisms and implications for chromosome structure. In *The Plant Genome,* Proc. 4th John Innes Symp., ed. D. R. Davies and R. A. Hopwood, pp. 61–72. Norwich: John Innes Charity.

Schweizer, D. (1981). Counterstain enhanced chromosome banding. *Hum. Genet., 57,* 1–14.

Schweizer, D., and Loidl, J. (In press). A model for heterochromatin dispersion and the evolution of C band patterns. *Chromosomes Today, 9.*

Sentis, C., Santos, J., and Fernandez Piqueras, J. (1986). C heterochromatin polymorphism in *Baetica ustulata:* intraindividual variation and fluorescence banding patterns. *Chromosoma, 94,* 65–70.

Sessions, S. K. (1980). Evidence for a highly differentiated sex chromosome heteromorphism in the salamander *Necturus maculosus* (Rafinesque). *Chromosoma, 77,* 157–168.

Sessions, S. K., and Wiley, J. E. (1985). Chromosome evolution in salamanders of the genus *Necturus. Brimeleyana, 10,* 37–52.

Sharma, T., and Raman, R. (1973). Variation of constitutive heterochromatin in the sex chromosomes of the rodent *Bandicota bengalensis bengalensis* (Gray). *Chromosoma, 41,* 75–84.

Sherwood, S. W., and Patton, J. L. (1982). Genome evolution in pocket gophers (genus *Thomomys*) II. Variation in cellular DNA content. *Chromosoma, 85,* 163–179.

Sieger, M., Pera, F., and Schwarzacher, H. G. (1970). Genetic activity of heterochromatin and heteropycnosis in *Microtus agrestis. Chromosoma, 29,* 349–364.

Simi, S. and Tursi, F. (1982). Polymorphisms for human chromosomes 1, 9, 16, Y: variations, segregation and mosaicism. *Hum. Genet., 62,* 217–220.

Sims, S. H., Macgregor, H. C., Pellatt, P. S., and Horner, H. A. (1984). Chromosome 1 in crested and marbled newts *(Triturus).* An extraordinary case of heteromorphism and independent chromosome evolution. *Chromosoma, 89,* 169–185.

Sinclair, D. A. R., Mottus, R. C., and Grigliatti, T. A. (1983). Genes which suppress position effect variegation in *Drosophila melanogaster* are clustered. *Mol. Gen. Genet., 191,* 326–333.

Singer, M. F. (1982). Highly repeated sequences in mammalian genomes. *Int. Rev. Cytol., 76,* 67–112.

Singh, L., Purdom, I. F., and Jones, K. W. (1976). Satellite DNA and evolution of sex chromosomes. *Chromosoma, 59,* 43–62.

Singh, L., Purdom, I. F., and Jones, K. W. (1980a). Sex chromosome associated DNA: evolution and conservation. *Chromosoma, 79,* 137–157.

Singh, L., Purdom, I. F., and Jones, K. W. (1980b). Conserved sex chromosome associated nucleotide sequences in eukaryotes. *Cold Spring Harbor Symp. Quant. Biol., 45,* 805–814.

Singh, L., Phillips, C., and Jones, K. W. (1984). The conserved nucleotide sequences of Bkm, which define Sxr in the mouse, are transcribed. *Cell, 36,* 111–120.

Sinha, A. K., Kakati, S., and Pathak, S. (1972). Exclusive localization of C bands with opposum sex chromosomes. *Exp. Cell Res., 75,* 265–268.

Smith, G. P. (1976). Evolution of repeated DNA sequences by unequal crossing over. *Science, 191,* 528–535.

Southern, D. I. (1967). Chiasma distribution in truxaline grasshoppers. *Chromosoma, 22,* 164–191.

Southern, D. I. (1980). Chromosome diversity in tsetse flies. In *Insect Cytogenetics,* Symp. Royal Entomol. Soc., Lond., *Vol. 10,* ed. R. L. Blackman, G. M. Hewitt, and M. Ashburner, pp. 225–243. Oxford: Blackwell Scientific.

Southern, E. M. (1984). DNA sequences and chromosome structure. *J. Cell. Sci.* [*Suppl.*], *1,* 31–41.

Spofford, J. B. (1976). Position effect variegation in *Drosophila.* In *Genetics and Biology of Drosophila, Vol. 1C,* ed. M. Ashburner and E. Novitski, pp. 954–1018. London: Academic Press.

Spradling, A., Penman, S., and Pardue, M. L. (1975). Analysis of *Drosophila* mRNA by in situ hybridization: sequences transcribed in normal and heat shocked cultured cells. *Cell, 4,* 395–404.

Steinemann, M. (1978). Coreplication of satellite DNA of *Chironomus melanotus* with main band DNA during polytenization. *Chromosoma, 66,* 127–139.

Steinemann, M. (1982). Multiple sex chromosomes in *Drosophila miranda:* a system to study the degeneration of a chromosome. *Chromosoma, 86,* 59–76.

Sueoka, N. (1961). Variation and heterogeneity of base composition of deoxyribonucleic acids: a compilation of old and new data. *J. Mol. Biol., 3,* 31–40.

Swartz, F. J., Henry, M., and Floyd, A. (1967). Observations on nuclear differences in *Ascaris, J. Exp. Zool., 164,* 297–308.

Szauter, P. (1984). An analysis of regional constraints on exchange in *Drosophila melanogaster* using recombination defective meiotic mutants. *Genetics, 106,* 45–71.

Tanaka, R. (1965). ^{3}H-thymidine autoradiographic studies on the heteropycnosis, heterochromatin and euchromatin in *Spiranthes sinensis. Bot. Mag., 78,* 50–62.

Tanaka, R. (1969). Deheterochromatinization of the chromosomes in *Spiranthes sinensis. Jpn. J. Genet., 44,* 291–296.

Tartof, K. D., Hobbs, C., and Jones, M. (1984). A structural basis for variegating position effects. *Cell, 37,* 869–878.

Therman, E. (1983). Mechanisms through which X chromosome constitutions affect the phenotype. In *Cytogenetics of the Mammalian X Chromosome, Part B: X Chromosome Anomalies and Their Clinical Manifestations,* ed. A. A. Sandberg, pp. 159–173. New York: Alan R. Liss.

Therman, E., and Sarto, G. E. (1983). Inactivation center on the human X chromosome. In *Cytogenetics of the Mammalian X Chromosome, Part A: Basic Mechanisms of X Chromosome Behavior,* ed. A. A. Sandberg, pp. 315–325. New York: Alan R. Liss.

Tone, M., Sakaki, Y., Hashiguchi, T., and Mizuno, S. (1984). Genus specificity and extensive methylation of the W chromosome specific repetitive DNA sequences from the domestic fowl, *Gallus gallus domesticus. Chromosoma, 89,* 228–237.

Torre de la, J., Lopez Fernandez, L., Nichols, R., and Gosálvez, J. (1986). Heterochromatin readjusting chiasma distribution in two species of the genus *Arcyptera:* the effects among individuals and populations. *Heredity, 56,* 177–184.

Traut, W., Winking, H. and Adolph, S. (1984). An extra segment in chromosome 1 of wild *Mus musculus:* a C band positive homogeneously staining region. *Cytogenet. Cell Genet., 38,* 290–297.

Vandeberg, J. L. (1983). Developmental aspects of X chromosome inactivation in eutherian and metatherian mammals. *J. Exp. Zool., 288,* 271–286.

Varley, J. M., Macgregor, H. C., Nardi, I., Andrews, C., and Erba, H. P. (1980). Cytological evidence of transcription of highly repeated DNA sequences during the lampbrush stage in *Triturus cristatus carnifex. Chromosoma, 80,* 289–307.

Venolia, L., and Gartler, S. M. (1983). Comparison of transformation efficiency of human active and inactive X chromosomal DNA. *Nature, 302,* 82–83.

Venolia, L., Cooper, D. W., O'Brien, S. A., Millette, C. F., and Gartler, S. M. (1984). Transformation of the *Hprt* gene with DNA from spermatogenic cells: implications for the evolution of X chromosome inactivation. *Chromosoma, 90,* 185–189.

Verma, R. S., and Dosik, H. (1980). Human chromosomal heteromorphisms: nature and clinical significance. *Int. Rev. Cytol., 62,* 361–385.

Verma, R. S., Jacob, J. P., and Babu, A. (1986). Heterochromatin organization in the nucleus of Indian muntjac *(Muntiacus muntjak). Can. J. Genet. Cytol., 28,* 628–630.

Vig, B. K. (1982). Sequence of centromere separation: role of centromeric heterochromatin. *Genetics, 102,* 795–806.

Vig, B. K., and Zinkowski, R. P. (1985). Sequence of centromere separation: influence of pericentromeric heterochromatin (repetitive DNA) in *Mus. Genetics, 67,* 153–159.

Virrki, N. (1974). The bouquet. *J. Agric. Univ. Puerto Rico, 58,* 338–350.

Vistorin, G., Gamperl, R., and Rosenkranz, W. (1977). Studies on sex chromosomes of four hamster species: *Cricetus cricetus, Cricetulus griseus, Mesocricetus auratus* and *Phodopus sungorus. Cell Genet., 18,* 24–32.

Voelker, R. A., and Kojima, K-I. (1971). Fertility and fitness of XO males in *Drosophila.* I. Qualitative study. *Evolution, 25,* 119–128.

Vogt, P., and Hennig, W. (1983). Y chromosomal DNA of *D. hydei. J. Mol. Biol., 167,* 37–56.

Wahlström, J., Kyllerman, M., Hansson, A., and Taranger, J. (1985). Unequal mitotic sister chromatid exchange and different length of Y chromosomes. *Hum. Genet., 70,* 186–188.

Wahrman, J., Richler, C., Neufeld, E., and Friedmann, A. (1983). The origin of multiple sex chromosomes in the gerbil, *Gerbillus gerbillus* (Rodentia: Gerbillidae). *Cytogenet. Cell Genet., 35,* 161–180.

Walker, P. M. B. (1971). Origin of satellite DNA. *Nature, 229,* 306–308.

Walter, L. (1973). Syntheseprozesse an den Riesenchromosomen von Glytoptendipes. *Chromosoma, 41,* 327–360.

Wargent, J. M., and Hartmann Goldstein, I. J. (1976). Replication behavior and morphology of a rearranged chromosome region of *Drosophila. Chromosomes Today, 5,* 109–116.

Weith, A. (1985). The fine structure of euchromatin and centromeric heterochromatin in *Tenebrio molitor* chromosomes. *Chromosoma, 91,* 287–296.

Wheeler, L. L., Arrighi, F., Cordeiro Stone, M., and Lee, C. S. (1978). Localization of *Drosophila nasutoides* satellite DNAs in metaphase chromosomes. *Chromosoma, 70,* 41–50.

White, M. J. D. (1940). The heteropycnosis of sex chromosomes and its interpretation in terms of spiral structure. *J. Genet., 40,* 67–81.

White, M. J. D. (1970). Asymmetry of heteropycnosis in tetraploid cells of a grasshopper. *Chromosoma, 30,* 51–61.

White, R., Pasztor, L. M., and Hu, F. (1975). Mouse satellite DNA in noncentromeric heterochromatin of cultured cells. *Chromosoma, 50,* 275–282.

Wilson, A. C., White, T. J., Carlson, S. S., and Cherry, L. M. (1977). Molecular evolution and cytogenetic evolution. In *Human Molecular Cytogenetics,* ed. R. W. Sparkes, D. E. Comings, and C. F. Fox, pp. 375–393. New York: Academic Press.

Wilson, E. B. (1925). *The Cell,* 3rd ed. New York: Macmillan.

Wolf, B. E. (1961). Y Chromosom und überzählige Chromosomen in den polytänen Somakernen von *Phryne cincta* Fabr. (Diptera, Nematocera). *Verhand. Deutsch. Zoolog. Gesell. in Saarbrücken,* pp. 110–123.

Wolf, B. E. (1970). Structure and function of alpha and beta heterochromatin: results on *Phryne cincta. The Nucleus,* Suppl. Vol., pp. 145–160.

Wolf, B. E. (1973). Chromosome structure and exchange frequency. *Chromosomes Today, 4,* 169–180.

Wolf, B. E., and Wolf, E. (1969). Adaptive value and interaction between alpha heterochromatin and autosomal polymorphism in *Phryne cincta. Chromosomes Today, 2,* 44–55.

Wolf, B. E., and Sokoloff, S. (1973). Migration of α heterochromatin from the giant X chromosome of the gnat *Phryne cincta. Chromosomes Today, 4,* 129–148.

Wolf, U., Flinspach, G., Böhm, R., and Ohno, S. (1965). DNS Reduplikationsmuster bei den Riesen Geschlectschromosomen von *Microtus agrestis. Chromosoma, 16,* 609–617.

Wrigley, J. M., and Marshall Graves, J. A. (in press). Sex chromosome homology and incomplete, tissue specific X inactivation in monotremes suggests a model for the evolution of the mammalian sex chromosomes and of X inactivation. *Nature.*

Wu, Z., Murphy, C., and Gall, J. G. (1986). A transcribed satellite DNA from the bullfrog *Rana catesbiana. Chromosoma, 93,* 291–297.

Wurster, D. H., Benirschke, K., and Noelke, H. (1968). Unusually large sex chromosomes in the Sitatunga *(Tragelaphus spekei)* and the Blackbuck *(Antilope cervicapra). Chromosoma, 23,* 317–323.

Wurster, D. H., Snapper, J. R., and Benirschke, K. (1971). Unusually large sex chromosomes: new methods of measuring and description of karyotypes of six rodents (Myomorpha and Hystricommorpha) and one agomorph (Ochotonidae). *Cytogenetics, 10,* 153–176.

Yamamoto, M. (1977). Cytological studies of heterochromatin function in the

Drosophila melanogaster male: autosomal meiotic pairing. *Chromosoma, 72,* 293–328.

Yamamoto, M. (1979). Interchromosomal effects of heterochromatic deletions on recombination in *Drosophila melanogaster. Genetics, 93,* 437–448.

Yamamoto, M., and Miklos, G. L. G. (1977). Genetic dissection of heterochromatin in *Drosophila:* the role of basal X heterochromatin in meiotic sex chromosome behavior. *Chromosoma, 60,* 283–296.

Yamamoto, M., and Miklos, G. L. G. (1978). Genetic studies on heterochromatin in *Drosophila melanogaster* and their implication for the functions of satellite DNA. *Chromosoma, 66,* 71–98.

Yoon, J. S., and Richardson, R. H. (1978). A mechanism of chromosomal rearrangements: the role of heterochromatin and ectopic pairing. *Genetics, 88,* 305–316.

Young, B. S., Pession, A., Traverse, K. L., French, C., and Pardue, M. L. (1983). Telomere regions in *Drosophila* share complex DNA sequences with pericentric heterochromatin. *Cell, 34,* 85–94.

Yunis, J. J., and Yasmineh, W. G. (1971). Heterochromatin, satellite DNA and cell function. *Science, 174,* 1200–1209.

Zacharias, H. (1979). Underreplication of a polytene chromosome arm in the chironomid *Prodiamesa olivacea. Chromosoma, 72,* 23–51.

Zacharias, H. (1981). Sex linked differences in DNA content of a polytene chromosome in *Prodiamesa* (Chironomidae). *Chromosoma, 82,* 657–672.

Zacharias, H. (1986). Tissue-specific schedule of selective replication in *Drosophila nasutoides. Roux's Arch. Dev. Biol., 195,* 378–388.

Zacharias, H., Hennig, W., and Leoncini, O. (1982). Microspectrophotometric comparison of the genome sizes of *Drosophila hydei* and some related species. *Genetics, 58,* 153–157.

Zhimulev, I. F., Semeshin, V. F., Kulichkov, V. A., and Belyaeva, E. S. (1982). Intercalary heterochromatin in *Drosophila.* I. Localization and general characteristics. *Chromosoma, 87,* 197–228.

Zuk, J. (1969). Analysis of Y chromosome heterochromatin in *Rumex thyrsiflorus. Chromosoma, 27,* 338–353.

Zuk, J. (1970a). Structure and function of sex chromosomes in *Rumex thyrsiflorus. Acta Soc. Bot. Pol., 39,* 539–562.

Zuk, J. (1970b). Function of Y chromosomes in *Rumex thyrsiflorus. Theor. Appl. Genet., 40,* 124–129.

II
Evolution of satellite DNA sequences in *Drosophila*

ALLAN LOHE AND PAUL ROBERTS

1. Introduction

A feature of genomes in higher eukaryotes is that certain portions show remarkably different physical properties from the bulk of the DNA. These fractions are called satellite DNAs and are characterized by their unusual base composition and highly reiterated nature. A satellite DNA arises from the repetition of a nucleotide sequence, from several base pairs (bp) up to several kilobases (kb) in length, into a long tandem array of repeats (see Brutlag, 1980, and Singer, 1982, for reviews). It is not uncommon to find a large number of different satellite DNAs in the one species, and for the combined amounts of satellites to represent 25–50% of the entire genome. Chromosome mapping in diverse species has shown that this class of DNA is largely confined to regions that are defined cytologically as heterochromatic. Further, purification of the major satellite DNAs in *Drosophila melanogaster* and mapping to mitotic chromosomes have shown that the vast majority of heterochromatin in this species is composed of these highly repeated DNAs (Peacock et al., 1976, 1977). Thus, heterochromatin in *Drosophila* can be partitioned into several classes of repeated DNA, each class corresponding to a particular satellite DNA fraction. The available evidence suggests that a similar conclusion may apply to other species of eukaryotes.

In this review, we will concentrate on satellite DNA, the chief physical component of constitutive heterochromatin. We refer those interested in the genetic makeup of heterochromatin in *Drosophila* to the review of Hilliker, Appels, and Schalet (1980). The subject of chromosome pairing and heterochromatin has been reviewed by John and Miklos (1979), and further details of the arrangement of satellite DNA sequences in *Drosophila* and other species can be found in Appels and Peacock (1978) and Brutlag (1980).

Paul Roberts is on sabbatical from the Department of Zoology, Oregon State University, Corvallis, Oregon.

1.1. Behavior of satellite DNA in evolution

An especially striking property of satellite DNA is its apparent lability in evolution, both in a qualitative and quantitative sense. Different species show satellite DNA profiles which are characteristic for that species only, because of variations in satellite buoyant densities in cesium chloride gradients (Walker, 1968; Hennig and Walker, 1970). The species specificity of satellite DNA profiles holds even for species that can give fertile hybrids (Hennig, Hennig, and Stein, 1970). Although some satellite DNAs of these species show identical buoyant densities, there are usually quantitative differences which distinguish the species (Mazrimas and Hatch, 1972; Gall and Atherton, 1974). However, individuals of a given species from different geographic isolates share the same DNA profiles, showing that alterations to satellite DNAs occur over those time periods required for the emergence of new species.

Given the rapid evolutionary changes in satellite DNA profiles, one might not predict that distantly related species would share the same satellite DNA sequences, but this is precisely what Salser et al. (1976) showed for two species of rodent. They noticed that the 6-bp repeating sequence of the Hs-α satellite from the kangaroo rat, *Dipodomys ordii,* is identical to the α satellite repeat from guinea pig, *Cavia porcellus,* reported previously by Southern (1970). The sequencing of satellite DNAs from other species of rodent identified the same repeat in these species as well (Fry and Salser, 1977). This demonstrates that tandem arrays of the same satellite sequence span the three major suborders of rodent, which may have separated from each other as long as 40–50 million years ago. To explain these results, Salser et al. (1976) proposed that related species share a "library" of conserved satellite sequences, any of which could be amplified into a major satellite DNA in evolution. Although unlikely in probability, de novo origin of the same satellite repeats by chance in different species, and horizontal transfer of repeats between species, are two possibilities also consistent with the results.

The extremes in evolutionary behavior of satellite DNAs have served to deepen the mystery surrounding this class of DNA and underline the complexities of the problem. An obvious question is, what function or functions do such large amounts of monotonously repeating DNA sequences serve in the genome? Hypotheses for function of satellite DNA usually involve generalized chromosomal functions such as chromosome organization, pairing, or recombination. One recent proposal is that changes in amounts of satellite DNA can affect cell proliferation and rates of development (Macgregor and Sessions, 1986). However, another possibility is that satellite DNA serves no purpose whatsoever, despite its ubiquity in the genomes of higher eukaryotes.

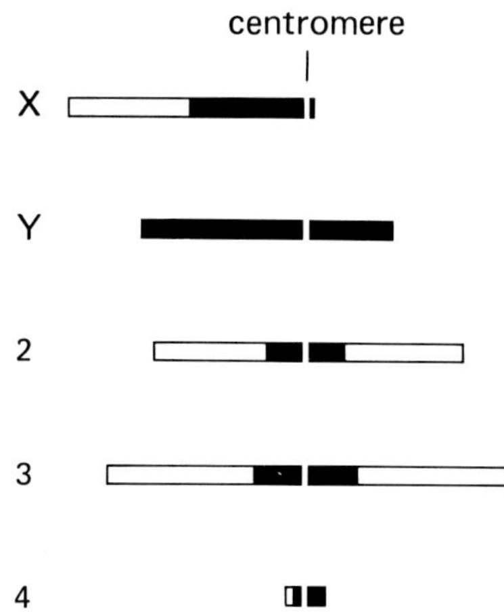

Figure 2.1. Locations of heterochromatin in mitotic chromosomes of *D. melanogaster*. Regions of heterochromatin are shown solid, whereas euchromatin is shown by clear areas within the bars. The position of the centromeres is indicated and is drawn as a small gap. (Modified after Cooper, 1950, and Hannah, 1951.)

1.2. Heterochromatin in *D. melanogaster*

Drosophila melanogaster is well suited to a detailed investigation of heterochromatin and associated DNA sequences. It contains large blocks of heterochromatin which extend for considerable distances from the centromeres of all chromosomes (see Figure 2.1), is amenable to genetic manipulation and cytological analyses, and has a relatively small genome size of about 170,000 kb (Rasch, Barr, and Rasch, 1971). Despite these advantages, genetic approaches to the question of function of heterochromatin have been handicapped by the inability to manipulate heterochromatin on a fine scale. For example, a characteristic of constitutive heterochromatin is the absence of crossing over (Brown, 1940), even in the presence of heterologous inversions, which strongly stimulate crossing over in proximal euchromatic regions (Roberts, 1965).

The existence of important loci and regions having genetic functions in heterochromatin (see Hilliker, Appels, and Schalet, 1980) also complicates genetic analyses. These functions include the centromere of each chromosome, ribosomal RNA genes on the X and Y chromosomes, six fertility factors on the heterochromatic Y chromosome (Brosseau, 1960; Kennison, 1981; Hazelrigg, Fornili, and Kaufman, 1982; Gatti and Pimpinelli, 1983), sites involved in pairing of the X and Y chromosomes during male meiosis (Cooper, 1964), and vital loci in the heterochromatin of chromosome 2 (Hilliker, 1976). Thus, any mutant effect which maps to a region of heterochromatin may not be attributable to a specific satellite sequence located at that site. However, even if deletion of a defined region of heterochromatic DNA were possible, one might expect little or no phenotypic effect on other grounds. Our current knowledge of the multichromosome distribution for many satellite sequences suggests that single deletions would not remove all satellite repeats of any one class from the genome, and also that the deliberate elimination of them would be an impossible task. Thus, although genetic functions can be mapped to heterochromatin, we do not know how these functions relate to satellite DNA, the principal component of heterochromatin.

Molecular approaches may provide information from which one might deduce constraints or other properties operating during the evolution of satellite DNAs. For example, the question of species specificity of satellite DNAs is open to experimental investigation. Does such specificity arise from qualitative differences in evolution, or can it be explained simply by quantitative changes to a common "library" of satellite sequences, as proposed by Salser et al. (1976)? We will focus on two evolutionary aspects of the *Drosophila* satellite DNAs. One is to compare the types of satellite DNAs found within one species, *D. melanogaster,* and to determine what relationships, if any, these satellites have with each other at the level of nucleotide sequence, genomic amount, and chromosomal location. The other aspect is to trace the evolution of specific satellite sequences between the closely knit group of sibling species in the *melanogaster* subgroup, coupling sensitive assays for sequence homologies with chromosomal mapping.

2. Evolution of satellite DNA sequences within a species

2.1. Satellite DNAs in *D. melanogaster*

Analysis of DNA from diploid nuclei has shown that *D. melanogaster* contains four well-defined satellite DNAs that can be easily purified from each other and main-band DNA sequences in CsCl–antibiotic gradients (Peacock et al., 1973). A fifth satellite, banding at 1.679 g/cm^3 in CsCl gradients, consists of ribosomal DNA sequences and other complex repeated DNA (Peacock et al., 1977). These satellites amount to about 20% of the genome and can be divided into two groups based on sequence complexity. One group contains tandem repeats of a simple sequence, only 5, 7, or 10 bp in length, and these satellites band at 1.672 g/cm^3, 1.686 g/cm^3, and 1.705 g/cm^3 in CsCl (Peacock et al., 1973, 1977; Brutlag and Peacock, 1975; Endow, Polan, and Gall, 1975; Sederoff and Lowenstein, 1975; Endow, 1977). The other group is represented by the 1.688

satellite, most of which consists of tandem repeats of a longer sequence, 359 bp in length (Hsieh and Brutlag, 1979). Although each of these satellites is AT rich, no obvious sequence homologies exist between the simple and complex satellite groups. However, there are clear relationships between satellite sequences within a group.

2.1.1. Simple satellites

Molecular cloning of the three simple satellite DNAs has been impeded by low efficiency in cloning these DNAs and instability of the cloned segment during propagation in *Escherichia coli* (Brutlag, Fry, Nelson, and Hung, 1977). These difficulties were overcome, in part, by cloning short satellite fragments 300 to 600 bp in length (Lohe and Brutlag, 1986). Analysis of these clones has provided a more refined view of the satellite DNA composition in this species. Nucleotide sequencing showed that each satellite consists predominantly of tandem repeats of a simple sequence in long, homogeneous arrays. In addition to the major sequences, numerous other repeats were also found in clones from any one satellite preparation. Quantitation showed them to constitute only a small proportion of the genome (Table 2.1). These minor repeated sequences were not interspersed among other DNA sequences or satellite repeats but were also organized in long, tandem arrays. They banded as satellites in CsCl gradients, and the homogeneity of repeats was similar to that shown by the major satellites.

To date, 11 simple repeated sequences have been identified in *D. melanogaster.* This large number of different tandemly repeated sequences was unexpected, given that only four satellite peaks are clearly visible in buoyant density analyses. However, the number of satellite sequences in *D. melanogaster* may be more extensive than suggested by the present cloning data, because only a limited number of satellite clones have been examined in detail, and even the least abundant satellite repeats amount to about 100 kb of the genome, or 20,000 copies. Many satellite DNA sequences can be present in much smaller amounts in a species. For example, in *D. erecta,* 10 of 11 satellite sequences detected by hybridization vary in copy number from about 350 to 19,000 copies (see section 3, Evolution of satellite DNA sequences between species). Therefore, instead of four repeated sequences which correspond to the major satellite DNAs of *D. melanogaster,* we should consider the possibility of a large number of different satellite sequences, perhaps 100 or more, in this one species.

The nucleotide sequences of ten simple repeats in *D. melanogaster* conforms to a formula, $(AAN)_m(AN)_n$, where N is any nucleotide (Endow

Table 2.1. Abundance of satellite DNAs in three sibling species of *Drosophila*

Sequence of repeat unit, 5′ → 3′	CsCl buoyant density (g/cm^3)	Amount in genome (%) of:		
		D. melanogaster[a]	*D. simulans*	*D. erecta*
AATAT	1.672	3.1	1.9	0.0088
AATAG	1.693	0.23	2.4	0.041
AATAC	1.680	0.52	0.0065	0.0018
AAGAC	1.689, 1.701	2.4	—[b]	0.011
AAGAG	1.705	5.6	0.71	0.055
AACAA	1.663	0.06	—	—[c]
AATAAAC	1.669	0.23	0.10	0.0015
AATAGAC	1.688	0.23	0.036	0.0016
AAGAGAG	—	1.5	0.074	0.0070
AATAACATAG	1.686	2.1	—[b]	0.0091
359 bp	1.688	5.1	0.11	0.24

[a]From Lohe and Brutlag, 1986.
[b]Repeats of this sequence could not be detected in *D. simulans*.
[c]Repeats of this sequence amount to less than 0.001% of the genome.

et al., 1975; Lohe and Brutlag, 1986). Since a single 5-bp repeated sequence, 5′ GGAGA 3′, was also cloned from this species but does not fit this rule, the more general formula $(RRN)_m(RN)_n$, where R is A or G, was proposed. A restriction on the nucleotide sequence of satellite repeats is also found in *D. virilis,* although these repeats fit the expression $(AAN)_m(NA)_n$ (Gall and Atherton, 1974). The similarities in structure of simple satellite sequences in *Drosophila* suggest that the expressions describing different repeating units are rules specifying form, rather than sequence, of a repeat.

At first glance, the close sequence relationships between the simple satellites in any one species would appear to support a hypothesis for the origin of one satellite sequence from a preexisting satellite. However, the changes in nucleotide sequence of the repeat unit are not random, as can be seen by comparison of sequences for six 5-bp repeats (Table 2.1). These differ in only two of five nucleotide positions, with three positions remaining invariant. If simple satellite sequences have evolved from a common sequence, we must conclude that not all sequence variations are generated, or else, if generated, they do not persist for long periods in evolution. These constraints on sequences may reflect either their mode of origin, or selective constraints on a function such as DNA conformation or specific satellite binding proteins. For example, the D1 protein binds tightly to repeats of 5′ AATAT 3′ (1.672 satellite) but poorly to 5′ AAGAG 3′ repeats (1.705 satellite; Levinger, 1985).

Unusual satellite DNA segments were also found in clones of the simple satellites. Most satellite clones showed about one nucleotide alteration per kilobase of repeats, but a small number of clones exhibited greater than two orders of magnitude more sequence variations (Figure 2.2; Lobe and Brutlag, 1987b). Several interpretations are possible to account for both homogeneity in some satellite arrays and heterogeneity in other arrays.

Repeated rounds of unequal crossover between sister chromatids can, in theory, result in homogenization of mutationally diversified sequences or even generate perfect tandem repeats from a random sequence (Smith, 1973, 1976). However, this model does not explain satisfactorily the small number of aberrant satellite clones showing sequence heterogeneity, because other clones containing the same repeated sequence showed no sequence alterations. To remain consistent with this mode of origin, one must propose that a highly mutated segment was located in a chromosomal region not conducive to the operation of unequal exchange, thereby preventing sequence homogenization. This region could be close to the end of a tandem array of repeats, where some heterogeneity in sequence may be expected (Smith, 1976), or in chromosomal regions isolated from major blocks of the same satellite. The recurrence of particular nucleotide alterations along the heterogeneous satellite segments that we sequenced is evidence for at least some unequal exchanges. Without these, the heterogeneity may have been even greater.

Sudden, large-scale amplification of a sequence into a satellite, as in saltatory replication (Britten and Kohne, 1968; Sutton and McCallum, 1972), would also result, at least initially, in sequence homogeneity of repeats. Those repeats with an ancient evolutionary history could be expected to show an accumulation of mutations (Southern, 1970), and this could account for the heterogeneous satellite segments. This possible mode of origin of the simple satellites could provide an alternative explanation for the homogeneity of repeats and explain the major quantitative changes occurring suddenly in satellite evolution (see section 3).

Figure 2.2. Homogeneous and heterogeneous segments of simple satellite DNA containing the same repeating sequence. The entire sequence of the cloned DNA is presented, and the sequence of the repeat is indicated. Adjacent repeats are aligned beneath one another and shown by a line. Cloned satellite fragments in (a) plasmid 1.686-171 and (b) plasmid 1.686-9 are shown. Both segments are comprised of the same tandemly repeated, 10-bp sequence, but differ in the number of nucleotide alterations. Homogeneous and heterogeneous satellite segments of the same 5-bp repeating sequence are shown for (c) plasmid 1.672-349 and (d) plasmid 1.672-97.

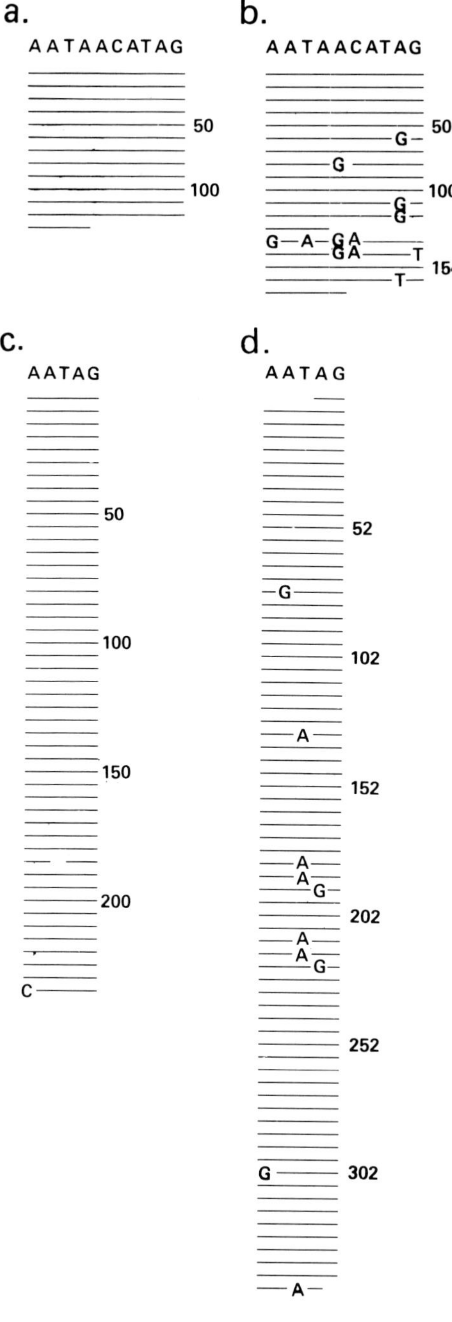

a.
AATAACATAG
50
100
b.
AATAACATAG
50
G
G
100
G
G
G A G A
G A T
154
T
c.
AATAG
50
100
150
200
C
d.
AATAG
52
G
102
A
152
A
A G
202
A
A G
252
G 302
A

Other cloned segments of satellite DNA were also unusual since they contained tandem arrays of two repeated sequences which were juxtaposed (Lohe and Brutlag, 1987b). In two clones, repeats of a simple sequence were joined to moderately repeated DNA. These more complex DNA sequences were shown, by Southern and in situ hybridizations, to be present at many sites in the genome. The junctions may have arisen by the integration of a mobile element into an array of simple satellite repeats. Six other clones were junctions between different simple repeats, and each showed similarities in structure at the region of the junction. In no cases were the tandem arrays of simple repeats separated by transitional sequences or other DNA, but arrays were always immediately adjacent to each other (Figure 2.3). Repeats on either side of a junction were usually extremely homogeneous, and the two repeat types in each array were closely related in sequence. Perhaps most surprising was that in five cases, the periodicity of repeats was maintained in phase across the junction.

Three conclusions that are pertinent to the evolution of satellite DNAs can be drawn from these observations. First, the conservation of repeat-unit periodicity across a junction is not predicted by the model of unequal exchange. It might be expected that one of the two terminal repeats at a junction would be incomplete in length, since adjacent tandem arrays may have repeat units with different nucleotide sequences and repeat lengths, and therefore would evolve independently in the unequal exchange model. Second, closely related satellite repeats are found in immediate proximity at the molecular level. This finding may be relevant to models of satellite evolution, because related repeats may derive from each other. Also, a prediction of chromosomal locations of satellites would be that closely related satellite sequences show similar chromosomal distributions. Third, the number of junctions between different satellite segments (eight junction clones in an unselected sample of 26 clones) was unexpectedly high, since physical and chromosome mapping data suggest an average segment length of about 750 kb for blocks of a given satellite (Brutlag et al., 1977b; Peacock et al., 1976, 1977). In contrast, our analysis of satellite fragments 0.5 kb in length suggests that segment length of one repeat type can be quite small, and in extreme examples is less than 100 bp (Figure 2.3b). These results may be reconciled when the arrangement of one simple satellite sequence with respect to another can be determined in molecules between 0.5kb and 750 kb in length.

The close relationship in nucleotide sequence of repeats in neighboring satellite arrays may not be coincidental. In the junction molecule shown in Figure 2.3a, the repeat unit of the 5-bp array, 5′ ATAGA 3′, forms part of the longer 10-bp repeat 5′ ATAAC<u>ATAGA</u> 3′ (the 5-bp sequence is

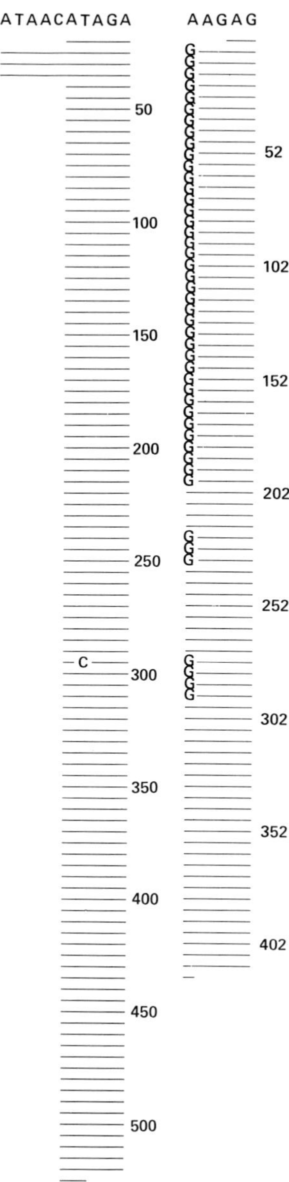

Figure 2.3 Junctions between tandem arrays of different simple satellite repeats: (a) Complete sequence of insert in plasmid 1.686-201, showing a tandem array of 10-bp repeats (35 bp in length) joined to a tandem array of 5-bp repeats (487 bp in length). (b) Complete sequence of insert in plasmid 1.705-1232, showing multiple junctions between the same two 5-bp repeated sequences.

underscored). This suggests that the 5-bp array may have arisen from part of a 10-bp repeat unit. Since similarities in nucleotide sequence of neighboring satellite arrays are a common feature in junction molecules involving the simple satellites, a model of de novo satellite generation has been proposed based on these patterns (Lohe and Brutlag, 1987b). If, during a satellite amplification event, a repeat unit were synthesized enzymatically using the preceding repeat as template, long homogeneous arrays would be formed. However, a simple error in copying such as misreading the repeat length of the template copy, or a base change during synthesis, would result in an abrupt switch in satellite arrays. The template amplification model requires only that changes in the repeat sequence conform to the "form rule," which is $(RRN)_m(RN)_n$ in *D. melanogaster.*

This mechanism of DNA amplification differs from other models for sequence amplification (see Bostock, 1986), including the slippage–replication–slippage cycle for de novo synthesis of simple repeats in vitro (Kornberg et al., 1964). This model does not provide for synthesis of a repeat using the preceding repeat as template, and repeat units longer than 2 bp are not formed by this mechanism. However, the mode of satellite generation we have proposed may be analogous to the terminal transferase-like synthesis in *Tetrahymena,* which adds homogeneous arrays of 6-bp repeats to telomeres (Greider and Blackburn, 1985).

2.1.2. Complex satellites

Most of the 1.688 satellite consists of tandem repeats of a 359-bp sequence (Hsieh and Brutlag, 1979). Unlike the simple satellite DNAs, different repeat units are not identical and show about 4–5% sequence variation. Two variant repeats in arrays separate from the 359-bp repeats have also been isolated from the 1.688 satellite. Both are minor in amount and are about 80% homologous to each other and to the 359-bp repeats (Carlson and Brutlag, 1979; Lohe and Brutlag, 1986). Despite the close relationship of these repeats at the nucleotide sequence level, the length of the repeat unit can differ markedly. One variant repeat is 254 bp in length, primarily due to a single 98-bp deletion relative to the 359-bp unit. The other variant has a 353-bp repeat sequence and differs from the 359-bp repeats by a 6-bp deletion. Apart from these deletions, most sequence variation between the variant arrays involves single base substitutions. The repeats may have derived from an ancestral repeat but have drifted apart through mutations at random positions of the nucleotide sequence. Separate amplification events of mutated copies, involving either a deleted copy or part of a 359-bp copy, could have generated the different tandem arrays.

2.1.3. Chromosomal locations of satellites

In situ hybridization using bulk satellites as probes has shown that each is located at the chromocenter of polytene chromosomes in *Drosophila* (Gall, Cohen, and Polan, 1971; Peacock et al., 1973) and in the heterochromatic portion of mitotic chromosomes (Peacock et al., 1976, 1977). Our mapping studies have extended these data and have avoided cross-contamination of satellite sequences by the use of cloned DNA templates for synthesis of complementary RNA probes. Each of seven simple satellite DNAs is restricted to the chromocenter of polytene chromosomes, and, with minor exceptions (see below), no labeling of euchromatic bands was observed, even with the 1000-fold lateral amplification afforded by polytene chromosomes. Since no hybridization was detected in those regions in the euchromatin termed *intercalary heterochromatin* (see Hannah, 1951; Zhimulev, Semeshin, Kulichkov, and Belyaeva, 1982), the characteristic properties of these regions (such as ectopic pairing, high frequency of x-ray breakage) may be mediated by other DNA sequences.

Although the simple satellite repeats show a remarkable restriction in their distribution along a *Drosophila* chromosome (to the chromocenter of polytene nuclei, to proximal or centromeric heterochromatin of mitotic cells), there are several exceptions involving minor amounts of satellite sequences. Repeats of the sequence 5′ AAGAG 3′ are present in a sufficient number of copies to be detected by in situ hybridization, in some cells at the telomeres (Figure 2.4). Under relaxed conditions of hybridization, this same repeated sequence also hybridizes to band 21D on chromosome arm 2L, but the labeling is due to cross-hybridization (Steffensen, Appels, and Peacock, 1981). Whether the sequences at band 21D are satellite repeats but are not represented among the 11 satellite sequences cloned (Table 2.1), or are repeats that have no counterparts in heterochromatin, must await the molecular cloning of these sequences. Other known examples of limited localization of satellite sequences to euchromatin involve a simple and a complex satellite. Eight tandem repeats of the satellite sequence 5′ AATAC 3′ are found in front of the s38 chorion gene, which is located on the X chromosome (Spradling et al., 1987). Although the 359-bp repeats of the 1.688 satellite are found predominantly in the chromocenter, related sequences are present near band 3C on the X chromosome (Tartof, Hobbs, and Jones, 1984). The discovery of short satellite segments interspersed among the limited number of single copy genes that have been sequenced suggests that more exceptions may be found.

The restriction of simple satellite DNA sequences to heterochromatin may be a consequence of several factors. For example, these may include infrequent transposition of satellite sequences from heterochromatin into

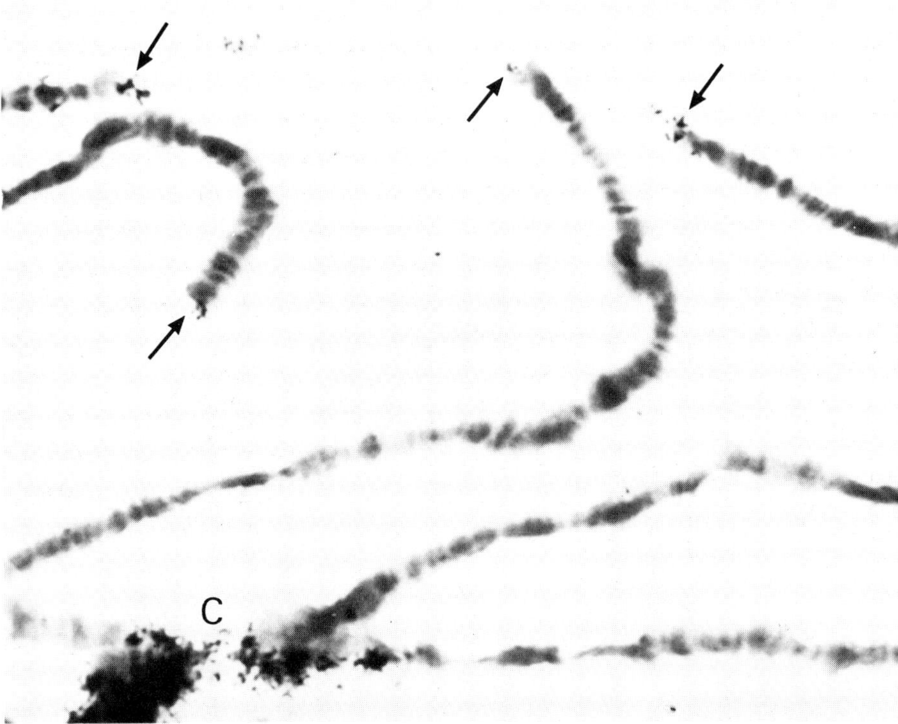

Figure 2.4. Localization of 5′ AAGAG 3′ repeats to telomeres of polytene chromosomes in *D. melanogaster,* strain Canton S. Four telomeres are labeled (arrows), and chromocentral labeling (C) is also shown. The probe was single-stranded RNA, synthesized with tritium-labeled guanosine 5′-triphosphate ([^{3}H]GTP) and SP6 polymerase.

euchromatin, or transposition followed by rapid removal. There could also be adverse effects on gene function from neighboring satellite sequences, such as occurs in the phenomenon of position-effect variegation (Spofford, 1976). To investigate these hypotheses, we introduced a small fragment of simple satellite DNA (0.3 kb) into euchromatin using the P-element-mediated transformation system of Rubin and Spradling (1982). The satellite fragment was linked to the *white*$^+$ (i.e., wild-type) gene which acted as a marker. Two independent transformants were obtained, each with a different satellite sequence adjacent to the *white* gene. As judged by the loss of red eye color phenotype to give white eyes, the transposons are about 30 times more unstable than similar transposons lacking the satellite segment (Lohe and Moran, in preparation).

Many of these white derivatives show no evidence for the transposon at the original site of integration, by in situ hybridization. Further, the instability is independent of the presence of authentic P-element transposase since the helper "wings-clipped" (Karess and Rubin, 1984) was used in the injections, and no complete P-element can be detected in the transformants by molecular or genetic criteria. If these experiments can be repeated with several other satellite sequences, it suggests that small segments of heterochromatic DNA would be rapidly eliminated from euchromatic locations by an unknown mechanism.

In situ hybridizations of bulk satellite fractions to mitotic chromosomes demonstrate that a satellite often has a multichromosome distribution in heterochromatin (Peacock et al., 1976, 1977). Detailed localizations have been presented for the 1.672 and 1.705 satellites (Steffensen et al., 1981). These should be considered as composite maps for the several closely related repeats cobanding in these satellites. However, the use of cloned satellite probes has refined the maps and has enabled the comparison of chromosomal locations for closely related satellite sequences, such as the 5-bp and 7-bp arrays which coband in the 1.705 satellite (Roberts and Lohe, in preparation). A striking result is that four satellite repeats with closely related sequences (5′ AATAG 3′, 5′ AAGAG 3′, 5′ AAGAC 3′, 5′ AAGAGAG 3′) have extremely similar chromosomal locations. These results are not due to cross-hybridization between the different repeats since stringent hybridization conditions had been defined for each probe by nitrocellulose filter assays before the in situ hybridizations were carried out. Further, although there was a concordance in the chromosomal positions of these repeats, there were also quantitative differences for some sites. For example, repeats of 5′ AAGAG 3′ (5.6% of the genome) are found on all chromosomes but, in particular, on chromosome 2 (multiple sites) and the Y chromosome (at least three sites). The 7-bp sequence, 5′ AAGAGAG 3′ (1.5% of the genome), is repeated in these same chromosomal regions, with chromosome 2 and the Y (two sites) showing most labeling, followed by the 3, 4, and X chromosomes in order of decreasing labeling intensity (Figure 2.5a). The two repeated sequences 5′ AAGAC 3′ and 5′ AATAG 3′, which differ from repeats of 5′ AAGAG 3′ by a single nucleotide change at different positions, also label predominantly chromosomes 2 and the Y. Again, there are differences at a quantitative level. The 5′ AAGAG 3′ repeats, amounting to 2.4% of the genome, map to the Y chromosome (3–4 sites) and chromosome 2 (Figure 2.5b). As determined by grain counts of autoradiographs, there is four times the amount of this satellite sequence on the Y than on each chromosome 2. The distribution of 5′ AATAG 3′ sequences (Figure 2.5c) includes a subset of sites shown by the other three closely related repeats.

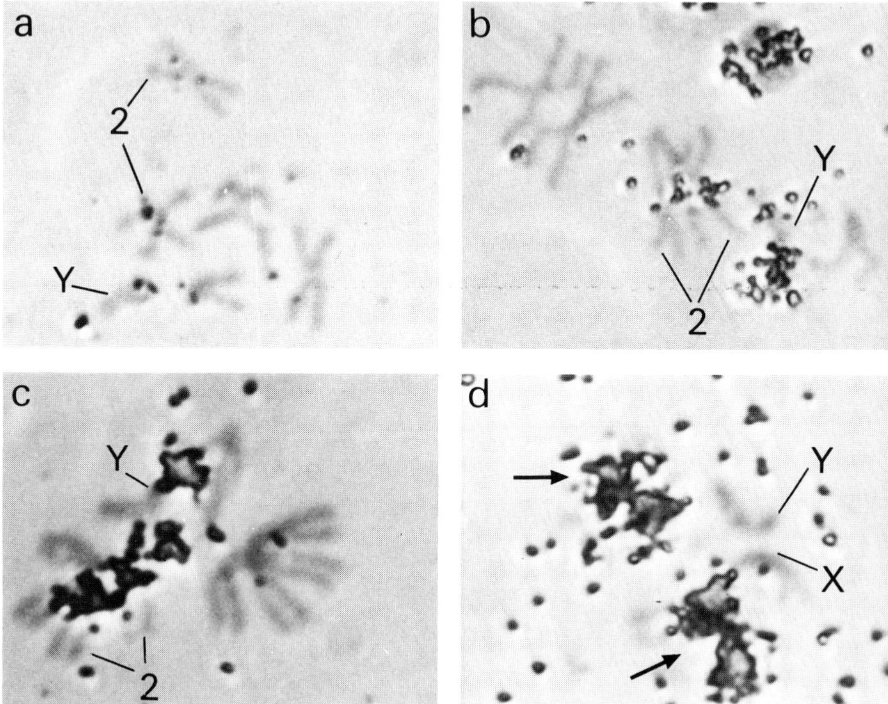

Figure 2.5. In situ hybridization of simple satellite DNAs to larval neuroblast cells of *D. melanogaster.* (a) The probe used is 5′ AAGAGAG 3′ repeats, and there is hybridization to chromosomes 2 and the Y (indicated). (b) The probe is 5′ AAGAC 3′ repeats. The Y chromosome is labeled at one site on the short arm and at least two sites, saturated with silver grains, on the long arm. Both chromosomes 2 are also labeled. (c) The probe is 5′ AATAG 3′ repeats. The Y chromosome is heavily labeled at one site, and chromosomes 2 are labeled extensively. (d) The probe is 5′ AATAACATAG 3′ repeats, the major repeating sequence of the 1.686 satellite. This autoradiograph was overexposed for the sites on chromosomes 2 and 3 (arrows). No grains are seen on the Y chromosome, and the X shows a single grain, possibly due to autoradiographic scatter from labeling of the autosomes.

In contrast to these overlapping distributions, other satellite sequences show quite different localizations. Repeats of the 10-bp sequence, the major component of the 1.686 satellite, are restricted to chromosomes 2 and 3 and are not present on the sex chromosomes or chromosome 4 (Figure 2.5d). The sequence 5′ AATAC 3′ is repeated at a single site, near the tip of the Y chromosome long arm. This restriction to a single site is reminiscent of the 359-bp repeats comprising the 1.688 satellite, which are located at one site on the X chromosome (Brutlag, unpublished observations; Hilliker and Appels, 1982). Our localizations for repeats of 5′ AATAT 3′, the major component of the 1.672 satellite, were to the Y and chromosome 4, with smaller amounts on the remaining chromosomes. This distribution is similar to that obtained by Steffensen et al. (1981), who used total 1.672 satellite as a probe. These results show that closely related repeats, such as those differing by a single nucleotide in their repeating sequence, often have similar chromosomal distributions. Other satellites can have different distributions and are sometimes confined to the sex chromosomes (5′ AATAC 3′ and 359-bp repeats) or the autosomes (10-bp repeats). The significance of this will be considered below.

2.2. Satellite DNAs in other species of *Drosophila*

2.2.1. D. simulans

This sibling species is the closest known relative of *D. melanogaster* and is therefore of special importance in evolutionary studies involving *D. melanogaster.* Molecular data based on the sequencing of alcohol dehydrogenase genes in the two species (Bodmer and Ashburner, 1984; Cohn, Thompson, and Moore, 1984) suggest an evolutionary separation of about 10 million years (Easteal and Oakeshott, 1985). *Drosophila melanogaster* and *D. simulans* have different satellite DNA profiles (Lohe, 1977, 1981; Barnes, Webb, and Dover, 1978), and only one of eight satellites in *D. simulans,* the 1.672 satellite, has identical physical properties to a satellite in *D. melanogaster.*

The nature of four satellites in *D. simulans* was, therefore, investigated by molecular cloning and sequencing (Lohe and Brutlag, 1987a). As would be expected from the physical criteria, clones of the *D. simulans* 1.672 satellite contain the repeated sequence 5′ AATAT 3′. The 1.695 satellite of *D. simulans* consists of 5-bp repeats with the sequence 5′ AATAG 3′, a repeat also present in *D. melanogaster* (Table 2.1). However, the reiteration frequency of this sequence is high enough in *D. simulans* so that a satellite band is visible. Molecular cloning has defined three other repeated sequences in *D. simulans* (Table 2.2). Although not identical to satellite repeats cloned from *D. melanogaster,* each is closely related in sequence, including the 240-bp repeats (Lohe and Roberts, unpublished data). An unusual repeating sequence is the 15-bp repeat of the 1.696 satellite which is comprised of three 5-bp subrepeats (5′ GAACA 3′), two being arranged as direct repeats with the third in an inverted orientation. As with other simple repeated sequences in *D. simulans,* this 5-bp subrepeat (or a permutation, 5′ AACAG 3′) conforms to the "form rule" of the *D. melanogaster* satellites, $(RRN)_m(RN)_n$.

Table 2.2. Sequences of major satellite DNAs in *D. simulans*

CsCl buoyant density (g/cm^3)	Sequence of repeat unit (5′ → 3′)
1.672	AATAT
1.694	240 bp
1.695	AATAG
1.696	AACAGAACATGTTCG
—	AACAAAC

We can conclude that the major satellites in *D. simulans* have sequences that either are identical to, or are variants of, the *D. melanogaster* type of sequences. Specificity in satellite DNA profiles between these sibling species therefore results from *quantitative* differences for satellite sequences held in common, and also from amplification of variant sequences that may not be held in common. Others have also proposed that quantitative changes in repeated DNA content rather than qualitative differences can mimic rapid evolutionary divergence (Klein et al., 1978; Moore, Scheller, Davidson, and Britten, 1978; Schmookler Reis, Timmis, and Ingle, 1981). More rigorous data in support of this conclusion, involving the hybridization of *D. melanogaster* satellite sequences to *D. simulans* DNA, are discussed in section 3.

2.2.2. D. virilis *and other species*

The three well-defined satellite DNAs in *D. virilis* were among the first *Drosophila* satellites to be sequenced. Each satellite is comprised of tandem repeats of a 7-bp sequence which conforms to the form rule $(AAN)_m(NA)_n$ (Gall and Atherton, 1974), a pattern very similar to that shown by the *D. melanogaster* satellite sequences. A fourth satellite, minor in amount, has repeats of the 7-bp sequence 5′ AATATAG 3′ (Mullins and Blumenfeld, 1979). This sequence does not fit the rule shown by the three major *D. virilis* satellites, but rather conforms to the *D. melanogaster* form rule. The coexistence in the one species of repeats obeying different sequence formulas shows that these rules are not constraints limiting the diversity of simple satellite sequences in a *Drosophila* species.

Three closely related species that belong to the *virilis* phylad of the *virilis* group contain the same satellite sequence. Nucleotide sequencing of the *D. americana americana* 1.692 satellite confirmed that it is identical to *D. virilis* satellite I, which also bands at 1.692 g/cm^3 in CsCl gradients (Gall and Atherton, 1974). Hybridization followed by thermal melting showed that the 1.692 satellite in *D. lummei* also has the same repeating sequence as *D. virilis* satellite I (Cohen and Kaplan, 1982). However, the satellite DNAs of *D. kanekoi,* a species in the *montana* phylad of the *virilis* group, are different from the three major satellites in *D. virilis.* Since one of the *D. kanekoi* satellites also has a buoyant density of 1.692 g/cm^3 in CsCl, equal buoyant density in CsCl gradients is not a reliable criterion of sequence identity. This has also been shown for a complex satellite DNA banding at 1.690 g/cm^3 and common to three species of Hawaiian *Drosophila* (Miklos and Gill, 1981). Nucleotide sequencing of repeats cloned from each species showed that the repeating units, which are 185 bp in length, are closely related. Nevertheless, there are many nucleotide alterations in the sequences apparent in between-species comparisons, suggesting that the 185-bp sequence is not highly conserved in evolution.

The satellite DNAs of *D. nasutoides* have also been described (Cordeiro et al., 1975). This species belongs to the *immigrans* group and is unusual in that the dot chromosome 4 in polytene squashes corresponds to a large, heterochromatic chromosome in the metaphase karyotype. In situ hybridization of complementary RNAs from the four major satellites of this species localized each to the large heterochromatic chromosome (Wheeler, Arrighi, Cordeiro-Stone, and Lee, 1978).

3. Evolution of satellite DNA sequences between species

3.1. Satellite sequences of *D. melanogaster* in sibling species

Evolutionary studies of *D. melanogaster* satellite DNAs are made possible by the existence of eight sibling species in the *melanogaster* subgroup. These can be divided into two species complexes based on interspecies hybridizations, polytene chromosome banding patterns, and other data (see Lemeunier and Ashburner, 1976, 1984). Two representative species chosen for examination were *D. simulans,* the closest known relative of *D. melanogaster,* and *D. erecta,* one of two species within this group most distant to *D. melanogaster.* The chromosomes of *D. melanogaster* and *D. simulans* differ by three paracentric inversions as well as a number of band differences between polytene chromosomes of hybrids (Horton, 1939). These species form infertile hybrids. No hybrids form in crosses between *D. melanogaster* and *D. erecta.* Our impressions of the content of heterochromatin in the three species, from mitotic chromosome preparations, are that the amounts are essentially similar, with the exception of the sex chromosomes. The X and Y chromosomes in *D. simulans* and *D. erecta* are noticeably shorter than their counterparts in *D. melanogaster,* as a result of differences in the heterochromatic regions.

One question of interest is whether small amounts of the *D. melanogaster* satellite repeats can be found in the sibling species, despite the paucity of similar-density satellites. The availability of 12 satellite DNAs cloned from *D. melanogaster* has permitted such a study (Lohe and Brutlag, 1987a). In theory, the degree of sequence conservation for each simple satellite sequence could be investigated by molecular cloning and nucleotide sequencing of clones isolated from *D. simulans* and *D. erecta* DNAs. However, this approach was not feasible because of the extreme inefficiency of cloning simple satellite sequences in plasmid vectors and the instability of short, tandemly repeated *Drosophila* satellite sequences in phage vectors (Brutlag et al., 1977a; Lohe and Brutlag, 1986). The reasons for the inefficiency in cloning are not understood in molecular terms, but the result is that the cloning of particular satellite sequences which are rare and which therefore cannot be purified in abundance by CsCl gradient centrifugations is not practical at present.

Instead, the method adopted involved DNA–DNA hydbridizations of *D. melanogaster* satellite probes to *D. simulans* or *D. erecta* DNAs, immobilized on nitrocellulose filters, followed by thermal melting of hybrids. Although single copies of the short satellite sequences must appear by chance among unique DNA sequences, they would not be detected by this method. Significant levels of hybridization occur only if tandem arrays of these sequences are about 100 bp or more in length. An advantage in using simple satellite sequences in thermal melts is that the melting profile is sharp and distinctive, permitting the unambiguous identification of given satellite repeats by the mean melting temperature (T_m). Cross-hybridization between these closely related sequences occurs rarely if hybridization is carried out about 10°C below the T_m value, because a single mismatch per repeat unit results in 10–20% mismatching in hybrid molecules. Also, the use of probes of high specific activity permits small amounts of satellite repeats to be detected.

Ten simple satellite DNAs, cloned from *D. melanogaster,* were hybridized to DNAs of *D. melanogaster* or *D. simulans.* Each hybridization was carried out at a specific temperature to ensure high stringency, as first determined by hybridization of the cloned DNA to itself and measurement of the T_m value. In general, there was substantial hybridization of these *D. melanogaster* satellite probes to *D. simulans* DNA. When T_m values (the average of two or more independent experiments) were compared, heterologous T_m values using *D. simulans* DNA were less than 1°C different from the homologous *D. melanogaster* values for six satellite repeats (Table 2.3). A decrease of 1°C in the T_m value represents approx-

Table 2.3. Melting temperatures of satellite DNA–DNA hybrids in sibling species of *Drosophila*

	Homologous T_m (°C)		Heterologous T_m (°C)	
Sequence 5′ → 3′	Plasmid	*D. melanogaster*	*D. simulans*	*D. erecta*
AATAT	38.1	37.1	37.1	35.2
AATAG	47.2	45.1	45.7	46.0
AATAC	—	42.9	42.9	42.9
AAGAC	52.6	51.6	—[a]	50.2
AAGAG	57.0	55.9	55.9	56.2
AACAA	51.0	49.6	—[a]	49.5
AATAAAC	—	46.2	45.9	45.7
AATAGAC	52.1	50.1	47.4	48.3
AAGAGAG	58.5	56.9	56.7	56.5
AATAACATAG	46.4	43.6	—[a]	42.9
359 bp	52.7	46.9	44.5	44.9

[a] Less than 200 cpm was eluted from the filter, and the transition was continuous.

imately 1% nucleotide mismatching (Hutton and Wetmur, 1973; Britten, Graham, and Neufeld, 1974), and, therefore, tandem arrays of six satellite sequences cloned from *D. melanogaster* are present in *D. simulans.* The close correspondence in thermal stabilities also shows that tandem arrays of these repeats are just as homogeneous in *D. simulans* as their counterparts are in *D. melanogaster.*

Repeats of the 7-bp sequence 5′ AATAGAC 3′ were also detected in *D. simulans,* but here the T_m value was 2.7°C below the corresponding *D. melanogaster* value. Some cross-hybridization to extremely abundant yet closely related repeats can account for this result, since the melting profile represents cumulative radioactivity released and since early-melting hybrids contributed to the shape of the curve. The presence of a substantial late-melting component provides good evidence that many tandem repeats of this 7-bp sequence are also present in *D. simulans.* However, even if the slightly lowered T_m value were due entirely to sequence alterations, the difference would indicate that the 7-bp arrays in *D. simulans* are still about 97% homologous to those in *D. melanogaster.*

Repeats of two 5-bp sequences, 5′ AAGAC 3′ and 5′ AACAA 3′, and the 10-bp sequence 5′ AATAACATAG 3′, could not be detected in *D. simulans* DNA, because less than 200 cpm was eluted in each of several experiments and the melting transition was continuous. It is not clear whether this result can be explained by low copy number, since the limits of sensitivity of the method are about 1 kb per genome, representing 200 copies of the 5-bp repeats or 100 copies of 10-bp repeats. Therefore, autoradiography was used to improve the level of sensitivity for detection of 10-bp repeats, but no signal was observed in *D. simulans* DNA when 50 copies (0.5 kb per genome) were easily detected in dilutions of *D. melanogaster* DNA. It is unlikely that more than a few contiguous copies, if any, of the 10-bp sequence exist in *D. simulans.*

Each of the ten simple satellite sequences of *D. melanogaster* is represented in *D. erecta,* including the three repeated sequences probably absent from *D. simulans* (Table 2.3). Not all T_m values in *D. erecta* were within 1°C of their *D. melanogaster* counterparts, and three repeated sequences (repeats of 5′ AATAT 3′, 5′ AAGAC 3′, and 5′ AATAGAC 3′) melted within 3°C of the control value. The lowered T_m values are most likely explained by small amounts of cross-hybridization to more abundant, closely related satellite sequences, which contribute to radioactivity eluted just above the hybridization temperature. With the use of these satellites as probes, extensive cross-hybridization occurred when the hybridization temperature was lowered, supporting this interpretation. Alternative explanations such as short segment lengths of these repeats in *D. erecta* or sequence divergence are also possible. However, both the amount of radioactivity eluted and the high temperature of melting show that homologous sequences of the three satellite repeats exist in *D. erecta.*

Thermal melting was not an appropriate method to determine if exact copies of the complex satellite repeats exist in the sibling species, because of the broader T_m profile and ease of cross-hybridization with related repeat families, even at high stringencies of hybridization. Nevertheless, sequences similar to the 359-bp family of the 1.688 satellite can be detected in both *D. simulans* and *D. erecta* DNAs, suggesting that these repeats were present in a common ancestor of the sibling species. The data are in agreement with findings of Strachan and colleagues (1982, 1985), who detected and cloned sequences related to 359-bp repeats throughout the sibling species of the *melanogaster* subgroup. However, no complex repeats cloned from different species were identical in nucleotide sequence, suggesting that sequence conservation may apply only to the simple satellite sequences.

Conservation of simple repeated sequences (Table 2.3) is surprising, considering the many satellites with different buoyant densities in the three species. It is apparent that in species comparisons, different satellite profiles do not necessarily mean that satellite sequences are also different. Quantitation of the repeated sequences in common (Table 2.1) provides, in part, an explanation for the paradox of species specificity of satellite profiles yet simultaneous conservation of satellite sequences. The copy number of each sequence varies widely among species and can be extremely low, sometimes as few as several hundred 5-bp copies in *D.*

erecta. These satellite sequences amount to 21% of the genome in *D. melanogaster,* but most would not be visible optically as satellite peaks in CsCl gradients of *D. simulans* or *D. erecta* DNAs, where together they amount to 5% or 0.4% of the genome, respectively. Thus, species-specific satellite profiles in *Drosophila* result primarily from changes in the copy number of repeats during evolution and not from alterations to the satellite sequences. Some of the differences can also be attributed to the amplification of a sequence present in one species but absent in another species.

Since there is a correlation between abundance of most satellites (if they are present) and the phylogenetic relationships of these three species, quantitative changes in satellite repeats during evolution may not be independent events. In general, the copy number of the *D. melanogaster* satellites is reduced in a coordinate manner in *D. simulans,* and is further reduced in *D. erecta.* The trend of lower abundance with evolutionary distance shown by these 11 satellite sequences is an example of the nonrandom evolutionary behavior of satellite DNAs which is not understood at present.

The hybridization results show that most satellite sequences, as tandem arrays, not only are common to three sibling species but also are identical, despite long periods of separation in evolution. This period is approximately 100 million years for the divergence of present-day *D. melanogaster* and *D. simulans* from a common ancestor. There are no comparable measurements for the period of separation of *D. melanogaster* and *D. erecta* from a common ancestor, but the value for *D. melanogaster* and *D. orena* is about 30 million years. Lines leading to modern *D. orena* and *D. erecta* may have appeared as distinct taxa at about the same time in evolution (Lemeunier and Ashburner, 1984). In contrast to this apparent extreme conservation of satellite sequences in tandem arrays, coding regions of the genome, such as the alcohol dehydrogenase genes, show many nucleotide changes over the same period of evolution. The conservation of the simple satellite sequences in *Drosophila* parallels the conservation of a 6-bp satellite repeat in rodents (Salser et al., 1976; Fry and Salser, 1977) and suggests that long evolutionary persistence of simple satellites may be a general phenomenon.

However, the conservation of some satellite sequences, as tandem arrays, during evolution does not preclude the loss of other sequences, since repeats of three *D. melanogaster* simple satellites could not be detected in *D. simulans.* Further, these same three satellite repeats are present in *D. erecta,* a more distant relative of *D. melanogaster* than is *D. simulans.* Thus, the distribution of satellite sequences in these sibling species does not match the phylogenetic relationships. A similar evolutionary discontinuity of a simple satellite sequence occurs between *D. mela-*

nogaster and *D. varians,* a member of the *ananassae* subgroup. Polypyrimidine repeats of the 1.705 satellite of *D. melanogaster* were found in the sibling species *D. simulans* and *D. mauritiana,* as well as in *D. varians,* but in no other related species that were examined (Cseko et al., 1979). In order to understand more about the sequence conservation and evolutionary discontinuity shown by the simple satellites, we need to know the extent of sequence representation and conservation in other species, both closely and distantly related to *D. melanogaster.*

Any hypothesis involving the evolution of simple satellite sequences in *Drosophila* should consider the following properties: sequence conservation, rapid quantitative change, and evolutionary discontinuity. One explanation for the extreme sequence conservation in evolution of so many simple satellite sequences is that the repeats in sibling species are identical by descent from a common ancestor, and there has been positive selection for the nucleotide sequence of each repeat unit. Selection is the most obvious and usual cause of evolutionary stability, but implies a function whose nature is, however, difficult to envision.

The effect of proposed homogenization mechanisms on tandem repeats, such as unequal exchange between sister chromatids (Smith, 1973), would be to keep simple satellite repeats homogeneous within a species, and this mechanism could also account for quantitative changes of repeats between species. However, this model allows for the introduction and spread of variant repeats throughout the existing tandem arrays with time. These variants can differ in nucleotide sequence, repeat length, or both (Smith, 1976). Sequence conservation of many satellite repeats over the long evolutionary timespans separating species would not be predicted by the unequal exchange model alone, without additional postulates.

Another explanation takes into account the large quantitative differences observed for many satellite sequences between the sibling species, but assumes that qualitative similarities arose by chance. If most repeats of a satellite DNA had been duplicated recently in evolution, the age of the satellite would be much less than the age of the species. In itself, this could explain the extreme sequence homogeneity of the short tandem repeats within a species. Amplification of the same satellite sequence in different species could yield similar satellite DNA arrays, thereby giving the illusion of sequence conservation (Lohe and Brutlag, 1987a).

Amplification of the same satellite sequence independently in related species could occur if a repeated sequence were present in a common ancestor and had persisted during evolution of these species. Another possibility is de novo synthesis of the same satellite sequence in related species, although the large number of possible sequence combinations from the chance generation of short sequences would favor amplification of unrelated sequences into satellite DNAs. However, the fact that each of 11 simple satellite sequences in *D. melanogaster* obeys the expression $(RRN)_m(RN)_n$ suggests that chance could operate in conjunction with those factors, either enzymatic or selectional, that give rise to sequence conformity and the "form rule." Provided that an amplification mechanism exists in *D. melanogaster* to produce tandem copies of repeats according to this rule, all that would be required to generate the present results would be a similar mechanism and form rule in different species. If related species contain different repeats which, however, still conform to this rule, mistakes in copying the satellite sequence, as proposed in the template amplification model (Lohe and Brutlag, 1987b), could generate de novo the same satellite sequence in related species. Such an origin can explain the evolutionary discontinuities of satellites as seen between the *D. melanogaster* sibling species (Table 2.3) or between *D. melanogaster* and *D. varians* (Cseko et al., 1979).

3.2. Polypyrimidine and other repeats in related species

If a satellite array is derived from an adjoining array of repeats by simple mistakes, such as single nucleotide substitutions, during copying, certain predictions about the sequences of different satellites can be made. For example, if a satellite sequence that conforms to the form rule $(RRN)_m(RN)_n$ is present in a species, other satellite sequences fitting this expression may also be present. These would be produced by simple errors in copying the satellite sequence during amplification. A test of the prediction was made possible by the observation of Cseko et al. (1979) that polypyrimidine satellite repeats of *D. melanogaster* are present in a distant relative, *D. varians.* These repeats could not be found in almost 100 related species that were examined, and for this reason, the evolutionary discontinuity is not easily explained by proposing that the common ancestor of *D. melanogaster* and *D. varians* contained the same polypyrimidine satellite repeats. The observation seems to fit better the proposal of de novo synthesis of these repeats independently in evolution.

Our knowledge of the *D. melanogaster* satellite sequences indicates that the polypyrimidine probe used by Cseko et al. (1979) is derived from a mixture of two satellite repeats, 5′ AAGAG 3′ and 5′ AAGAGAG 3′, each in separate tandem arrays. Hybridization of these cloned repeats to *D. varians* DNA identified the sequence in common with *D. melanogaster* as only the 5-bp repeats. To investigate whether other satellite sequences of *D. melanogaster* also exist in *D. varians,* repeats cloned from *D. melanogaster* were used to probe *D. varians* DNA. We found that repeats of both 5′ AATAT 3′ and 5′ AATAG 3′ are present in *D. varians*

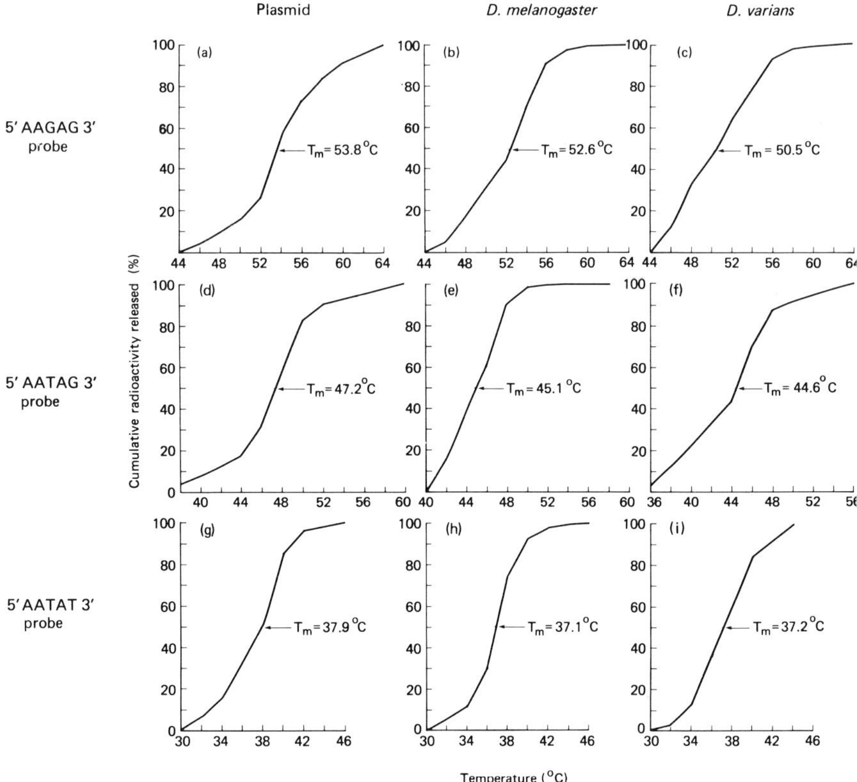

Figure 2.6. Three satellite sequences of *D. melanogaster* are also present in *D. varians.* Thermal melting profiles of a ^{32}p-labeled satellite DNA probe, excised from the plasmid vector with appropriate restriction endonucleases, and hybridized to itself (1 μg of plasmid DNA), *D. melanogaster* DNA (10 μg), or *D varians* DNA (10μg). In (a)–(c), the probe is repeats of 5′ AAGAG 3′. It was hybridized at 43°C to (a) plasmid DNA [49,100 total counts per minute (cpm) eluted], (b) *D. melanogaster* DNA (234,000 cpm eluted) and (c) *D. Varians* DNA (61,500 cpm eluted). In (d)–(f), the probe is repeats of 5′ AATAG 3′. It was hybridized at 35°C to (d) plasmid DNA (13,200 cpm eluted), (e) *D. melanogaster* DNA (9,300 cpm eluted), and (f) *D. varians* DNA (11,400 cpm eluted). In (g)–(i), the probe is repeats of 5′ AATAT 3′. It was hybridized at 30°C to (g) plasmid DNA (51,400 cpm eluted), (h) *D. melanogaster* DNA (31,600 cpm eluted), and (i) *D. varians* DNA (14,500 cpm eluted).

since in both cases there was substantial hybridization, and the heterologous hybrids melted close to the *D. melanogaster* control value (Figure 2.6). Full homology to other satellite sequences of *D. melanogaster,* including the complex 359-bp repeats, was not found, although some probes showed extensive cross-hybridization.

The cluster of *D. melanogaster*-type satellite sequences in the distantly related *D. varians* is consistent with the possibility that an evolutionary event has generated, de novo, a satellite sequence obeying the form rule $(RRN)_m(RN)_n$. The three *D. melanogaster*-type satellite sequences then may have derived from a copy of this ancestral satellite sequence by single base changes, or from each other through the series 5′ AAGAG 3′ ↔ 5′ AATAG 3′ ↔ 5′ AATAT 3′. Alternatively, these sequences may have been conserved and amplified recently, as discussed previously. Additional phylogenetic studies with other repeats should permit a choice between these possibilities.

The improvement in sensitivity of hybridizations, using cloned satellite probes labeled to high specific activity, has also provided evidence for the presence of a satellite sequence in a species that had been thought to lack it entirely. Although repeats of 5′ AAGAG 3′ were not found in *D. erecta* (Cseko et al., 1979), we have detected about 100 kb in this species (Table 2.1; Lohe and Brutlag, 1987a). Nine other simple satellite sequences of *D. melanogaster* are also present in *D. erecta* but are less abundant than repeats of 5′ AAGAG 3′ (Table 2.1), which suggests that they, too, would have been missed by these assaying procedures. Perhaps the *D. melanogaster* simple satellite sequences are widespread in the *melanogaster* species group, and even in the genus *Drosophila*, as indicated by the findings in *D. varians*, but are only rarely amplified into amounts sufficient to appear as a distinct buoyant density satellite. To answer this question, it will be necessary to extend the probing (using cloned *D. melanogaster* satellites, and methods to detect rare sequences in the range of 1–100 kb of satellite repeats per genome) to additional *Drosophila* species.

3.3. Chromosomal locations of *D. melanogaster* satellites in related species

Evolutionary studies of three simple satellite DNAs have been extended to the chromosome level by use of in situ hybridization. The most abundant simple satellite in *D. melanogaster* is the 1.705 satellite. From this satellite, the 5-bp repeats (sequence: 5′ AAGAG 3′) amount to 5.6% of the genome. These repeats are found on all chromosomes of the *D. melanogaster* complement, but particularly on chromosomes 2 and the Y (see section 2.1.3, Evolution of satellite DNA sequences within a species). In contrast, the same repeats are found only on the sex chromosomes of *D. simulans* (Figure 2.7a). The X chromosome showed one site of labeling, at the euchromatic–heterochromatic junction, while the Y showed three sites. The Y locations correspond well with the locations on the *D. melanogaster* Y chromosome. No labeling was detected on the *D. simulans* chromosome 2, in complete contrast to the heavy labeling of this chro-

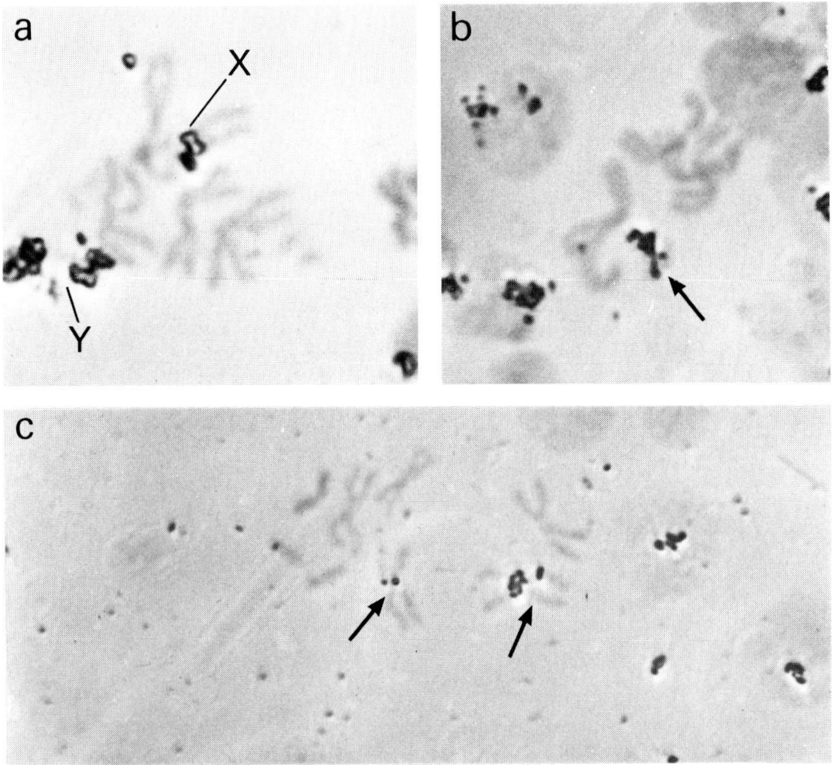

Figure 2.7. In situ hybridization of *D. melanogaster* 5′ AAGAG 3′ repeats to mitotic chromosomes of related species: (a) *D. simulans,* in which only the sex chromosomes (indicated) are labeled. The autoradiograph is overexposed for the Y chromosome, and the three sites appear as two in this cell. (b) *D. mauritiana,* with labeling restricted to a single site on the Y chromosome (arrow). (c) *D. varians,* in which labeling is concentrated on a single pair of autosomes (arrows). In this species, chromosome 4 appears as a rod because of a large heterochromatic region.

mosome in *D. melanogaster.* In *D. mauritiana,* these 5-bp repeats are found at a single location at the tip of the Y chromosome (Figure 2.7b). Since *D. simulans* and *D. mauritiana* can interbreed to form fertile hybrids (David, Lemeunier, Tsacas, and Bocquet, 1974), they are particularly closely related. Nevertheless, these species differ significantly in the content of 5′ AAGAG 3′ repeats at three of four chromosomal sites. We also mapped these repeats to the chromosomes of *D. varians* and found that they are present, instead, on a single large autosome (Figure 2.7c).

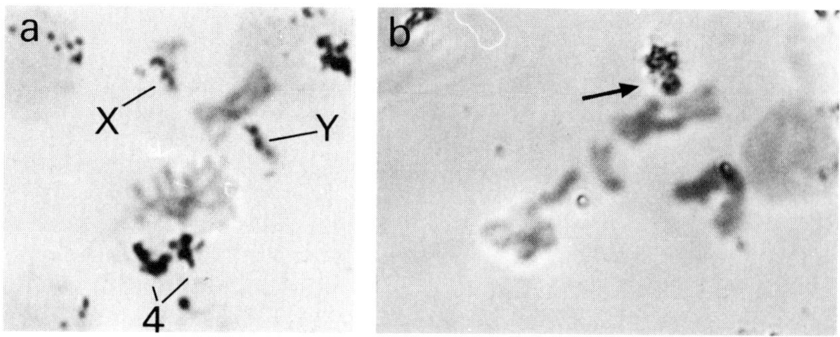

Figure 2.8. Chromosomal localizations of 5′ AATAG 3′ repeats in (a) *D. simulans,* which contains major blocks on these repeats on chromosomes 4 and the sex chromosomes (indicated); and (b) *D. mauritiana,* showing labeling of chromosomes 4 (arrow). A single grain is also present on a pair of large autosomes.

Repeats of the sequence 5′ AATAG 3′ in *D. melanogaster* (0.2% of the genome) are found on chromosome 2 and at one site on the Y chromosome. These repeats are about 10 fold more abundant in *D. simulans,* and, unlike *D. melanogaster,* they are located primarily on the *D. simulans* chromosome 4, the X, and the Y chromosome at two sites (Figure 2.8a). It is not clear whether the distributions are nonoverlapping between these species, or whether small amounts are present at the same chromosomal positions in both species but are masked by the abundance of repeats elsewhere. We found a difference in distribution for 5′ AATAG 3′ repeats between *D. simulans* and *D. mauritiana* (Figure 2.8b). Grain counts of autoradiographs showed that 93% of this satellite sequence is present on the *D. mauritiana* chromosome 4, with 4% present on the X, and less than 2% on each of the Y chromosome and autosomes. Therefore, the sex chromosomes of *D. mauritiana,* but not chromosome 4, are diminished in their relative content of 5′ AATAG 3′ repeats, compared with *D. simulans.* The chromosomal locations of 5′ AATAT 3′ repeats (1.672 satellite) were determined in five sibling species: *D. melanogaster, D. simulans, D. mauritiana, D. yakuba,* and *D. teissieri.* The distributions are remarkably similar in each species, with the Y and chromosome 4 being heavily labeled. The X chromosome was lightly labeled in most species.

The hybridization patterns of satellite sequences held in common by sibling species show that quantitative changes usually occur at many chromosomal sites during speciation. Nevertheless, there is usually a correspondence in locations of a given satellite between the sibling species, despite major quantitative differences. This concordance of position is particularly marked for the 5′ AATAT 3′ repeats, where a similar pattern of hybridization was observed in each of five species despite differences in abundance of this satellite. Thus, the location of DNA sequences in heterochromatin may remain stable over long periods of evolution. In this respect, the evolutionary rules of DNA arrangement in *Drosophila* heterochromatin may parallel those in euchromatin, where linkage groups can have long evolutionary histories (Sturtevant and Novitski, 1941). These satellite distributions are also inconsistent with a transposon-like propagation of most satellite sequences throughout heterochromatin, as might be predicted if they were yet another example of "selfish" DNA (Doolittle and Sapienza, 1980; Orgel and Crick, 1980). One exception, however, could be the 15-bp repeats in *D. simulans* (see below).

Multichromosome distribution is a general property of most satellites, and this dispersal could occur during a single step of sequence amplification. Alternatively, there may be a gradual spread of satellite repeats from one amplified site to other chromosomes, as suggested by Macgregor and Sessions (1986). An opportunity to deduce some properties of satellite amplification is provided by those satellite repeats that are major in one species but cannot be detected in a sibling species. These satellites may have arisen following speciation. Two repeated sequences, 5′ AAGAC 3′ and 5′ AATAACATAG 3′, are major satellites in *D. melanogaster* but could not be detected in *D. simulans.* Repeats of both sequences show multichromosome distributions in *D. melanogaster,* although the 10-bp repeats may be absent from the sex chromosomes and chromosome 4 (Figure 2.5d). Another repeated sequence that may have arisen since speciation of *D. melanogaster* and *D. simulans* is the 15-bp repeat of the 1.696 satellite in *D. simulans.* No homology was detected to *D. melanogaster* DNA in filter hybridizations of these repeats. We found that the 15-bp repeats also show a multichromosome distribution in *D. simulans,* with chromosomes 2 and 3 being the most heavily labeled.

If these three satellite DNAs arose by amplification events following the divergence of *D. melanogaster* and *D. simulans,* there may have been rapid dispersion of satellite repeats to some, although not all, chromosomes of the complement. Supporting this interpretation is the hybridization pattern of the 15-bp repeats of *D. simulans* to polytene chromosomes of this species. Although the chromocenter was heavily labeled, as would be predicted from hybridizations to mitotic chromosomes, at least four euchromatic bands were also labeled (Figure 2.9). Each of these sites of labeling was at the base of a chromosome arm, adjacent to the chromocenter. It would seem that during amplification of the 15-bp repeats in recent evolutionary history, the repeats have spread quickly to many locations in heterochromatin and have even invaded neighboring euchromatic regions. One possible explanation for this unusual evolutionary

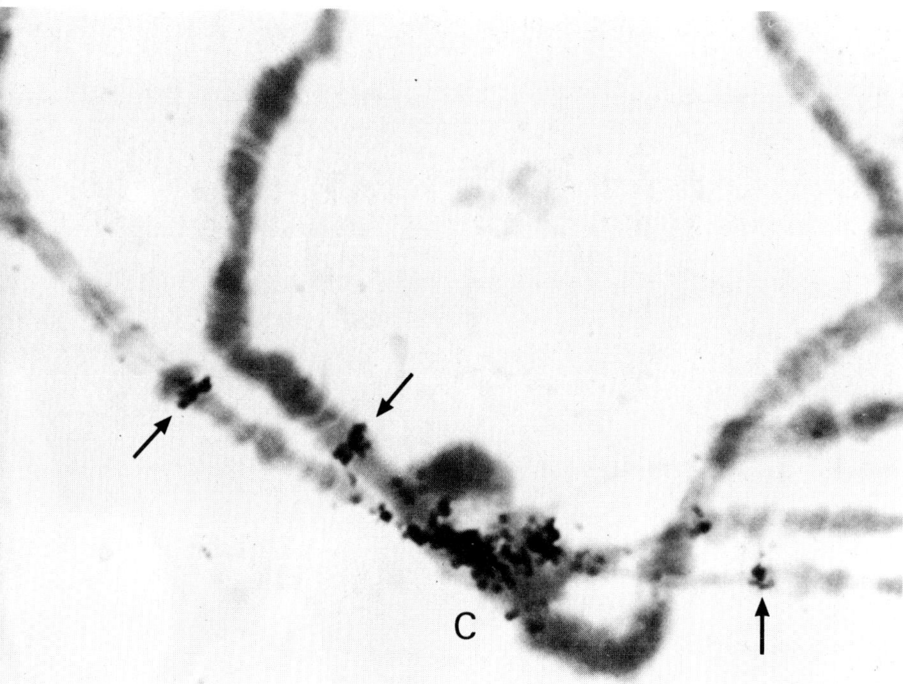

Figure 2.9. In situ hybridization of cloned 15-bp repeats (1.696 satellite) of *D. simulans,* to polytene chromosomes of this species. Note heavy labeling of the chromocenter (C) and three euchromatic bands in the chromosome arms (arrows).

behavior is that unlike other simple satellites in *Drosophila,* each 15-bp repeat contains a 10-bp palindrome (Table 2.2) which could make it resemble the termini of some transposable elements. Further information on the evolution of this satellite may be provided by examination of two sibling species of *D. simulans* – *D. mauritiana* and *D. sechellia* – for 15-bp sequences and, if present, their chromosomal distributions. Both species are more closely related to *D. simulans* than to *D. melanogaster.*

4. Discussion

It is now possible to give a detailed molecular description of the DNA sequences in the heterochromatin of *D. melanogaster,* and their chromosomal organization and evolutionary behavior. Most of the heterochromatin is comprised of tandem repeats of six sequences which are arranged in long, homogeneous stretches. These highly repeated sequences band as four large satellites in CsCl gradients. However, many other tandemly repeated sequences, minor in amount, have been discovered by molecular cloning. They have properties similar to those of the major repeated sequences. This indicates that the concept of a discrete number of satellite DNAs should be replaced by one of a multiplicity of related repeats which differ in their reiteration frequency. Most of the *D. melanogaster* repeated sequences are simple, with repeat units of 5, 7, or 10 bp in length, but several are complex, as exemplified by the 359-bp repeat family. There is no obvious relationship between the sequence of a satellite and its abundance in the genome.

The 11 simple satellite sequences cloned from *D. melanogaster* are related to nucleotide sequence and fit the expression $(RRN)_m(RN)_n$, where R is A or G, and N is any base. This rule is descriptive only and is unsatisfactory in that it gives no clue for its existence in a molecular framework. The rule need not be exclusive for all repeats within the one species of *Drosophila,* and should be considered as a rule of repeat structure rather than nucleotide sequence. Thus, of the four abundant satellite repeats in *D. virilis,* three obey the rule $(AAN)_m(NA)_n$ whereas one fits the *D. melanogaster* expression. Another exception may be the 15-bp repeats of the 1.696 satellite of *D. simulans.* Although the 5-bp subrepeats of this sequence fit the *D. melanogaster* rule, the basic 15-bp repeat conforms for only 10 bp. These rules, however, demonstrate constraints on the evolution of simple repeats, which are unexplained at present.

An additional constraint in evolution is evident from the chromosomal locations of these repeated DNAs. In situ hybridization to polytene chromosomes, where the lateral replication of euchromatic regions allows high sensitivity, shows that simple satellite sequences are almost never found in euchromatin. These repeats are, for the most part, confined to the chromocenter, a region in polytene chromosomes corresponding to the heterochromatic portion of mitotic chromosomes. In contrast, within the heterochromatin there is little or no restriction in their locations. Most satellites show a multichromosome distribution, and some are found at several sites on the one chromosome. The absence of crossing over in heterochromatin would prevent the formation of translocations by crossing over between homologous DNA sequences (such as 5′ AAGAG 3′ repeats) located on nonhomologous chromosomes.

Molecular cloning of satellite molecules containing different simple sequences shows that these arrays can be immediately juxtaposed without intervening, transitional sequences. Repeats flanking a junction usually differ by one nucleotide in their repeating sequence. In addition, the periodicity of repeats is maintained across the junction. The proximity of closely related satellite repeats extends to the chromosome level, where in situ mapping has provided evidence that, in general, satellite arrays differing by a single nucleotide change in their repeating sequence are found in neighboring chromosomal regions. Satellite repeats not as closely related in nucleotide sequence have quite different chromosomal distributions. Therefore, molecular and cytological evidence indicate that closely related satellite repeats occupy adjacent regions of the chromosomal DNA. Such topological constraints might arise from either functional restrictions or, more likely, from their mode of origin, one from the other.

The evolutionary histories of the *D. melanogaster* satellite sequences are seen to be much longer than would be expected from the different satellite DNA profiles in sibling species, which have few, if any, buoyant-density satellites in common. Clearly, molecular cloning has revealed that comparison of satellite position in a density gradient alone is inadequate as an indicator of similarity of simple-sequence DNA contained therein. Hybridization of cloned DNA probes shows that each of ten *D. melanogaster* simple satellite sequences is repeated in the sibling species *D. erecta,* as are seven of these same sequences in *D. simulans.* The satellite repeats would be easily missed by conventional analyses because in most cases the amounts are small, sometimes representing less than 10 kb per genome. However, the simple satellite repeats in *Drosophila* cannot be regarded as a conserved library of sequences common to related species, as proposed by Fry and Salser (1977) for satellites in rodents, since three satellite repeats of *D. melanogaster* could not be detected in the sibling species, *D. simulans.* Another satellite sequence cloned from *D. simulans* seems to be absent from *D. melanogaster.* Of course, the missing sequences might be present in a low number of copies (less than 50 for the 10-bp repeats in *D. simulans*), but this makes the library hypothesis unfalsifiable in the absence of complete sequencing of the genome.

These data illustrate the contrasting fates of simple satellite DNAs in evolution, namely persistence over evolutionary periods as long as about 30 million years since the divergence of present-day *D. melanogaster* and *D. erecta* from a common ancestor, or the apparent loss over the probably shorter period of time separating the divergence of *D. melanogaster* and *D. simulans* from a common ancestor. Some satellite repeats showed both properties simultaneously, thereby creating an evolutionary discontinuity. For example, three satellite repeats of *D. melanogaster* (5′ AACAA 3′, 5′ AAGAC 3′, and 5′ AATAACATAG 3′) are apparently absent from *D. simulans,* yet each can be detected readily in DNA from *D. erecta,* a species more distant to *D. melanogaster* than is *D. simulans.* The results also caution against the premature identification of a satellite sequence as being of "old" or "young" phylogenetic age. If only *D. mel-*

anogaster and *D. simulans* are considered, the three satellite sequences would appear to be "young" in evolution. When *D. erecta* is included in the comparison, these repeats show a longer evolutionary history and therefore are no different from seven other simple satellite sequences.

It would seem unlikely that simple loss of repeats during evolution can account for the evolutionary discontinuity of the three satellite sequences. One might expect that the distribution of satellite repeats in the present-day species would match the phylogenetic relationships, if this were the case. Another possible explanation is that none of the three satellite repeats existed in a common ancestor, but were created de novo in *D. melanogaster* and *D. erecta* independently, following the evolutionary separation of lines leading to the current species. The evolutionary discontinuity shown by repeats of 5′ AAGAG 3′, which span the even greater evolutionary distance between *D. melanogaster* and *D. varians,* leads us to favor the hypothesis of de novo synthesis and amplification. The probability that newly created satellite repeats have the same nucleotide sequence is increased if one repeat derives from the other, and if there are restrictions on generation or selective constraints on persistence of nucleotide sequences (the form rule), as described for the simple satellites in *D. melanogaster.* If de novo synthesis and amplification of simple satellite sequences are the mechanisms of their generation, then sequence homogeneity, sequence "conservation," and evolutionary discontinuity could follow as inevitable consequences.

The amount of heterochromatin in *D. erecta* is not significantly different from that in *D. melanogaster,* but the *D. melanogaster* satellites used as probes represent only 0.4% of the haploid genome of *D. erecta* compared with 21% in *D. melanogaster.* Thus, the bulk of heterochromatin in *D. erecta* must be comprised of other satellite sequences, but we do not have cloning and sequencing data available to define these differences. It seems that a balance between the processes of addition and deletion of satellite sequences is somehow achieved. This balance could be maintained directly by stabilizing selection if the satellite DNAs of constitutive heterochromatin are functional. These functions could be a supporting role in the spatial arrangement of the chromosomes within the nucleus, or in the differential compaction of chromatin with specific recognition proteins. Failure of compaction might interfere with chromosome movement, centromere activity, or both. But the balance of generation and deletion of satellite DNAs could also be maintained indirectly by stabilizing selection if much of the simple sequence DNA is neutral and if there are a sufficient number of vital loci and structures scattered throughout the heterochromatin. Large deletions would have a high probability of removing important functions such as centromeres, ribosomal

RNA genes, fertility factors, pairing sites, or vital loci, and would be selected against.

5. Conclusions

Large blocks of pericentromeric heterochromatin are found in the chromosomes of most *Drosophila* species. Constitutive heterochromatin persists, although, in a quantitative sense, the individual actors (short satellite repeats) that make up most of the heterochromatic blocks differ between related species. Individual actors come and go in the course of evolutionary time, but the characters they represent remain roughly constant, as indicated by the "form rule." The characters appear to be involved in a drama (function), but we are, at the present time, in the unenviable position of observing the drifting actors without understanding the significance of the characters, or the meaning of the drama. At the present stage of our knowledge, it is difficult, therefore, to decipher the meaning of the shifting structure of *Drosophila* satellite DNAs and their distributions and to relate these to possible functions. If satellite DNA were simple "parasitic" DNA (Doolittle and Sapienza, 1980; Orgel and Crick, 1980), or "junk" DNA, the nonrandom patterns of evolution shown by the simple satellites might not be expected. This nonrandom behavior in evolution is evident from the constraints on nucleotide sequence of repeats, from their chromosomal locations, and from quantitative differences in satellites between sibling species. Thus, in *D. melanogaster,* the repeating sequence of 11 simple satellites conforms to the expression $(RRN)_m(RN)_n$. In situ mapping of satellites to polytene chromosomes reveals limitations in their distribution. Apart from several exceptions involving minor amounts of some satellites, these highly repeated DNAs are restricted to pericentromeric regions of chromosomes and are absent from euchromatin. Quantitation of satellite sequences in three sibling species shows that amounts of different satellites do not vary independently in evolution. The ratios of amounts for some closely related repeats (for example, repeats of 5′ AAGAG 3′ and 5′ AAGAGAG 3′, or 5′ AATAAAC 3′ and 5′ AATAGAC 3′) are similar in *D. melanogaster, D. simulans,* and *D. erecta.* Individual amounts of these repeats change dramatically in the three sibling species. Nevertheless, the amount of heterochromatin in these species does not vary greatly despite the large quantitative changes accompanying satellite evolution. A plausible explanation for these nonrandom patterns of evolution is stabilizing selection, acting on constitutive heterochromatin, for important genetic, chromosomal, nuclear, or physiological function(s). Deletions or amplifications of satellite repeats appear to be tolerated only if they do not interfere with these functions. The role of constitutive heterochromatin in the cell, however, remains as one of the major unsolved problems in biology.

6. References

Appels, R., and Peacock, W. J. (1978). The arrangement and evolution of highly repeated (satellite) DNA sequences with special reference to *Drosophila. Int. Rev. Cytol.* [*Suppl.*], *8,* 69–126.

Barnes, S. R., Webb, D. A., and Dover, G. (1978). The distribution of satellite and main-band DNA components in the *melanogaster* species subgroup of *Drosophila. Chromosoma, 67,* 341–363.

Bodmer, M., and Ashburner, M. (1984). Conservation and change in the DNA sequences coding for alcohol dehydrogenase in sibling species of *Drosophila. Nature, 309,* 425–429.

Bostock, C. J. (1986). Mechanisms of DNA sequence amplification and their evolutionary consequences. *Philos. Trans. R. Soc. Lond.* [*B*], *312,* 261–273.

Britten, R. J., and Kohne, D. E. (1986). Repeated sequences in DNA. *Science, 161,* 529–540.

Britten, R. J., Graham, D. E., and Neufeld, B. R. (1974). In *Methods in Enzymology,* Vol. 29E, ed. L. Grossman and K. Moldave, pp. 363–418. New York: Academic Press.

Brosseau, G. E. (1960). Genetic analysis of the male fertility factors on the Y chromosome of *Drosophila melanogaster. Genetics, 45,* 257–274.

Brown, M. (1940). Studies in the genetics of *Drosophila.* II. Chiasma formation in the *bobbed* region of the X chromosome of *D. melanogaster.* The University of Texas Publication No. 4032, pp. 65–70.

Brutlag, D. L. (1980). Molecular arrangement and evolution of heterochromatic DNA. *Annu. Rev. Genet., 14,* 121–144.

Brutlag, D. L., and Peacock, W. J. (1975). Sequences of highly repeated DNA in *Drosophila melanogaster.* In *The Eukaryote Chromosome,* ed. W. J. Peacock and R. D. Brock, pp. 35–46. Canberra: Australian National University Press.

Brutlag, D., Fry, K., Nelson, T., and Hung, P. (1977a). Synthesis of hybrid bacterial plasmids containing highly repeated satellite DNA. *Cell 10,* 509–519.

Brutlag, D. L., Appels, R., Dennis, E. S., and Peacock, W. J. (1977b). Highly repeated DNA in *Drosophila melanogaster. J. Mol. Biol., 112,* 31–47.

Carlson, M., and Brutlag, D. (1979). Different regions of a complex satellite DNA vary in size and sequence of the repeating unit. *J. Mol. Biol., 135,* 483–500.

Cohen, E. H., and Kaplan, G. C. (1982). Analysis of DNAs from two species of the *virilis* group of *Drosophila* and implications for satellite DNA evolution. *Chromosoma, 87,* 519–534.

Cohn, V. H., Thompson, M. A., and Moore, G. P. (1984). Nucleotide sequence comparison of the *Adh* gene in three drosophilids. *J. Mol. Evol., 20,* 31–37.

Cooper, K. W. (1950). Normal spermatogenesis in *Drosophila.* In *Biology of Drosophila,* ed. M. Demerec, pp. 1–61. New York: John Wiley and Sons.

Cooper, K. W. (1964). Meiotic conjunctive elements not involving chiasmata. *Proc. Natl. Acad. Sci. USA, 52,* 1248–1255.

Cordeiro, M., Wheeler, L., Lee, C. S., Kastritsis, C. D., and Richardson, R. H. (1975). Heterochromatic chromosomes and satellite DNAs of *Drosophila nasutoides. Chromosoma, 51,* 65–73.

Cseko, Y. M. T., Dower, N. A., Minoo, P., Lowenstein, L., Smith, G. R., Stone,

J., and Sederoff, R. (1979). Evolution of polypyrimidines in *Drosophila. Genetics, 92,* 459–484.

David, J., Lemeunier, F., Tsacas, L., and Bocquet, C. (1974). Hybridation d'une nouvelle espèce, *Drosophila mauritiana* avec *D. melanogaster* et *D. simulans. Ann. Génét. (Paris), 17,* 235–241.

Doolittle, W. F., and Sapienza, C. (1980). Selfish genes, the phenotype paradigm and genome evolution. *Nature, 284,* 601–603.

Easteal, S., and Oakeshott, J. G. (1985). Estimating divergence times of *Drosophila* species from DNA sequence comparisons. *Mol. Biol. Evol., 2,* 87–91.

Endow, S. A. (1977). Analysis of *Drosophila melanogaster* satellite IV with restriction endonuclease *MboII. J. Mol. Biol., 114,* 441–449.

Endow, S. A., Polan, M. L., and Gall, J. G. (1975). Satellite DNA sequences of *Drosophila melanogaster. J. Mol. Biol., 96,* 665–692.

Fry, K., and Salser, W. (1977). Nucleotide sequences of HS-α satellite DNA from kangaroo rat *Dipodomys ordii* and characterization of similar sequences in other rodents. *Cell, 12,* 1069–1084.

Gall, J. G., and Atherton, D. D. (1974). Satellite DNA sequences in *Drosophila virilis. J. Mol. Biol., 85,* 633–664.

Gall, J. G., Cohen, E. H., and Polan, M. L. (1971). Repetitive DNA sequences in *Drosophila. Chromosoma, 33,* 319–344.

Gatti, M., and Pimpinelli, S. (1983). Cytological and genetic analysis of the Y chromosome of *Drosophila melanogaster.* I. Organization of the fertility factors. *Chromosoma, 88,* 349–373.

Greider, C. W., and Blackburn, E. H. (1985). Identification of a specific telomere terminal transferase activity in *Tetrahymena* extracts. *Cell, 43,* 405–413.

Hannah, A. (1951). Localization and function of heterochromatin in *Drosophila melanogaster. Adv. Genet., 4,* 87–125.

Hazelrigg, T., Fornili, P., and Kaufman, T. C. (1982). A cytogenetic analysis of x-ray induced male steriles on the Y chromosome of *Drosophila melanogaster. Chromosoma, 87,* 535–559.

Hennig, W., and Walker, P. M. B. (1970). Variations in the DNA from two rodent families (Cricetidae and Muridae). *Nature, 225,* 915–919.

Hennig, W., Hennig, I., and Stein, H. (1970). Repeated sequences in the DNA of *Drosophila* and their localization in giant chromosomes. *Chromosoma, 32,* 31–63.

Hilliker, A. J. (1976). Genetic analysis of the centromeric heterochromatin of chromosome 2 of *Drosophila melanogaster:* deficiency mapping of EMS-induced lethal complementation groups. *Genetics, 83,* 765–782.

Hilliker, A. J., and Appels, R. (1982). Pleiotropic effects associated with the deletion of heterochromatin surrounding rDNA on the X chromosome of *Drosophila. Chromosoma, 86,* 469–490.

Hilliker, A. J., Appels, R., and Schalet, A. (1980). The genetic analysis of *D. melanogaster* heterochromatin. *Cell, 21,* 607–619.

Horton, I. H. (1939). A comparison of the salivary gland chromosomes of *Drosophila melanogaster* and *D. simulans.* Genetics, *24,* 234–243.

Hsieh, T., and Brutlag, D. (1979). Sequence and sequence variation within the 1.688 g/cm^3 satellite DNA of *Drosophila melanogaster. J. Mol. Biol., 135,* 465–481.

Hutton, J. R., and Wetmur, J. G. (1973). Effect of chemical modification of the rate of renaturation of deoxyribonucleic acid: deaminated and glyoxalated deoxyribonucleic acid. *Biochemistry, 12,* 558–563.

John, B., and Miklos, G. L. G. (1979). Functional aspects of satellite DNA and heterochromatin. *Int. Rev. Cytol., 58,* 1–114.

Karess, R. E., and Rubin, G. M. (1984). Analysis of P transposable element functions in *Drosophila. Cell, 38,* 135–146.

Kennison, J. A. (1981). The genetic and cytological organization of the Y chromosome of *Drosophila melanogaster. Genetics, 98,* 529–548.

Klein, W. H., Thomas, T. L., Lai, C., Scheller, R. H., Britten, R. M., and Davidson, E. H. (1978). Characteristics of individual repetitive sequence families in the sea urchin genome studied with cloned repeats. *Cell, 14,* 889–900.

Kornberg, A., Bertsch, L. L., Jackson, J. F., and Khorana, H. G. (1964). Enzymatic synthesis of deoxyribonucleic acid. XVI. Oligonucleotides as templates and the mechanism of their replication. *Proc. Natl. Acad. Sci. USA, 51,* 315–323.

Lemeunier, F., and Ashburner, M. (1976). Relationships within the *melanogaster* species subgroup of the genus *Drosophila (Sophophora).* II. Phylogenetic relationships between six species based upon polytene chromosome banding sequences. *Proc. R. Soc. Lond.* [*B*], *193,* 275–294.

Lemeunier, F., and Ashburner, M. (1984). Relationships within the *melanogaster* species subgroup of the genus *Drosophila (Sophophora).* IV. The chromosomes of two new species. *Chromosoma, 89,* 343–351.

Levinger, L. F. (1985). D1 protein of *Drosophila melanogaster. J. Biol. Chem., 260,* 14311–14318.

Lohe, A. R. (1977). Highly repeated DNA in *Drosophila simulans.* Ph.D. thesis, Australian National University, Canberra.

Lohe, A. R. (1981). The satellite DNAs of *Drosophila simulans. Genet. Res., 38,* 237–250.

Lohe, A. R., and Brutlag, D. L. (1986). Multiplicity of satellite DNA sequences in *Drosophila melanogaster. Proc. Natl. Acad. Sci. USA, 83,* 696–700.

Lohe, A. R., and Brutlag, D. L. (1987a). Identical satellite DNA sequences in sibling species of *Drosophila. J. Mol. Biol., 194,* 161–170.

Lohe, A. R., and Brutlag, D. L. (1987b). Adjacent satellite DNA segments in *Drosophila:* structure of junctions. *J. Mol. Biol., 194,* 171–179.

Lohe, A. R., and Moran, C. Instability and repression of a *white* transposon is caused by blanking satellite DNA sequences. (In preparation)

Macgregor, H. C., and Sessions, S. K. (1986). The biological significance of variation in satellite DNA and heterochromatin in newts of the genus *Triturus:* an evolutionary perspective. *Philos. Trans. R. Soc. Lond.* [*B*], *312,* 243–259.

Mazrimas, J. A., and Hatch, F. T. (1972). A possible relationship between satellite DNA and the evolution of kangaroo rat species (genus *Dipodomys*). *Nature, 240,* 102–105.

Miklos, G. L. G., and Gill, A. C. (1981). The DNA sequences of cloned complex satellite DNAs from Hawaiian *Drosophila* and their bearing on satellite DNA sequence conservation. *Chromosoma, 82,* 409–427.

Moore, G. P., Scheller, R. H., Davidson, E. H., and Britten, R. J. (1978). Evolutionary change in the repetition frequency of sea urchin DNA sequences. *Cell, 15,* 649–660.

Mullins, J. I., and Blumenfeld, M. (1979). Satellite Ic: a possible link between the satellite DNAs of *D. virilis* and *D. melanogaster. Cell, 17,* 615–621.

Orgel, L. E., and Crick, F. H. C. (1980). Selfish DNA: the ultimate parasite. *Nature, 284,* 604–607.

Peacock, W. J., Brutlag, D., Goldring, E., Appels, R., Hinton, C. W., and Lindsley,

D. L. (1973). The organizaton of highly repeated DNA sequences in *Drosophila melanogaster* chromosomes. *Cold Spring Harbor Symp. Quant. Biol., 38,* 405–416.

Peacock, W. J., Appels, R., Dunsmuir, P., Lohe, A. R., and Gerlach, W. L. (1976). Highly repeated DNA sequences: chromosomal localization and evolutionary conservatism. In *International Cell Biology, 1976–1977,* ed. B. K. Brinkley and K. R. Porter, pp. 494–506. New York: Rockefeller University Press.

Peacock, W. J., Lohe, A. R., Gerlach, W. L., Dunsmuir, P., Dennis, E. S., and Appels, R. (1977). Fine structure and evolution of DNA in heterochromatin. *Cold Spring Harbor Symp. Quant. Biol., 42,* 1121–1135.

Rasch, E. M., Barr, H. J., and Rasch, R. W. (1971). The DNA content of sperm of *Drosophila melanogaster. Chromosome, 33,* 1–18.

Roberts, P. A. (1965). Difference in the behavior of eu- and heterochromatin: crossing over. *Nature, 205,* 725–726.

Roberts, P. A., and Lohe, A. R. Chromosomal distributions of satellite DNAs in *D. melanogaster* and sibling species. (In preparation)

Rubin, G. M., and Spradling, A. C. (1982). Genetic transformation of *Drosophila* with transposable element vectors. *Science, 218,* 348–353.

Salser, W., Bowen, S., Browne, D., El Adli, F., Federoff, N., Fry, K., Heindell, H., Paddock, G., Poon, R., Wallace, B., and Whitcome, P. (1976). Investigation of the organization of mammalian chromosomes at the DNA sequence level. *Fed. Proc., 35,* 23–35.

Schmookler Reis, R., Timmis, J. N., and Ingle, J. (1981). Divergence, differential methylation and interspersion of melon satellite DNA sequences. *Biochem. J., 195,* 723–734.

Sederoff, R., and Lowenstein, L. (1975). Polypyrimidine segments in *Drosophila melanogaster* DNA. II. Chromosome location and nucleotide sequence. *Cell, 5,* 183–194.

Singer, M. F. (1982). Highly repeated sequences in mammalian genomes. *Int. Rev. Cytol., 76,* 67–112.

Smith, G. P. (1973). Unequal crossover and the evolution of multigene families. *Cold Spring Harbor Symp. Quant. Biol., 38,* 507–513.

Smith, G. P. (1976). Evolution of repeated DNA sequences by unequal crossover. *Science, 191,* 528–535.

Southern, E. M. (1970). Base sequence and evolution of guinea-pig α-satellite DNA. *Nature, 227,* 794–798.

Spofford, J. B. (1976). Position-effect variegation in *Drosophila.* In *The Genetics and Biology of Drosophila,* ed. M. Ashjburner and E. Novitski, Vol. 1c, pp. 955–1018. London: Academic Press.

Spradling, A. C., de Cicco, D. V., Wakimoto, B. T., Levine, J. F., Kalfayan, L. J., and Cooley, L. (1987). Amplification of the X-linked *Drosophila* chorion gene cluster requires a region upstream from the s38 chorion gene. *EMBO J., 6,* 1045–1053.

Steffensen, D. M., Appels, R., and Peacock, W. J. (1981). The distribution of two highly repeated DNA sequences within *Drosophila melanogaster* chromosomes. *Chromosoma, 82,* 525–541.

Strachan, T., Coen, E., Webb, D., and Dover, G. (1982). Modes and rates of change of complex DNA families of *Drosophila. J. Mol. Biol., 158,* 37–54.

Strachan, T., Webb, D., and Dover, G. A. (1985). Transition stages of molecular drive in multiple-copy DNA families in *Drosophila. EMBO J., 4,* 1701–1708.

Sturtevant, A. H., and Novitski, E. (1941). The homologies of the chromosome elements in the genus *Drosophila. Genetics, 26,* 517–541.

Sutton, W. D., and McCallum, M. (1972). Related satellite DNAs in the genus *Mus. J. Mol. Biol., 71,* 633–656.

Tartof, K. D., Hobbs, C., and Jones, M. (1984). A structural basis for variegating position effects. *Cell, 37,* 869–878.

Walker, P. M. B. (1968). How different are the DNAs from related animals? *Nature, 219,* 228–232.

Wheeler, L. L., Arrighi, F., Cordeiro-Stone, M., and Lee, C. S. (1978). Localization of *Drosophila nasutoides* satellite DNAs in metaphase chromosomes. *Chromosoma, 70,* 41–50.

Zhimulev, I. F., Semeshin, V. F., Kulichkov, V. A., and Belyaeva, E. S. (1982). Intercalary heterochromatin in *Drosophila. Chromosoma, 87,* 197–228.

III

The mammalian kinetochore

DANIEL A. PEPPER

1. Introduction

Mitosis is an essential process for the equal distribution of genetic material between two daughter cells. During cell division, this genetic material is packaged in the form of chromosomes. The equal partitioning of the chromosomes to the daughter cells is dependent upon the orderly formation and function of a bipolar spindle apparatus which, in most eukaryotes, basically consists of two spindle poles and their associated spindle fibers. The use of electron microscopy has revealed that spindle fibers are composed of microtubules. For present purposes these microtubules can be grouped into two major classes. One class of microtubules, usually termed *interpolar* (or *continuous*) microtubules, traverses most, if not all, the distance between the two spindle poles. The other class, usually termed *kinetochore* microtubules, has one end terminating at the spindle pole and the other end terminating at a specialized locus on the centromeric region of the chromosome called the *kinetochore.*

The attachment of spindle microtubules to the chromosome via a specialized kinetochore is a feature of most higher eukaryotic cells. In most organisms studied, this microtubule attachment point is situated on the chromosome at the region of the centromere (primary constriction). Consequently the terms centromere and kinetochore have been used synonymously by many investigators for a number of years, resulting in considerable confusion. With the advent of electron microscopy and the discovery of a distinct structure for microtubule attachment to the chromosome, workers in the field have felt an increasing need to distinguish between this structure and the underlying chromosomal material of the primary constriction. Therefore, it has gradually become common to refer to the region of the primary constriction as the *centromere* and to use the term *kinetochore* to denote the specialized microtubule attach-

My thanks to many colleagues for their stimulating discussions and comments, and especially to Drs. B. R. Brinkley, Ron Balczon, and Ray Zinkowski for reviewing the manuscript. I also thank Dr. Sari Brenner for providing micrographs, and Dr. Janet Vitiello for her invaluable assistance with this paper.

ment point of the chromosome. It has been formally suggested that the two terms be standardly used in this fashion (Rieder, 1982); this terminology will be adopted in the present review because of its current popularity and in order to avoid possible further confusion.

As I hope to illustrate, the kinetochore is an interesting and singular structure which possesses properties likely to be of fundamental importance in normal chromosome distribution among progeny cells. Relatively little is known of the true nature and functional properties of this organelle. The purpose of this chapter will be to give an overview of the recent work in this specialized area of research, with an emphasis placed on the structure, composition, and function of the mammalian kinetochore. For broader or more historical treatments of work in this and related fields, the reader will be referred to previous reviews where appropriate.

2. Kinetochore structure and composition

2.1. Kinetochore structure in lower organisms

A few species among the lower plants and animals have been found to exhibit diffuse (holocentric) kinetochores distributed along the entire length of the chromosomes, and these chromosomes appear to lack a primary constriction (reviewed in Rieder, 1982). However, the kinetochores of most organisms are localized at the primary constriction, as mentioned above. These localized kinetochores vary in ultrastructural appearance among lower species and may appear as a ball of ill-defined fibrillar material embedded in a more electron-opaque chromatin cup (common in most plants). Alternatively a kinetochore structure may be microscopically absent, with spindle fiber microtubules apparently terminating on the centromeric deoxyribonucleoprotein (fungi, yeasts, etc.; for review see Rieder, 1982). The metaphase kinetochore of mammals, which appears ultrastructurally as a multilayered disk on the surface of the centromere of the chromosome, will be the primary subject of the remainder of this review.

2.2. Kinetochore structure in mammals

The mammalian kinetochore at metaphase is illustrated ultrastructurally in Figures 3.1 and 3.2. As revealed by conventional electron microscopy, the structure is a trilaminar disk and consists of an outer "plate" or layer of condensed fibrillar or granular material approximately 40 nm thick and an inner layer of approximately the same thickness which is closely

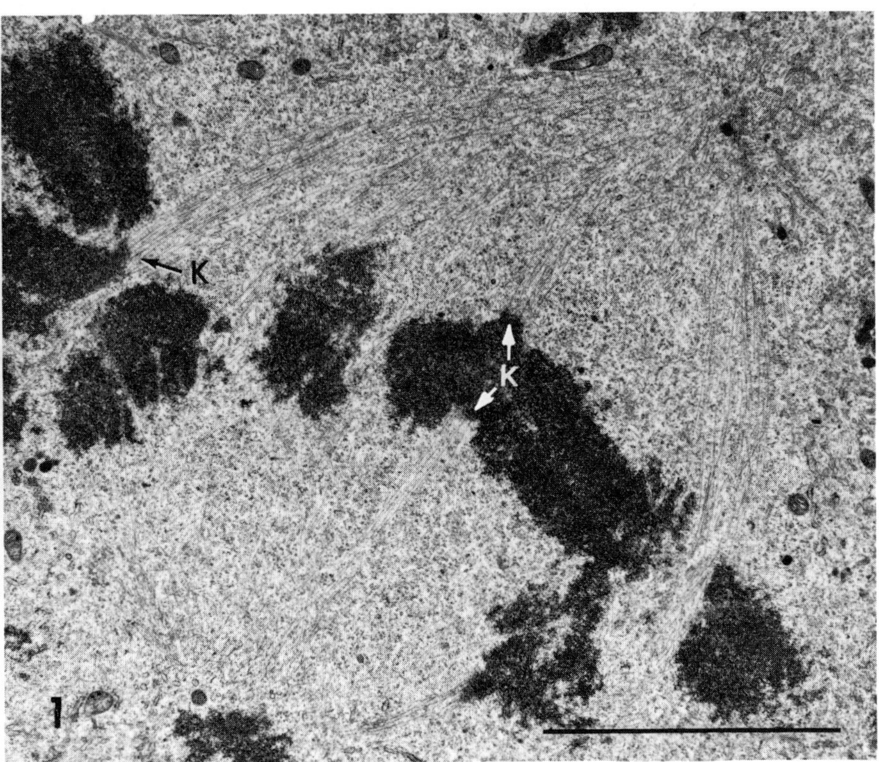

Figure 3.1. Routine transmission electron micrograph showing typical appearance of the metaphase spindle of a PtK_1 cell. Spindle microtubules and kinetochores (K) can be seen. Bar = 5 μm. (From Pepper and Brinkley, 1977.)

adjacent to the underlying centromeric chromatin. These two electron-opaque layers are separated by a more electron-translucent layer approximately 30 nm thick. It is interesting to note that in mitotic cells treated with spindle-disrupting drugs the kinetochore appears to retain only a single outer layer (Brinkley and Stubblefield, 1966; Roos, 1973) and has associated with it a "corona" of fibrillar material. The mechanism(s) of action of these drugs in altering the structure of the kinetochore has been the subject of much speculation, but is still essentially unknown. Also until recently, the composition of the kinetochore was largely speculative. However, as we shall see below, an increasing number of studies have recently been devoted to the chemical dissection of this organelle, especially since the advent of new techniques available to, and new approaches taken by researchers in this field.

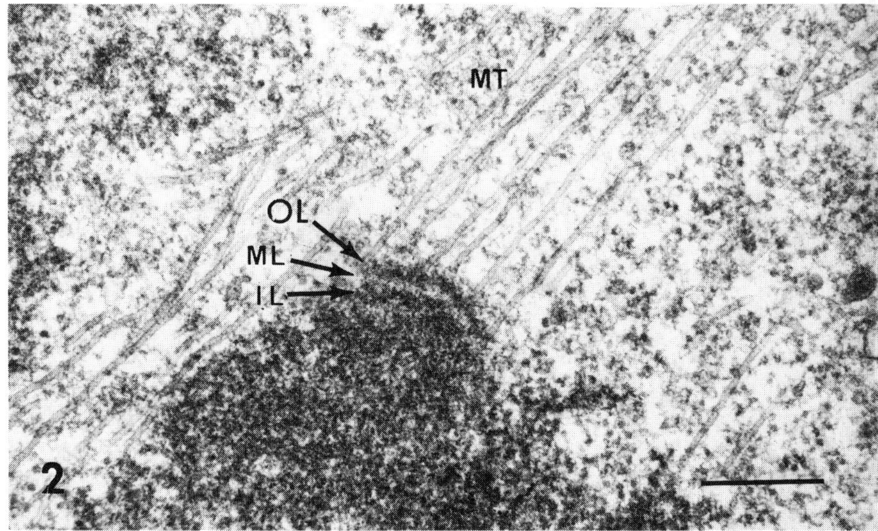

Figure 3.2. Higher magnification of a metaphase kinetochore as seen in routine electron-microscopic preparations, showing the outer-layer (OL), middle-layer (ML). and inner-layer (IL) kinetochore elements as well as kinetochore microtubules (MT). Bar = 0.5 μm. (From Pepper and Brinkley, 1977.)

2.3. Kinetochore composition

The kinetochore layers are differentiated from the underlying chromatin by electron microscopy using conventional heavy metal staining. This suggests that the composition of this organelle is somehow different from that of the chromosome, even though the kinetochore might be derived from (Luykx, 1970) or organized by (Holmquist and Dancis, 1979) the subjacent chromatin of the centromeric region.

The presence of DNA in the kinetochore has been suggested by several workers who found microtubules attached to decondensed centromeric chromatin in hypotonically treated mitotic cells (Brinkley, Cox, and Pepper, 1980; Ris and Witt, 1981). Also Pepper and Brinkley (1980) found that DNase treatment of lysed mitotic cells specifically decondensed the outer plate structure of the kinetochore. A study by Rattner (1986) has illustrated that the kinetochore may be digested with nucleases, although it is more resistant to digestion than the rest of the chromosome. Indirect evidence for the role of DNA in kinetochore function has been presented (Corces et al., 1980; Marx and Denial, 1985) and will be further discussed below. Earnshaw and Laemmli (1983) observed that isolated and dehistonized chromosomes retain differentiated regions that resemble kinetochores and are apparently derived from the chromatid axis. Results from earlier studies have indicated either that chromatin is lacking in the kinetochore outer plate (Roos, 1977) or that kinetochores are apparently not digested by DNase treatment (Rattner, Krystal, and Hamkalo, 1978). However, these observations were based either on conventional ultrastructural staining (Roos, 1977) or on analysis of whole-mount chromosome preparations that do not reveal kinetochore fine structure (Rattner et al., 1978). At present, the bulk of the evidence seems to indicate that the kinetochore does contain some form of chromatin and may possibly be derived from (or organized by) the centromeric DNA, although observations on the derivation and structural organization of the kinetochore are largely speculative at this point. However, some preliminary results obtained by Palmer and Margolis (1985) have indicated that specific kinetochore proteins interact with chromatin and may substitute for mononucleosome components (such as histone H_1), and that such an approach might be used for the isolation and sequencing of kinetochore-specific DNA. It is hoped that more studies along this line will soon yield definitive molecular data concerning putative kinetochore DNA sequences.

Evidence for the presence of RNA in kinetochores has been presented and is based largely on cytochemical studies (Bielek, 1978; Rieder, 1979). Bielek (1978) found in colcemid-treated HeLa cells a single outer plate structure which appeared to contain ribonucleoprotein (using a procedure that preferentially stains this type of material), and that this ribonucleoprotein was sensitive to RNase digestion. In a similar but more extensive study on PtK_2 cells, Rieder (1979) found evidence for the presence of ribonucleoprotein in the inner layer of untreated trilaminar kinetochores as well as the single kinetochore plate seen in colcemid-treated cells. (In addition to the possibility that kinetochores contain RNA, these studies combined raise an interesting question as to the derivation of the single kinetochore plate seen in cells treated with mitoclastic drugs.) In contrast to these studies, another group reported that RNase had no apparent effect on the ultrastructural appearance of kinetochores (Pepper and Brinkley, 1980), although the methodologies used were somewhat different and therefore do not rule out the presence of RNA in this structure.

The protein composition of the kinetochore has lately been the subject of intensified interest, which is not surprising in light of its apparent role in specific microtubule attachment to the chromosome. Ultrastructural immunocytochemical evidence was presented some years ago indicating the presence of tubulin in the kinetochore plate structures (Pepper and Brinkley, 1977). However, results from another group failed to localize

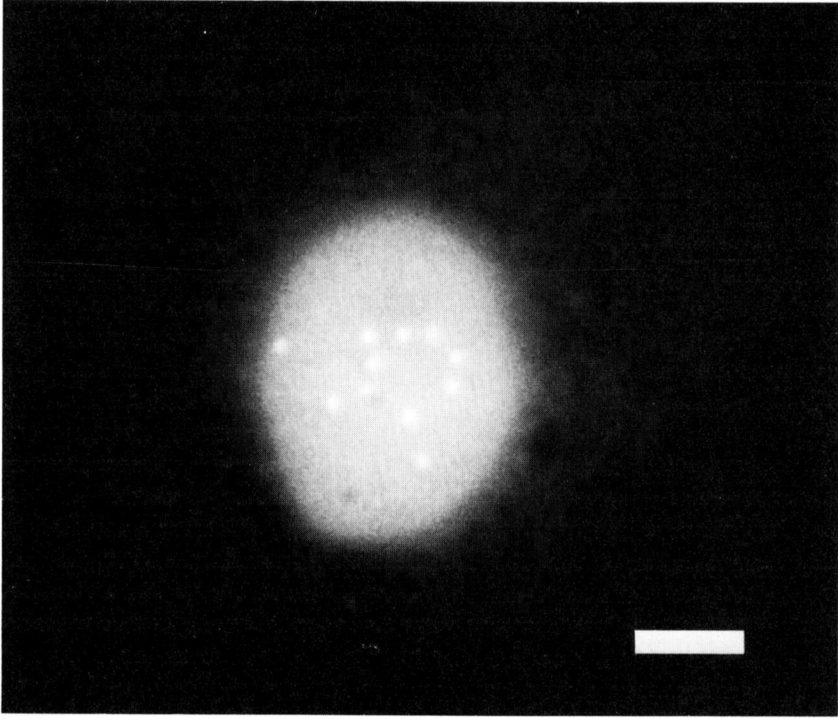

Figure 3.3. Immunofluorescent localization of scleroderma antikinetochore serum in interphase PtK_2 cells. Cells were fixed, indirectly labeled using a fluorescein-conjugated goat anti-human second antibody, and counterstained with propidium iodide to reveal DNA. Note the discrete localization of "prekinetochores" in the nucleus. Bar = 5 μm. (Micrograph courtesy of Dr. Sari Brenner.)

tubulin in this organelle (De Brabander, Geuens, De Mey, and Joniaiau, 1979), and the reason for this discrepancy is unknown. More recent studies from other authors (Mitchison and Kirschner, 1985a) have provided evidence that tubulin is associated with the kinetochores of colcemid-treated cells and with kinetochores of chromosomes incubated with tubulin in vitro, but not with kinetochores in cells treated with vinblastine. Thus the association of tubulin with kinetochores appears sensitive to the methods used in preparing the cells for study. The presence of other proteins in the kinetochore – for example, Ca-ATPase (Lyubskii, Buchwalow, and Raikhlin, 1979) and phosphoproteins (Vandre, Davis, Rao, and Borisy, 1984) – has been suggested. Although these results are potentially

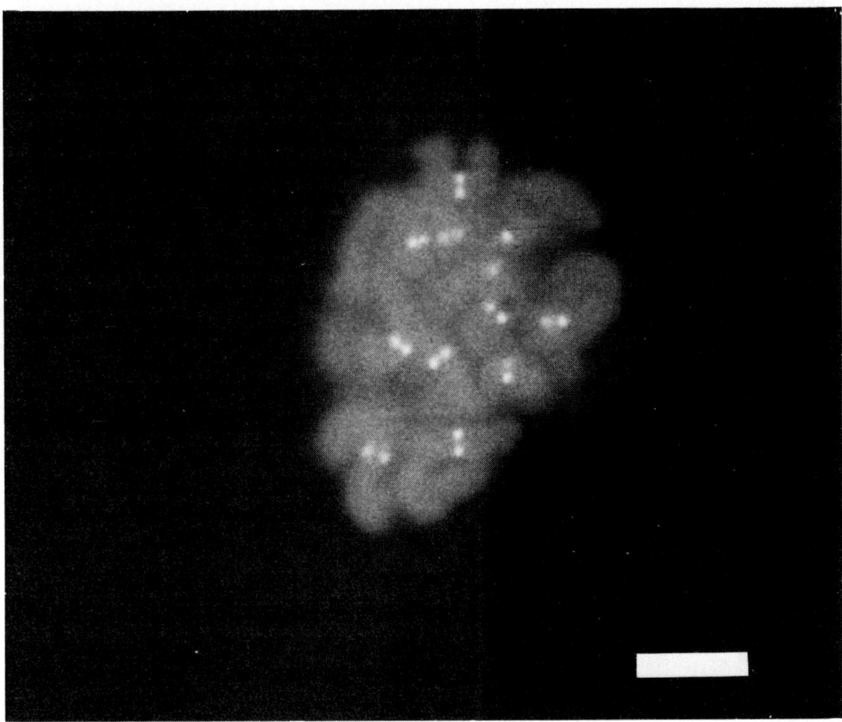

Figure 3.4. Immunofluorescent localization of kinetochores in mitotic PtK_2 cells with the use of scleroderma antiserum as in Figure 3.3. Note localization of discrete kinetochores on centromeres of chromosomes. Bar = 5 μm. (Micrograph courtesy of Dr. Sari Brenner.)

quite significant, these protein species need to be characterized in further detail.

The description by Moroi et al. (1980) of a naturally occurring autoantibody from scleroderma patients that specifically binds kinetochores has recently stimulated a significant amount of work focused on the characterization of kinetochores using this antibody. Brenner et al. (1981) found that this antibody localizes antigen(s) in the inner and outer kinetochore plates during mitosis, and that it localizes discrete foci in interphase nuclei that correspond to the chromosome numbers of the species studied (see Figures 3.3 and 3.4). These foci in nuclei were termed *presumptive kinetochores* or *prekinetochores,* and were found by immunofluorescence studies to double (or replicate) during G_2 of the cell cycle. Workers in this field have previously speculated on the existence of a kinetochore organizer region somewhat analogous to the nucleolar organizer (see Rieder, 1982, for review). The results of Brenner et al. (1981), although based solely on immunocytochemical studies, provide compelling evidence that such an organizer does exist within the genome.

Recent work using kinetochore-specific antibody from scleroderma patients has been focused on identifying or purifying the antigen(s) that this antibody recognizes. Several groups have used this type of antibody to label electrophoretically separated protein extracts from nuclei or chromosomes. Polypeptide species reportedly recognized by the antibody range in molecular weight from 14 to 140 (Cox, Schenk and Olmsted, 1983; Ayer and Fritzler, 1984; Earnshaw, Halligan, Cooke, and Rothfield, 1984; Guldner, Lakomek, and Bautz, 1984; Earnshaw and Rothfield, 1985; Nishikai, Okano, Yamashita, and Watanabe, 1985). Valdivia and Brinkley (1985) have isolated centromeres from human chromosomes and have identified in them kinetochore antigens with a mass of 18 and 80 kd. The reason for the molecular-weight discrepancies reported by these groups is unclear at present, but results from these studies could reflect differences in the antisera used or differences in the technical procedures employed, or they might indicate that there are multiple protein species involved in kinetochore structure. Undoubtedly, more definitive data on this subject will be forthcoming since the scleroderma antiserum described above is the first truly kinetochore-specific probe available and the dissection of the composition of the kinetochore is of fundamental importance in determining its molecular organization.

3. Kinetochore function

3.1. Kinetochore–microtubule interactions

It is well documented that mammalian kinetochores form on the condensing chromosomes during prophase of mitosis. The initiation of prometaphase is characterized by the breakdown of the nuclear envelope and the attachment of the kinetochores to the newly forming spindle. It has been observed that the primary function of the kinetochore is to anchor the chromosome to the spindle (Nicklas, 1971) and that attachment to the spindle (via microtubules) is a prerequisite for directed chromosome movement during mitosis (reviewed in Bajer and Mole-Bajer, 1972; McIntosh, 1979).

As discussed above, ultrastructural studies have revealed that kinetochore microtubules are usually seen terminating in the outer kinetochore plate. The number of microtubules associated with a kinetochore differs considerably from one organism to another (but less so between chromosomes within the same cell); the number is apparently not related to chromosome size, but instead may be related to the surface area of the kinetochore (Rieder, 1982). Cross sections of the kinetochore parallel to the plane of the outer plate show microtubules saturating the area of the outer plate and spaced an average of 60–90 nm apart in PtK_1 cells (Rieder, 1981) and Chinese hamster ovary cells (Witt, Ris, and Borisy, 1981). However, the molecular factors that determine the number of microtubules associated with a kinetochore are unknown.

In light of the tubulin assembly-promoting characteristics of microtubule associated proteins (MAPs), recent work has addressed the possibility that MAP-like molecules might be components of the kinetochore and may provide the structural link between the microtubule and the chromatin of the chromosome. Corces et al. (1980) and Marx and Denial (1985) have provided biochemical evidence of a preferential association between isolated MAPs and satellite DNA. Whether the associations seen in these studies were specific or fortuitous is unclear, since the MAPs used are not known to have kinetochore-specific functions and the DNA employed was from the total satellite fraction (and therefore may or may not have contained kinetochore-specific DNA sequences). De Brabander et al. (1981), using an antibody against a 210-kd MAP, failed to immunocytochemically localize this molecule in the kinetochores of mitotic cells. Nevertheless, in view of the intimate association between kinetochore microtubules and the kinetochore outer plate, the presence of MAP-like molecules in the kinetochore would not be surprising, but the question awaits further study.

3.2. Kinetochore vs. centrosomal microtubule nucleation

The orderly formation of the mammalian mitotic spindle is apparently facilitated by the microtubule-nucleating capacity of the centrosomes and possibly the kinetochores. An overview of current models of spindle formation and function will not be presented here; instead the reader is referred to Pickett-Heaps, Tippit, and Porter (1982) and Mitchison, Evans, Schulze, and Kirschner (1986) for review and critique.

It is established that kinetochores can nucleate the assembly of tubulin subunits into microtubules in vitro by adding exogenous tubulin to lysed cells (McGill and Brinkley, 1975; Snyder and McIntosh, 1975) and isolated chromosomes (Telzer, Moses, and Rosenbaum, 1975; Gould and Borisy, 1978). Evidence that this occurs in vivo as well has also been presented (Witt, Ris, and Borisy, 1980; De Brabander, Geuens, De Mey, and Jonaiau, 1981; Ris and Witt, 1981). Putative molecular components involved in the capacity of the kinetochore to nucleate microtubule assembly include tubulin (Pepper and Brinkley, 1979), DNA (Pepper and Brinkley, 1980), and an undefined component which is apparently identical with an antigen that binds the scleroderma-associated kinetochore antibody described above (Cox et al., 1983). However, definitive data on the components responsible for the ability of kinetochores to nucleate microtubule assembly (and their molecular arrangement within the kinetochore) are certainly incomplete at present.

In an attempt to elucidate the molecular architecture of the mitotic spindle, workers have addressed the question of the polarity of kinetochore microtubules, that is, whether the kinetochore is associated with the tubulin-assembling or the tubulin-disassembling end ("plus" or "minus" end, respectively) of kinetochore microtubules. (For a review of the methods employed in microtubule polarity determinations, see McIntosh and Euteneuer, 1984.) Polarity determinations of mitotic cells lysed in situ indicate that it is usually the plus ends of microtubules that are associated with kinetochores (Euteneuer and McIntosh, 1981; Telzer and Haimo, 1981). This and other considerations have led some authors to suggest that kinetochore microtubules are derived from those nucleated from the centrosomes and are simply "captured" by kinetochores in prometaphase (Pickett-Heaps et al., 1982; Rieder, 1982), since tubulin would not be expected to assemble onto a microtubule end that is already anchored in the kinetochore plate (Borisy, 1978). Thus the significance of microtubule nucleation by the kinetochore in vivo (especially with regard to its putative role in spindle formation) is currently the subject of debate (Pickett-Heaps et al., 1982; Rieder, 1982; Mitchison and Kirschner, 1985b; Sluder, Rieder, and Miller, 1985). Studies by Mitchison and Kirschner (1985a, 1985b) show that tubulin can be incorporated into microtubules of mixed polarities at the kinetochore in vitro. This group of researchers also microinjected labeled tubulin into metaphase cells and found in vivo evidence for kinetochore-mediated tubulin assembly at metaphase and disassembly at anaphase (Mitchison et al., 1986). How these results relate to current models of spindle function (reviewed in Pickett-Heaps et al., 1982; Rieder, 1982) is unclear at present. However, such an approach applied to prometaphase cells might provide valuable information as to the role of kinetochores in the formation of the mitotic spindle apparatus.

In light of the studies described above, it becomes apparent that the characteristics and function of kinetochore-mediated microtubule assembly (and possibly disassembly) is a complex issue. Its significance must await further experimentation, which likely will be soon forthcoming.

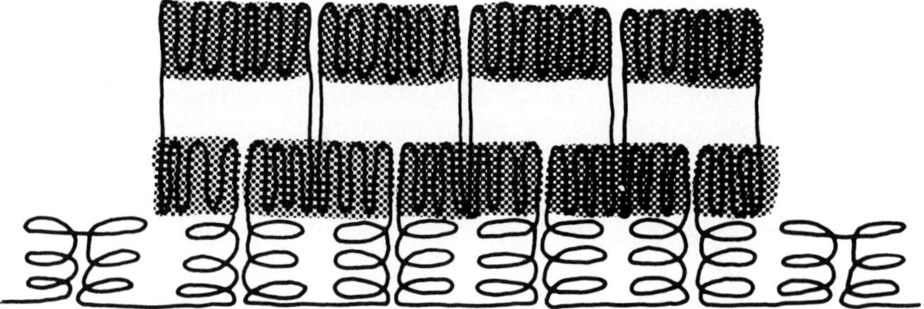

Figure 3.5. Diagrammatic representation of a hypothetical model for the structure of the mammalian kinetochore. The DNA scaffold is designated as a single black line. Horizontal loops represent the packed DNA of the centromeric chromatin underlying the kinetochore. Vertical loops represent the DNA scaffold of the inner and outer kinetochore plates. Kinetochore-associated proteins are designated by the shaded (stippled) areas of the inner and outer layers.

4. Cytogenetic aspects

4.1. Mammalian kinetochore evolution

A detailed structural model of the trilaminar kinetochore was first presented by Roos (1977), which described the organelle as being composed of a series of repeating units derived from a looping out of discrete repeated segments of the centromeric DNA. Other similar models have been suggested, and inherent in them is the implication that these loops contain DNA sequences that are responsible for the specific binding and structural organization of kinetochore components (protein, RNA, etc.). Such a hypothetical model is illustrated in Figure 3.5. Regardless of accuracy of details, an attractive feature of such models is that they provide a general explanation of why kinetochores appear structurally consistent on a particular chromosome from one cell generation to the next.

Robertsonian rearrangements can lead to abnormally large centromeres with a concomitant increase in the number of kinetochore-associated microtubules. This has led some investigators to the hypothesis that the relatively large complex kinetochores seen in higher eukaryotes may have evolved from simpler units (such as the one-microtubule-per-chromosome arrangement seen in yeasts) by a series of genomic reorganizations and linear fusions. In this regard, an interesting system for study has involved the genus *Muntiacus* or Asiatic muntjac deer. The Indian muntjac *M. muntjak vaginalis* has a low chromosome number [$2N$ = 6 (female), 7 (male)] with large chromosomes and associated kinetochores (especially on the X chromosome), whereas the Chinese muntjac *M. m. reevesi* has a large number of smaller chromosomes ($2N = 46$). Shi, Ye, and Duan (1980) have studied the chromosomal banding patterns of these two species as well as F_1 hybrids between them, and on this basis have suggested that the Indian muntjac evolved from the Chinese muntjac concomitant with a series of centric and repeated tandem translocations among the chromosomes. This study has been supported and elaborated upon by Brinkley, Valdivia, Tousson, and Brenner (1984), who studied the chromosomes of these two species with the use of kinetochore-specific antibody (from scleroderma patients, as described above in section 2, Kinetochore structure and composition) in immunofluorescence experiments. This group interpreted their results to indicate that distinct homologies exist between the interphase prekinetochore staining patterns of the two species (despite the vast difference in their diploid chromosome numbers), and suggested that the kinetochores of the Indian muntjac evolved by linear fusion of the smaller kinetochores of the Chinese muntjac. This is an interesting idea, although the test of such an hypothesis awaits the construction of an appropriate experimental system.

4.2. Active vs. inactive centromeres

Robertsonian translocations may generate stable di- and multicentric chromosomes possessing apparently "inactive" centromeres. Workers have recently begun to study the nature of these inactive centromeres in more detail, especially with regard to kinetochore formation. Whether such centromeres are inactive because of the lack of kinetochore structures is difficult to determine, since electron microscopic identification of individual chromosomes in thin section is hardly possible in most cases.

More recently, investigators have taken advantage of kinetochore-specific scleroderma-associated antibodies (as described above) as immunofluorescent probes for the presence of kinetochores in spread chromosome preparations. Anti-kinetochore immunofluorescence of a stable isodicentric X chromosome in humans (Earnshaw and Migeon, 1985) showed positive labeling at the active centromere only. However, in a similar study of human fibroblasts exhibiting a 9;11 translocation (Merry, Pathak, Hsu, and Brinkley, 1985), it was seen that both centromeres were labeled with the antibody, although the inactive centromere (No. 9) was labeled to a lesser degree. Zinkowski, Vig, and Broccoli (1986) carried out anti-kinetochore staining of stable rat and mouse cell lines which exhibited di- and multicentric chromosomes, and found that rat cells displayed as many kinetochores as there were centromeres, but that mouse cells showed only one pair of kinetochores on di- and multicentric chromosomes.

Taken together, these studies indicate that kinetochores may be present on inactive centromeres but, if so, are not necessarily functional. If probes for individual components of the kinetochore can be constructed, these multicentric systems may provide a means for the identification of kinetochore components specifically responsible for binding to spindle microtubules (i.e., those components that may *not* be present in inactive kinetochores).

4.3. Aneuploidy and the kinetochore

Aneuploidy in mammals is most often associated with developmental abnormalities, benign tumors, and malignant neoplasias. Even though aneuploid conditions are suspected of often having dire consequences for the individual, cause-and-effect relationships between aneuploidy and clinical abnormalities are poorly understood at the molecular level.

Equal chromosome distribution to daughter cells requires the formation of a bipolar spindle and the proper attachment of chromosomes to this spindle (i.e., attachment of sister kinetochores to opposite poles of the spindle via kinetochore microtubules). Nicklas (1985) has reviewed various general aspects of mitotic and meiotic spindle malfunction that may lead to aneuploidy. These include (1) the failure to form a bipolar spindle, which may result in nondisjunction (in the case of monopolar spindles) or many different types of chromosomal distributions (in the case of multipolar spindles); (2) the absence of a functional kinetochore, which eventually results in chromosome loss; and (3) the absence of the usual constraints on kinetochore position and polarity, which may result in chromosome breakage or any of a number of distributional abnormalities. Certainly other mechanisms such as effects of mutagens and clastogens could play a role here (see Liang and Brinkley, 1985, for review), and it will be a formidable task to sort these out.

Many workers have directed their efforts toward understanding the control (and the absence of control) of cell division. Many of these efforts have, in addition, begun to shed light on possible mechanisms involved in the generation of abnormal genomic content and arrangements. Cytogeneticists have become interested in the structural and molecular aspects of centromere and kinetochore function. Thus cytogeneticists and cell biologists are now combining their efforts to understand the molecular mechanisms that result in both normal cell division and aneuploidy. It is hoped that this association will prove fruitful in elucidating these important fundamental mechanisms, which are the basis of cellular inheritance.

5. References

Ayer, L. M., and Fritzler, M. J. (1984). Anti-centromere antibodies bind to trout testis histone 1 and a low molecular weight protein from rabbit thymus. *Mol. Immunol., 21,* 761–770.

Bajer, A. S., and Mole-Bajer, J. (1972). Spindle dynamics and chromosome movements. *Int. Rev. Cytol.* [*Suppl.*], *3,* 1–271.

Bielek, E. (1978). Structure and ribonucleoprotein staining of kinetochores of colchicine-treated HeLa cells. *Cytobiologie, 16,* 480–484.

Borisy, G. G. (1978). Polarity of microtubules in the mitotic spindle. *J. Mol. Biol., 124,* 565–570.

Brenner, S., Pepper, D., Berns, M. W., Tan, E., and Brinkley, B. R. (1981). Kinetochore structure, duplication, and distribution in mammalian cells: analysis by human autoantibodies from scleroderma patients. *J. Cell Biol., 91,* 95–102.

Brinkley, B. R., and Stubblefield, E. (1966). The fine structure of the kinetochore of a mammalian cell in vitro. *Chromosoma, 19,* 28–43.

Brinkley, B. R., Cox, S. M., and Pepper, D. A. (1980). Structure of the mitotic apparatus and chromosomes after hypotonic treatment of mammalian cells in vitro. *Cytogenet. Cell Genet., 26,* 165–174.

Brinkley, B. R., Valdivia, M. M., Tousson, A., and Brenner, S. L. (1984). Compound kinetochores of the Indian muntjac: evolution by linear fusion of unit kinetochores. *Chromosoma, 91,* 1–11.

Corces, V. G., Manso, R., De La Torre, J., Avila, J., Nasr, A., and Wiche, G. (1980). Effects of DNA on microtubule assembly. *Eur. J. Biochem., 105,* 7–16.

Cox, J. V., Schenk, E. A., and Olmsted, J. B. (1983). Human anticentromere antibodies: distribution, characterization of antigens, and effect on microtubule organization. *Cell, 35,* 331–339.

De Brabander, M., Geuens, G., De Mey, J., and Jonaiau, M. (1979). Light microscopic and ultrastructural distribution of immunoreactive tubulin in mitotic mammalian cells. *Biol. Cell., 34,* 213–226.

De Brabander, M., Geuens, G., De Mey, J., and Jonaiau, M. (1981). Nucleated assembly of mitotic microtubules in living PtK_2 cells after release from nocodazole treatment. *Cell Motil., 1,* 469–483.

Earnshaw, W. C., and Laemmli, U. K. (1983). Architecture of metaphase chromosomes and chromosome scaffolds. *J. Cell Biol., 96,* 84–93.

Earnshaw, W. C., and Migeon, B. R. (1985). Three related centromere proteins are absent from the inactive centromere of a stable isodicentric chromosome. *Chromosoma, 92,* 290–296.

Earnshaw, W. C., and Rothfield, N. (1985). Identification of a family of human centromere proteins using autoimmune sera from patients with scleroderma. *Chromosoma, 91,* 313–321.

Earnshaw, W. C., Halligan, N., Cooke, C., and Rothfield, N. (1984). The kinetochore is part of the metaphase chromosome scaffold. *J. Cell Biol., 98,* 352–357.

Euteneuer, U., and McIntosh, J. R. (1981). Structural polarity of kinetochore microtubules. *J. Cell Biol., 89,* 338–345.

Gould, R. R., and Borisy, G. G. (1978). Quantitative initiation of microtubule assembly by chromosomes from Chinese hamster ovary cells. *Exp. Cell Res., 113,* 369–374.

Guldner, H. H., Lakomek, H.-J., and Bautz, F. A. (1984). Human anti-centromere sera recognise a 19.5 kD non-histone chromosomal protein from HeLa cells. *Clin. Exp. Immunol., 58,* 13–20.

Holmquist, G. P., and Dancis, B. (1979). Telomere replication, kinetochore organizers, and satellite DNA evolution. *Proc. Natl. Acad. Sci. USA, 76,* 4566–4570.
Liang, J. C., and Brinkley, B. R. (1985). Chemical probes and possible targets for the induction of aneuploidy. In *Aneuploidy,* ed. V. L. Dellarco, P. E. Voytek, and A. Hollaender, pp. 491–505. New York: Plenum.
Luykx, P. (1970). Cellular mechanisms of chromosome distribution. *Int. Rev. Cytol.* [*Suppl.*], *2,* 1–173.
Lyubskii, S. L., Buchwalow, I. B., and Raikhlin, N. T. (1979). Ultrastructural localization of ATPase and 5-prime nucleotidase activity in Chinese hamster mitotic cells. *Acta Histochem. Cytochem., 12,* 1–6.
Marx, K. A., and Denial, T. (1985). Chromosome segregation, kinetochores and DNA-microtubule interaction. A preferential satellite DNA–MAP interaction may be conserved in evolution. In *Molecular Basis of Cancer,* Part B: *Macromolecular Recognition, Chemotherapy, and Immunology,* ed. R. Rein, pp. 65–75. New York: Alan R. Liss.
McGill, M., and Brinkley, B. R. (1975). Human chromosomes and centrioles as nucleating sites for the in vitro assembly of microtubules from bovine brain tubulin. *J. Cell Biol., 67,* 189–199.
McIntosh, J. R. (1979). Cell division. In *Microtubules,* ed. K. Roberts and J. S. Hyams, pp. 428–441. London: Academic Press.
McIntosh, J. R., and Euteneuer, U. (1984). Tubulin hooks as probes for microtubule polarity: an analysis of the method and an evaluation of data on microtubule polarity in the mitotic spindle. *J. Cell Biol., 98,* 525–533.
Merry, D. E., Pathak, S., Hsu, T. C., and Brinkley, B. R. (1985). Anti-kinetochore antibodies: use as probes for inactive centromeres. *Am. J. Hum. Genet., 37,* 425–430.
Mitchison, T., and Kirschner, M. W. (1985a). Properties of the kinetochore in vitro. I. Microtubule nucleation and tubulin binding. *J. Cell Biol, 101,* 755–765.
Mitchison, T., and Kirschner, M. W. (1985b). Properties of the kinetochore in vitro. II. Microtubule capture and ATP-dependent translocation. *J. Cell Biol, 101,* 766–777.
Mitchison, T., Evans, L., Schulze, E., and Kirschner, M. (1986). Sites of microtubule assembly and disassembly in the mitotic spindle. *Cell, 45,* 515–527.
Moroi, Y., Peebles, C., Fritzler, M. J., Steigerwald, J., and Tan, E. M. (1980). Autoantibody to centromere (kinetochore) in scleroderma sera. *Proc. Natl. Acad. Sci. USA, 77,* 1627–1631.
Nicklas, R. B. (1971). Mitosis. *Adv. Cell Biol., 2,* 225–297.
Nicklas, R. B. (1985). Mitosis in eukaryotic cells: an overview of chromosome distribution. In *Aneuploidy,* ed. V. L. Dellarco, P. E. Voytek, and A. Hollaender, pp. 183–195. New York: Plenum Press.
Nishikai, M., Okano, Y., Yamashita, H., and Watanabe, M. (1984). Characterization of centromere kinetochore antigen reactive with sera of patients with a scleroderma variant CREST syndrome. *Ann. Rheum. Dis., 43,* 819–824.
Palmer, D. K., and Margolis, R. L. (1985). Kinetochore components recognized by human autoantibodies are present on mononucleosomes. *Mol. Cell. Biol., 5,* 173–186.
Pepper, D. A., and Brinkley, B. R. (1977). Localization of tubulin in the mitotic apparatus of mammalian cells by immunofluorescence and immunoelectron microscopy. *Chromosoma, 60,* 223–235.
Pepper, D. A., and Brinkley, B. R. (1979). Microtubule initiation at kinetochores and centrosomes in lysed mitotic cells. Inhibition of site-specific nucleation by tubulin antibody. *J. Cell Biol., 82,* 585–591.

Pepper, D. A., and Brinkley, B. R. (1980). Tubulin nucleation and assembly in mitotic cells: evidence for nucleic acids in kinetochores and centrosomes. *Cell Motil., 1,* 1–15.
Pickett-Heaps, J. D., Tippit, D. H., and Pcrter, K. R. (1982). Rethinking mitosis. *Cell, 29,* 729–744.
Rattner, J. B. (1986). Organization within the mammalian kinetochore. *Chromosoma, 93,* 515–520.
Rattner, J. B., Krystal, G., and Hamkalo, B. A. (1978). Selective digestion of mouse metaphase chromosomes. *Chromosoma, 66,* 259–268.
Rieder, C. L. (1979). Localization of ribonucleoprotein in the trilaminar kinetochore of PtK_1. *J. Ultrastruct. Res., 66,* 109–119.
Rieder, C. L. (1981). The structure of the cold-stable kinetochore fiber in metaphase PtK_1 cells. *Chromosoma, 84,* 145–158.
Rieder, C. L. (1982). The formation, structure, and composition of the mammalian kinetochore and kinetochore fiber. *Int. Rev. Cytol., 79,* 1–58.
Ris, H., and Witt, P. L. (1981). Structure of the mammalian kinetochore. *Chromosoma, 82,* 153–170.
Roos, U.-P. (1973). Light and electron microscopy of rat kangaroo cells in mitosis. II. Kinetochore structure and function. *Chromosoma, 44,* 195–220.
Roos, U.-P. (1977). The fibrillar organization of the kinetochore and kinetochore region of mammalian chromosomes. *Cytobiologie, 16,* 82–90.
Shi, L., Ye, Y., and Duan, X. (1980). Comparative cytogenetic studies on the red muntjac, Chinese muntjac, and their F_1 hybrids. *Cytogenet. Cell Genet., 26,* 22–27.
Sluder, G., Rieder, C. L., and Miller, F. (1985). Experimental separation of pronuclei in fertilized sea-urchin eggs. Chromosomes do not organize a spindle in the absence of centrosomes. *J. Cell Biol, 100,* 897–903.
Snyder, J. A., and McIntosh, J. R. (1975). Initiation and growth of microtubules from mitotic centers in lysed mammalian cells. *J. Cell Biol., 67,* 744–760.
Telzer, R. B., and Haimo, L. T. (1981). Decoration of spindle microtubules with dynein: evidence for uniform polarity. *J. Cell Biol., 89,* 373–378.
Telzer, B. R., Moses, M. J., and Rosenbaum, J. L. (1975). Assembly of microtubules onto kinetochores of isolated mitotic chromosomes of HeLa cells. *Proc. Natl. Acad. Sci. USA, 72,* 4023–4027.
Valdivia, M. M., and Brinkley, B. R. (1985). Fractionation and initial characterization of the kinetochore from mammalian metaphase chromosomes. *J. Cell Biol., 101,* 1124–1134.
Vandre, D. D., Davis, F. M., Rao, P. N., and Borisy, G. G. (1984). Phosphoproteins are components of mitotic microtubule organizing centers. *Proc. Natl. Acad. Sci. USA, 81,* 4439–4443.
Witt, P. L., Ris, H., and Borisy, G. G. (1980). Origin of kinetochore microtubules in Chinese hamster ovary cells. *Chromosoma, 81,* 483–505.
Witt, P. L., Ris, H., and Borisy, G. G. (1981). Structure of kinetochore fibers: microtubule continuity and inter-microtubule bridges. *Chromosoma, 83,* 523–540.
Zinkowski, R. P., Vig, B. K., and Broccoli, D. (1986). Characterization of kinetochores in multicentric chromosomes. *Chromosoma, 94,* 243–248.

IV
The organization of the centromere and centromeric heterochromatin

JEROME B. RATTNER AND CHYI-CHYANG LIN

1. Introduction

The question of how chromatin is organized during the cell cycle is central to understanding the molecular basis of gene expression. Since the mid-1970s, our understanding of chromatin organization has undergone a profound change so that it is now possible to describe the arrangement and reorganization of chromatin in some detail. In this chapter, we shall consider the ultrastructural organization of various levels of chromatin, its higher-order structure, and the relation of these hierarchies to the structure of the centromere and to the arrangement of centromeric heterochromatin in the metaphase chromosome.

2. Organization of chromatin

Chromatin is organized into a uniform repeating structure that is folded to form a 30-nm fiber.

Studies over the past decade have established that DNA is complexed with histone to form the basic subunit of chromatin organization, the *nucleosome* (see McGhee and Felsenfeld, 1980, for review). Electron microscopy of chromatin from interphase nuclei and metaphase chromosomes has confirmed that the basic chromatin fiber (the 10-nm fiber) has a repeated structure appearing as a string of beads (Olins and Olins, 1974; Rattner, Branch, and Hamkalo, 1975). Digestion of chromatin for brief periods with micrococcal nuclease yields nucleosomes containing about 200 base pairs (bp) of DNA and two each of the four core histone molecules H_{2A}, H_{2B}, H_3, and H_4. Further digestion yields a core particle which consists of 146 bp of DNA wrapped around the octameric histones.

This work has been supported by the Medical Research Council of Canada (C.C.L.), the Alberta Cancer Board (C.C.L. and J.B.R.), and NSERC (J.B.R.). J.B.R. is a scholar of the Alberta Heritage Foundation for Medical Research. We wish to thank Mrs. Theresa Wang for her excellent technical assistance during the course of this work.

The structure of the core particle has now been studied in great detail (Richmond et al., 1984; Burlingame et al., 1984). There is some evidence that the core histones in heterochromatin are modified. However, the role these modifications may play in heterochromatin organization is unclear (Halleck and Gurley, 1982). Although the majority of the DNA associated with the nucleosome is in a right-handed or B conformation, there is some in vitro experimental evidence suggesting that alternate conformations such as the left-handed or Z conformation can also be accommodated within this structure (Miller, Rattner, and van de Sande, 1982).

Adjacent nucleosomes along the 10-nm fiber are separated by a portion of histone-free DNA. This "linker DNA" is of an average length that is both species and tissue specific. However, in some cells such as those of yeast and *Physarum* (Rattner, Saunders, Davie, and Hamkalo, 1982; Scheer, Zentgraf, and Sauer, 1981), the linker length may vary considerably along a length of chromatin, and there is some suggestion that some degree of variability exists in all cell types. This variability may have ramifications for the regularity of the next level of chromatin packing, the 30-nm fiber described below.

Whereas thin sections of intact nuclei have been unable to provide a detailed picture of chromatin organization, recent advances in whole-mount electron microscopic (EM) techniques have made considerable contributions to our understanding of chromatin organization. Since whole-mount electron microscopy has provided many of the results that will be discussed here, we shall briefly describe one of the most informative protocols. In 1969, Miller and Beatty reported a whole-mount EM procedure that allowed the visualization of transcriptional complexes. Subsequently, Rattner et al. (1975) modified this procedure to visualize the basic level of metaphase chromosome structure. In this modified method, metaphase chromosomes are released from mitotic cells by detergent lysis at low ionic strength. The chromosomes are deposited onto coated EM grids by centrifugation through a sucrose cushion containing 10% formalin at pH 9.5. The low ionic strength coupled with the formalin at a high pH relaxes the higher-order structure of chromatin, allowing the observation of the beaded nature of both euchromatin and heterochromatin. The beaded nature of heterochromatin has also been detected in preparations of meiotic prophase chromosomes of the moth, *Ephestia kiehniella* (Weith and Traut, 1980). Support for the ultrastructural observations of the nucleosomal organization of euchromatin has also come from biochemical studies (see Brutlag, 1980; Igo-Kemenes and Zachau, 1982).

Whole-mount EM preparations of chromosomes from organisms, such as the mouse, that have extensive centromeric heterochromatin illustrate several interesting features of this class of chromatin. First, this type of chromatin can be distinguished from the remainder of the chromosome by its increased electron density (Figure 4.1). Since this feature is also apparent in unstained specimens, it is not based solely on the affinity of electron-dense stains for this type of chromatin. Second, this class of chromatin is highly resistant to mechanical and/or chemical disruption. This latter feature has made it possible to enrich for centromeric heterochromatin by use of micrococcal nuclease or DNase I. Thus, under appropriate conditions it is possible to digest the euchromatin while leaving the centromere and centromeric heterochromatin intact (Rattner, Krystal, and Hamkalo, 1978). In addition, in some organisms it is possible to enrich for centromeric heterochromatin by using restriction enzymes that do not recognize sites in this class of chromatin but cut frequently in euchromatic regions (Lica and Hamkalo, 1983). This discovery of differential sensitivity to nucleases has opened the way to the further biochemical characterization of this chromosomal domain.

The presence of histone H_1 or histone H_1-like proteins seems essential for the folding of the basic beaded fiber into the next level of structure, the 30-nm fiber (Figure 4.2). There is some evidence that histone H_1 in heterochromatin may be modified. For example, satellite DNA-containing heterochromatin of *Drosophila virilis* has been shown to be highly phosphorylated. It has been suggested that this modification provides the basis for the compact nature of centromeric heterochromatin and determines its genetic inactivity (Blumenfeld et al., 1978; Billings et al., 1979; Weith, 1985).

The presence of a 30-nm fiber class within interphase nuclei and metaphase chromosomes was revealed by early EM studies (see Ris and Korenberg, 1979, for review). However, unlike the basic 10-nm fiber, the internal organization of this fiber class remains controversial. To date, there are three basic types of models that have been presented, which attempt to describe the manner in which the nucleosomal fiber is folded to form the 30-nm fiber (for a review, see Felsenfeld and McGhee, 1986). These models are the solenoid model, the superbead model, and the twisted ribbon model.

The first model is based on the morphology and conformation of chromatin fragments at increasing salt concentration (Finch and Klug, 1976; Thoma, Koller, and Klug, 1979). This model suggests that the basic 10-nm fiber is folded in a helical manner to form a solenoid with six nucleosomes per turn. Support for this type of arrangement has been provided by Widom and Klug (1985) using x-ray diffraction of oriented samples. However, similar studies by Williams and co-workers (1986) have provided very different results which have been interpreted to support an alternative arrangement of the nucleosomes within the fiber. The second model, also initially suggested by EM studies, proposed that nucleosomes

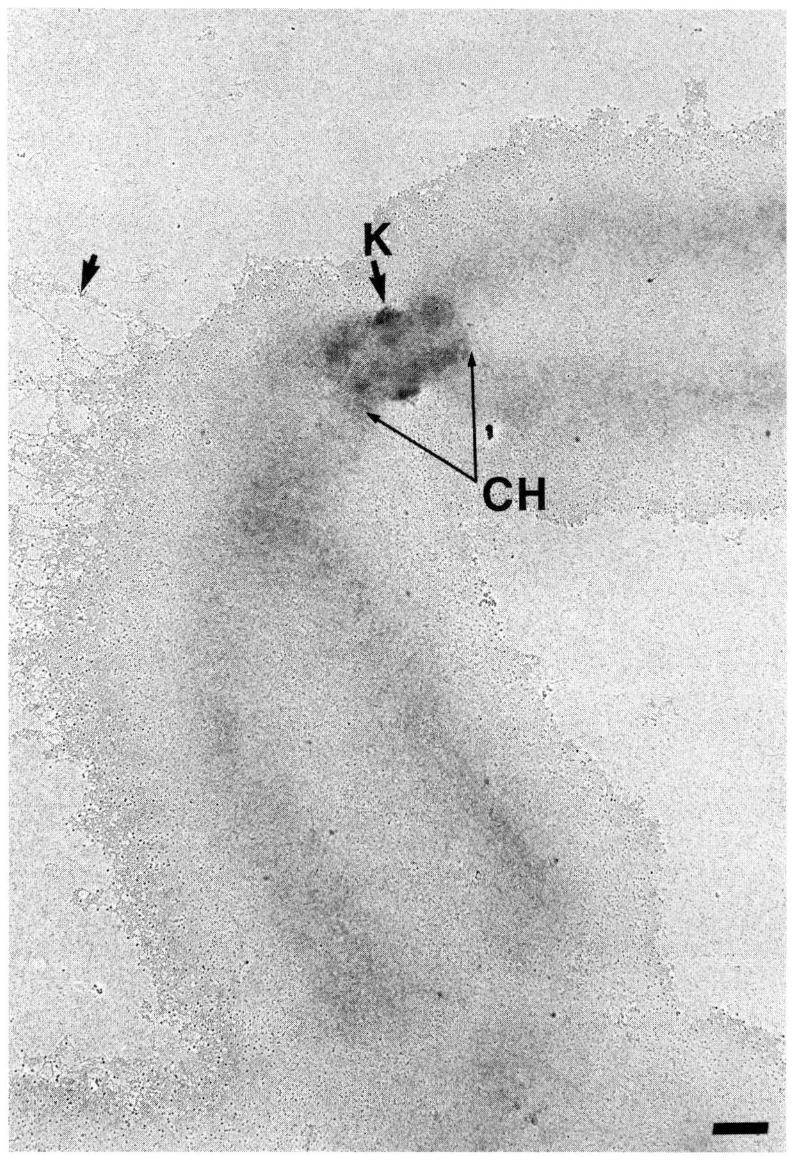

Figure 4.1. Whole-mount electron micrograph of a biarmed mouse L929 metaphase chromosome. The kinetochore (K) is associated with the electron-dense centromeric heterochromatin (CH). The euchromatin has been relaxed to the beads-on-a-string level of structure (small arrow). Bar = 1 μm.

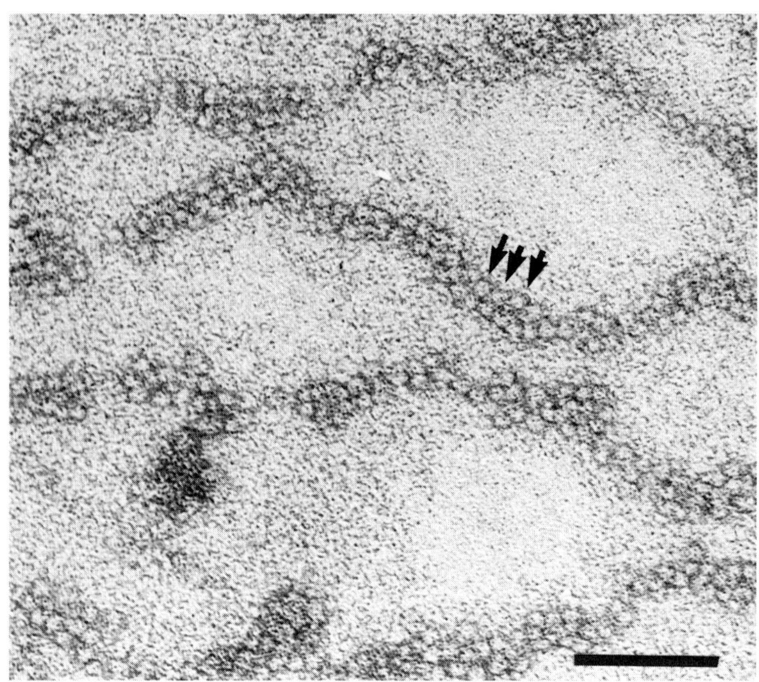

Figure 4.2. A portion of several 30-nm chromatin fibers from a mouse L929 metaphase chromosome. The close apposition of nucleosomes (arrows) of the basic 10-nm fiber can be seen within the fiber. Bar = 0.1 μm.

are aggregated into "superbead" structures containing 8–48 nucleosomes, depending on the source of the chromatin (Renz, Nehls, and Hozier, 1977; Zentgraf, Miller, and Franke, 1980). Proponents of the solenoid model, however, think that the discontinuities reflect breakdown and rearrangements of a continuous helically folded fiber. The third model suggests that the nucleosomal fiber is folded into a zig-zag ribbon, which is in turn folded in a helical manner to form the compact 30-nm fiber (Woodcock, Frado, and Rattner, 1984). Less complex arrangements have also been proposed, in which the zig-zag ribbon is simply twisted along the long axis of the fiber (Worcel, Strogartz, and Riley, 1981). The existence of such divergent models illustrates the latitude allowed by the existing biochemical and ultrastructural data. A definitive understanding of this level of structure must await further study. However, it is clear that this level of structure provides the basis for further levels of organization within interphase and metaphase cells and probably represents the

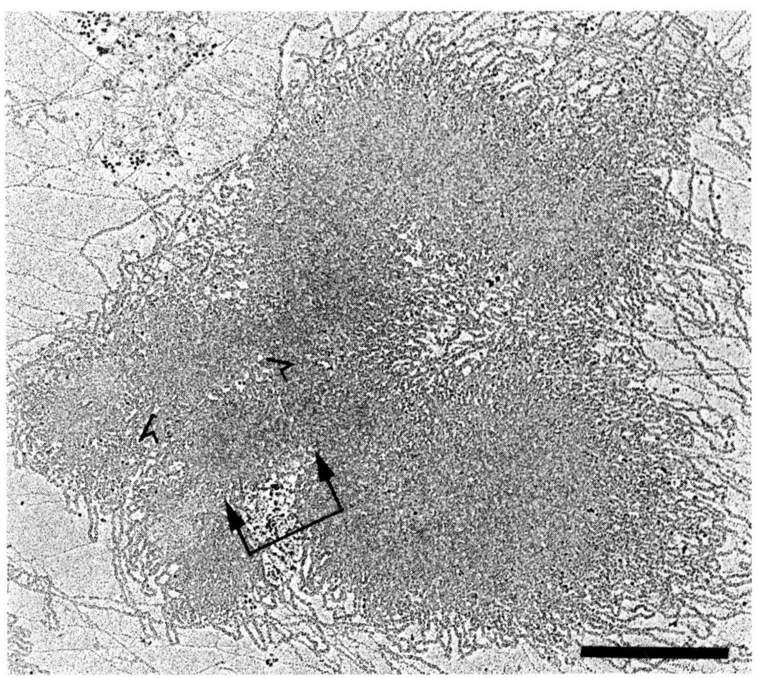

Figure 4.3. Electron micrograph of a metaphase chromosome, illustrating the 30-nm fibers that constitute both the euchromatic and heterochromatic (brackets) regions. Bar = 1 μm.

conformation of most of the chromatin in vivo, with the extended 10-nm conformation existing only in short localized regions involved in replication or transcription.

It is possible to visualize the 30-nm fiber in spread preparation of both interphase and metaphase chromosomes by using a modification of the Miller procedure described above. In this case, the maintenance of physiological ionic strength and pH throughout the preparative procedure and the elimination of formalin fixation allows the preservation of this level of structure. Under these conditions it has been possible to visualize this fiber class in the centromere region and the centromeric heterochromatin of the mouse as well as in chromatin of the noncentromeric region (Figure 4.3). Because of the highly condensed nature of the centromere region, the 30-nm fibers are more readily seen in chromosome preparations in which the euchromatin has been solubilized by brief enzyme digestion. This method also facilitates the identification and visualization of interphase heterochromatic regions and their underlying 30-nm fiber conformation (Rattner, unpublished observations). In such preparations, no difference in the diameter of this fiber class between heterochromatic and euchromatic regions is apparent. However, there is a report that the diameter of heterochromatic fibers are larger in pachytene chromosomes of the insect *Tenebrio molitor* prepared in 0.6 mM magnesium (Weith, 1985).

3. Structure of centromeric heterochromatin in metaphase chromosomes

Centromeric heterochromatin is folded into radial loops in metaphase chromosomes.

Several models have been proposed to explain how the 30-nm fiber is folded to form the next level of high-order structure of chromatin in the heterochromatin and euchromatic region of the metaphase chromosome. Since the 1960s, the most popular model of chromosome organization has been the folded fiber model of Du Praw (1965). In this model, the 30-nm fiber folds back on itself randomly as it transverses the chromatid. Recently, a radial loop arrangement of the 30-nm fiber has received considerable attention in light of the visualization of the scaffold and loop architecture of histone-depleted chromosomes (Adolph, Cheng, and Laemmli, 1977; Paulson and Laemmli, 1977). As originally proposed, the chromatid is seen as a series of loops of the 30-nm fiber which originate at the axis of the chromatid. Along this axis is a scaffold, composed of a subset of nonhistone chromosomal proteins, that functions to associate the loops together. Several studies have suggested that in the compact state each loop is collapsed or folded toward the central axis (Marsden and Laemmli, 1979; Adolph, Kreisman, and Kuehn, 1986).

The presence of a discrete scaffold element within the native chromosome has been the subject of considerable debate. It now seems likely that the scaffold may not be as well defined as first thought, and the structure visualized in histone-depleted preparations may represent aggregation toward the axis of the chromatid. In vivo, the scaffold proteins probably form a diffuse network within the chromosome. The recent identification of topoisomerase II as a major component of the scaffold supports the notion that the proteins that generate this structure in vitro may be involved in the organization and function of chromatin loops in vivo (Earnshaw and Heck, 1985).

Whereas the existence and form of the scaffold remain controversial, the presence of chromatin loops within both meiotic and mitotic chromosomes has been well documented. One of the clearest illustrations of the formation and organization of chromatin loops is found in the EM

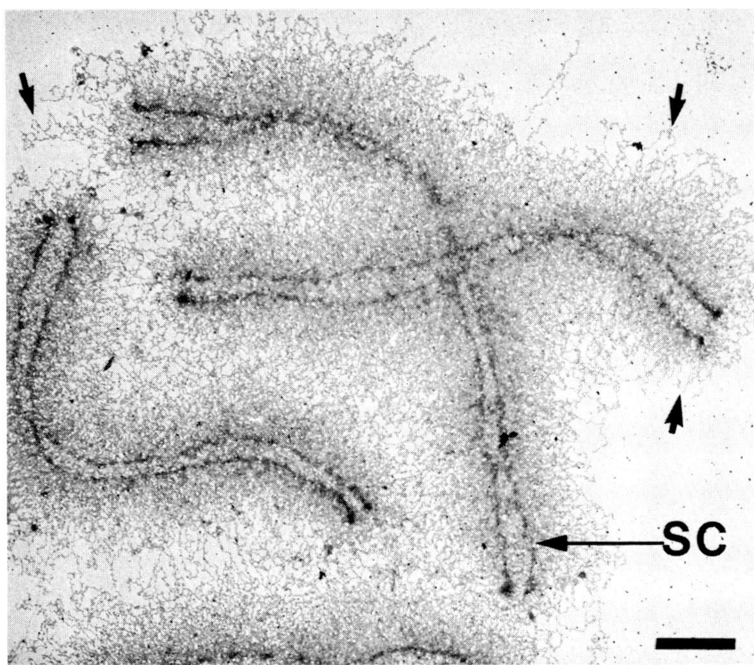

Figure 4.4. Three pachytene chromosomes of the silkmoth, *Bombyx mori*. The chromatin is arranged into a series of radial loops (arrows) about the synaptonemal complex (SC). Bar = 1 μm.

study of male meiosis in the silkmoth *Bombyx mori* (Rattner, Goldsmith, and Hamkalo, 1980, 1981). At the onset of meiotic prophase, the 30-nm fibers within the nucleus appear to have an ill-defined organization. With time, discrete chromatin loops can be seen forming within chromatin adjacent to the nuclear envelope. Subsequently, adjacent loops become arranged into clusters or foci, forming a linear array which defines the future axis of the meiotic prophase chromatid. Within the central region of the clusters, elements of the synaptonemal complex form so that further chromatin-loop aggregation in concert with the maturation of the synaptonemal complex results in the formation of a pachytene chromosome. In a relaxed state, these chromosomes have a lampbrush appearance, with the synaptonemal complex located along the long axis of the structure and chromatin loops extending radially about this structure (Figure 4.4). It is interesting to note that the chromosomes at this stage are considerably longer than are the identical chromosomes in the subsequent first meiotic division (Rattner et al., 1980). This arrangement

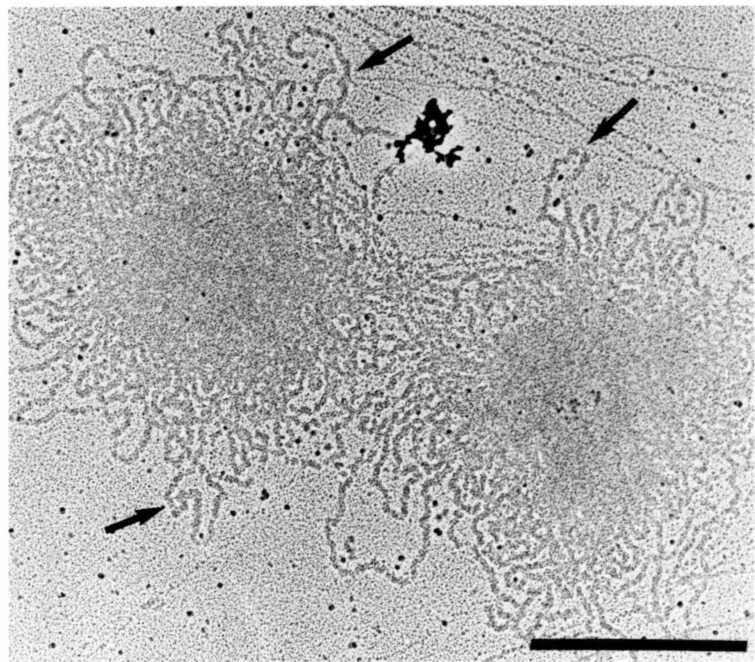

Figure 4.5. A double minute chromosome from a human colon carcinoma cell line COLO 320. Radial loops (arrows) extend from the central region of the chromosome. Bar = 1 μm.

suggests that this linear array of chromatin loops (including the centromeric heterochromatin region) must fold into additional packaging hierarchies in order to generate the typical metaphase chromosome form. The same type of preparation that reveals the loop nature of meiotic chromosomes also allows the visualization of loop structures in mitotic chromosomes (Figure 4.5). However, it has not been possible to visualize the base of individual loops.

4. The final level of organization within the centromere

The final form of the centromere is a fiber 200–300 nm in diameter.

Are radial loops the final level of organization within the centromere? To answer this question, it is necessary to reexamine several concepts of chromosome organization. Historically, the first model of chromosome structure to gain wide consideration employed a number of hierarchies

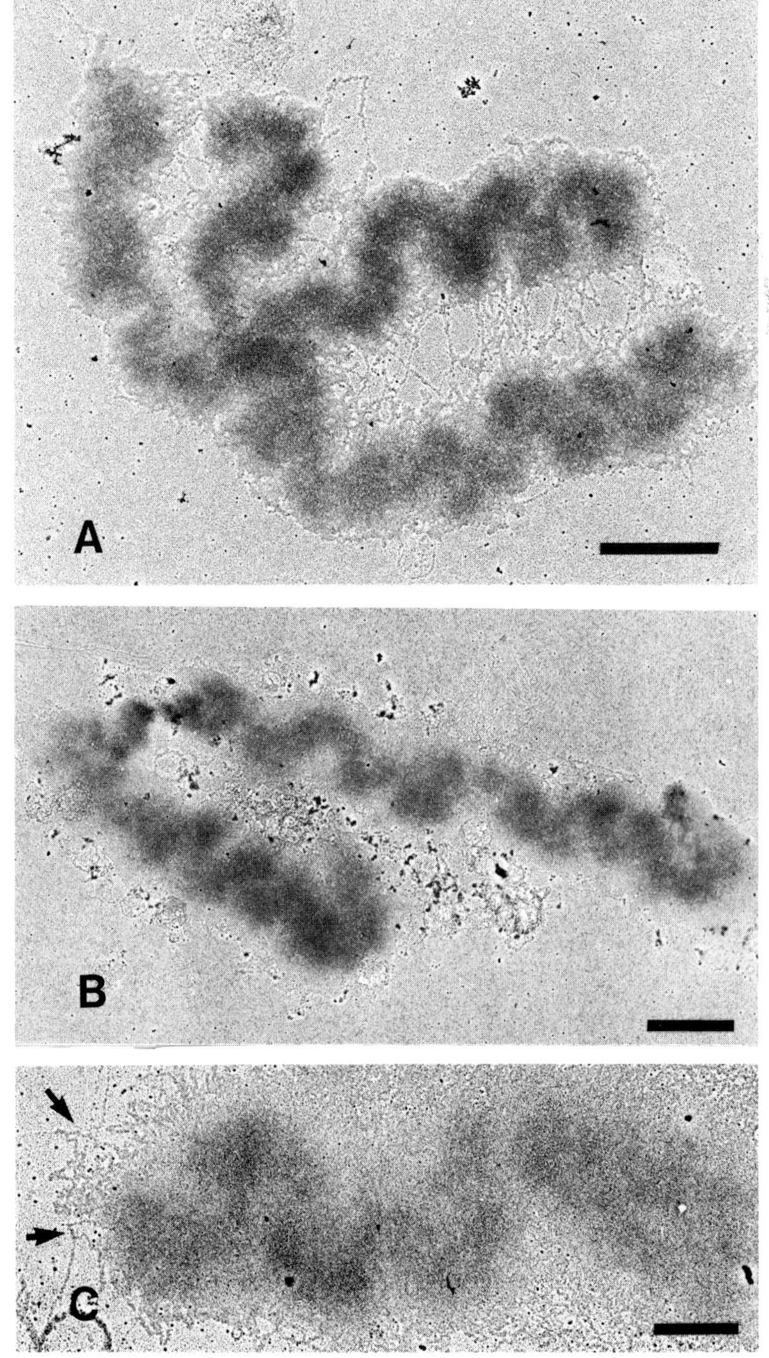
A
B
C

of helical coiling to account for final metaphase form. This model was based on observations, by classical cytologists, of the structure of plant chromosomes primarily during the meiotic cycle (for review, see White, 1973). This level of structure has also been observed in animal cells but has usually been restricted to preparations that have been subjected to rather harsh pre- or postfixation treatments (see Rattner and Lin, 1985b, and references therein). Perhaps the most convincing of these studies is that of Onnuki (1968), who demonstrated clearly that the chromatids of human cells can appear as a spiral or zig-zag fiber. With the popularization of the scaffold model the relevance of these classical observations was unclear although there was some suggestion that the base of the radial loops may be arranged in a helical manner about the scaffold (Adolph et al., 1977).

Recently, it has become possible to confirm the coiled nature of untreated human chromosomes by use of transmission (T) and scanning (S) EM procedures and to resolve the apparent paradox between the scaffold and helical coil models (Rattner and Lin, 1985b). When we examined untreated metaphase chromosomes from a cell line derived from a human colon carcinoma, we found that it was possible to visualize a fiber approximately 200–300 nm in diameter within the metaphase chromatid (Figure 4.6A,B). This fiber followed a helical path along the length of the chromatid, and close apposition of successive coils was responsible for the compact form of the chromosome (Figure 4.6B). Observations under conditions that relaxed the internal substructure reveal that the fiber is composed of numerous chromatin loops (Figure 4.6C). Thus both helical coils and radial loops coexist in metaphase chromosomes. The coiled substructure of human chromosome is frequently seen in standard cytological preparations. However, such chromosomes are generally ignored since they do not lend themselves to good karyotyping and show no differential staining after treatment with various banding protocols (Rattner and Lin, unpublished observations).

One striking finding in the study described above was that within the centromere the 200- to 300-nm fiber did not undergo the final helical packing found in the rest of the chromosome. Rather, the fiber followed a straight course through the centromere (Figure 4.7). This organization provides the structural bases for the constricted appearance of this chromosomal location. Since the degree of compaction of the radial loops

Figure 4.6. (A) Transmission electron micrograph of an untreated human chromosome from a cell line derived from a human colon carcinoma, illustrating the coiled substructure of metaphase chromosomes. (B) The close apposition (lower chromosome arm) of successive coils results in the compact form of the chromosome. (C) Relaxation of this coiled fiber reveals its radial loop (arrows) substructure and results in the increase in fiber diameter. Bar = 1 μm.

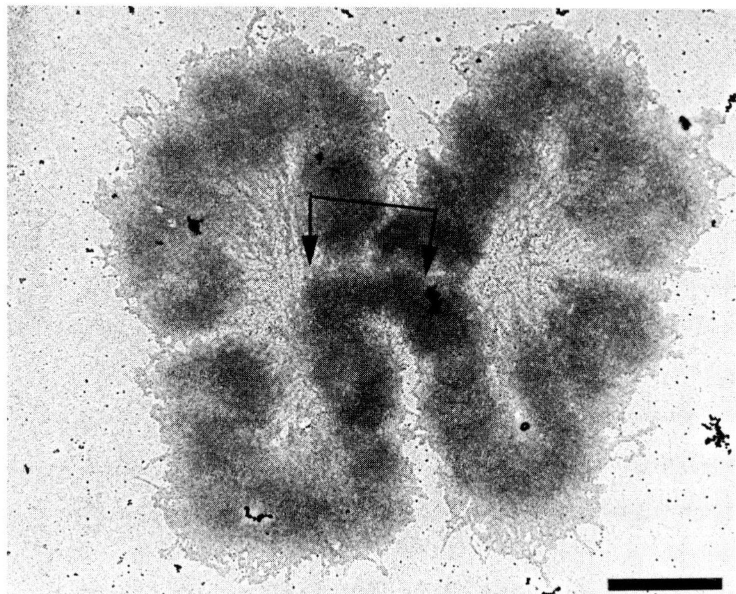

Figure 4.7. The 200- to 300-nm fiber follows a straight path in the centromere region (brackets). Bar = 1 μm.

within the 200- to 300-nm fiber is dependent on preparative conditions, and since the form of the centromere also varies with the experimental method, the variations in centromere diameter reported in our study and in the literature probably reflect this relationship. The same relationship is also seen in the noncentromeric chromatin.

5. Arrangement of radial loops within the centromeric fiber

Radial loops follow a complex path with the centromere: The 200- to 300-nm centromere fiber has a helical substructure.

Attempts to study the manner in which radial loops are arranged within the centromeric fiber are complicated by the highly condensed nature of this region. To circumvent this problem, we have used agents that selectively affect the condensation of the primary construction as probes to elucidate the organization of this region.

In the mouse, several types of drugs affect centromere condensation; these include Hoechst (H) 33258 (Marcus, Greitein, and Grapp, 1979) and 5-azacytidine (Viegas-Pequignot and Dutrillaux, 1981). When the

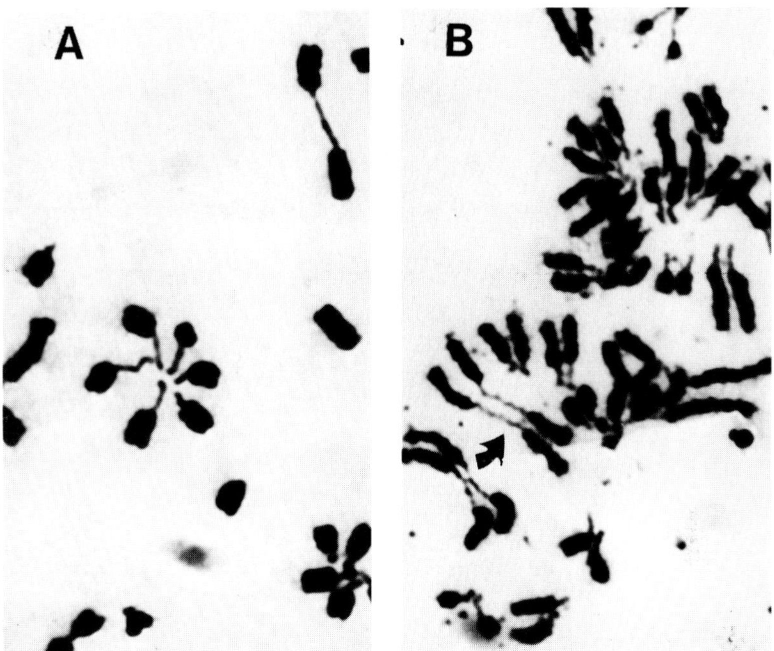

Figure 4.8. Light micrograph of a mouse L929 chromosome from a cell grown in the presence of Hoechst 33258 (A) and 5-azacytidine (B). The extended centromere fiber occasionally displays a prominent coiled morphology (B, arrow). Bar = 1 μm.

mouse tissue culture cell line L929 is grown in the presence of H33258 during the G_2 period of the cell cycle and is visualized by standard light microscopy, after acid fixation the centromere region in a majority of the chromosomes displays an extended morphology. A similar conformation can be seen after growth in the presence of 5-azacytidine. These drugs seem to affect specifically the centromere region while leaving intact the remainder of the chromosome, which appears to condense in a normal manner (Figure 4.8A,B).

The degree of decondensation induced by drug treatment may be determined by taking advantage of a marker chromosome in the L929 karyotype that contains multiple centromeres (Figure 4.9). In situ hybridization indicates that each of these sites contains the 1.691 g/cm^3 satellite, and TEM studies illustrate that the kinetochore is confined to the central primary constriction (Rattner and Lin, 1985b). This is also the only region that will bind anti-kinetochore antisera (Rattner, unpublished observations). These regions are particularly prominent after drug treat-

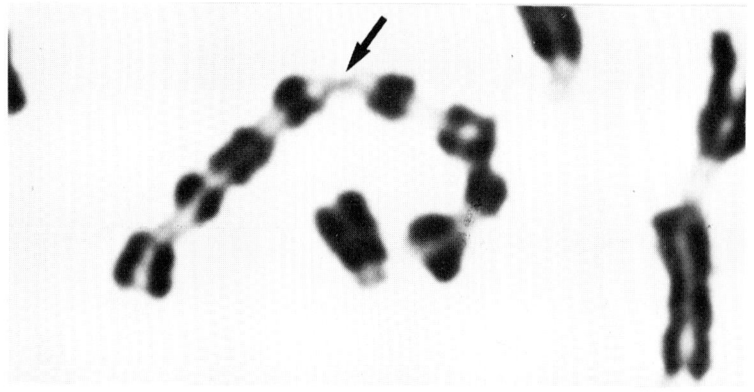

Figure 4.9. Marker chromosome from a L929 karyotype. The regions containing centromeric heterochromatin display an extended morphology after growth in 5-azacytidine. The primary centromere is indicated by an arrow. Bar = 1 μm.

ment so that eight major and two minor sites are easily identifiable. We have compared the length of the central primary centromere of this chromosome in the karyotypes of cells grown in the presence and absence of the drugs. These measurements indicate a four- to fivefold increase in the total centromere length as a result of H33258 treatment, and a three- to fourfold increase in length after 5-azacytidine treatment.

Scanning EM examination of these acid-fixed preparations allows the three-dimensional visualization of the extended centromere fiber. This fiber has a roughly uniform diameter, and fibrous (radial loop) substructure is apparent in both telocentric and metacentric chromosomes (Figure 4.10A,B). Frequently, the fiber has the appearance of a relaxed coil, a morphology that was occasionally detectable in light micrographs (Figure 4.8A). When the extended fiber is examined in whole-mount TEM preparations from cells that have been grown in the presence of H33258 for 30 hours or more, chromatin loops are seen forming the substructure of the fiber (Figure 4.11).

The relationship between the extended fiber found in drug-treated cells and the native centromere has been elucidated by the discovery of conditions that allow the visualization of the sequential folding of the extended fiber. When cells grown in the presence of 5-azacytidine are held in mitotic arrest in the presence of the drug, the extended centromere region can be seen in various stages of condensation in subsequent cytological preparations (Figure 4.12). With time, the centromere assumes a compact form, identical to that found in the centromere of untreated chromosomes. This feature implies that the folding patterns observed in the presence of the drug probably reflect stages of the natural process, and

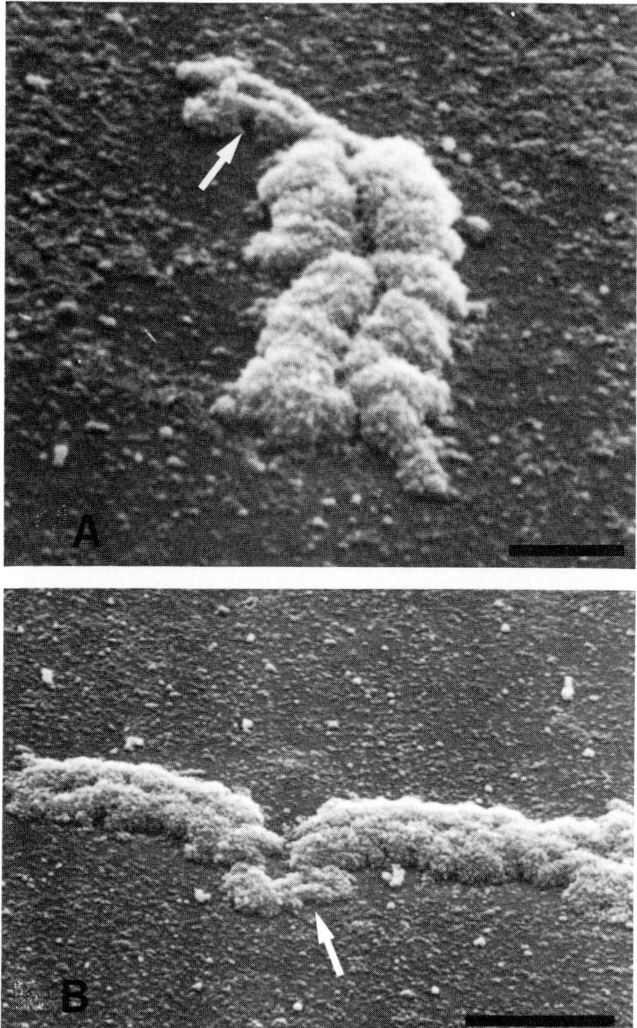

Figure 4.10. Scanning electron micrograph of a mouse L929 chromosome from a cell grown in the presence of H33258: (A) telocentric chromosome; (B) metacentric chromosome. The fibrous nature and cylindrical profile of the extended fiber (arrow) are apparent in these preparations. Bar = 1 μm.

that the configurations of drug-treated centromeres may reflect aspects of native organization.

The pathway of centromeric condensation has been followed in SEM preparations of chromosomes showing different degrees of condensation after 5-azacytidine treatment. Particularly prominent in the partially con-

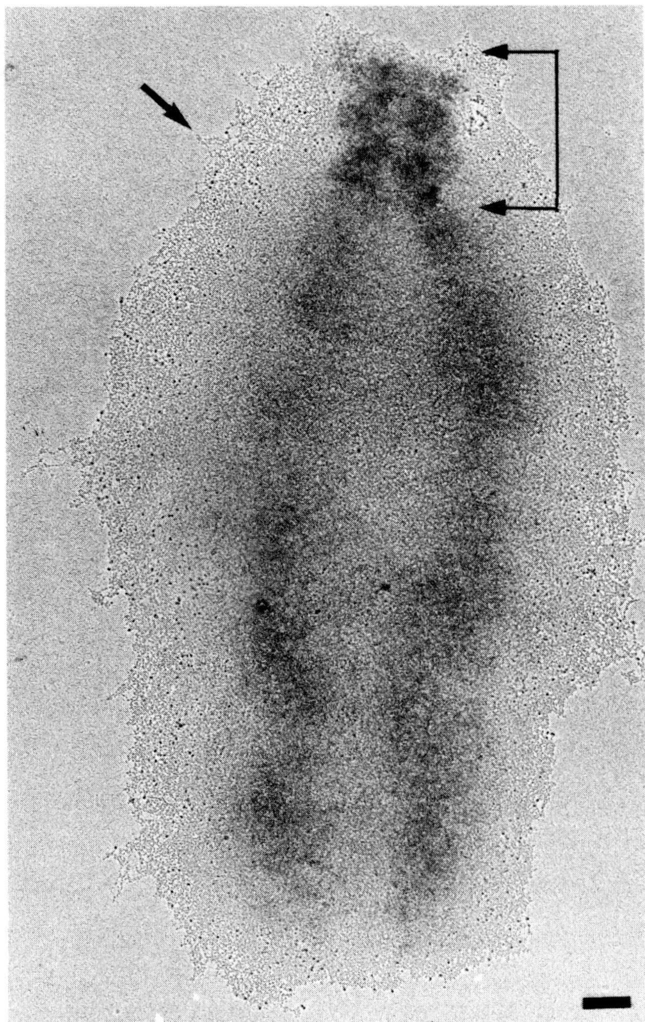

Figure 4.11. Transmission electron micrograph of a mouse L929 chromosome from a cell grown in the presence of H33258 for 30 hours. The extended centromere (brackets) is composed of numerous chromatin loops (arrows). Bar = 1 μ m.

densed centromeres are a number of cross-striations with diameters approximating that of the extended fiber described above (Figure 4.13). As condensation proceeds, the boundaries between some of the striations are lost so that the fiber appears to be composed of striations of variable dimensions (Figure 4.14). Throughout condensation, the fiber has a cylin-

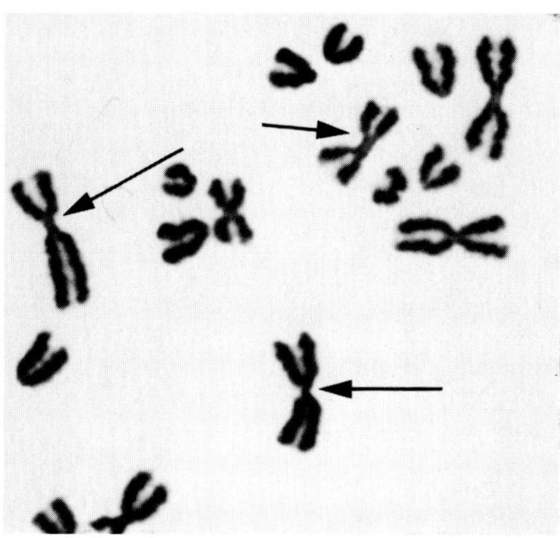

Figure 4.12. Light micrograph of mouse chromosome from a cell that had been grown in the presence of 5-azacytidine through the G_2 period and then held in mitotic arrest for several hours prior to fixation. Under these conditions, the centromere begins to condense so that various stages of condensation can be seen in cytological preparations (arrows). Bar = 1 μm.

drical profile, as does the fully condensed fiber. The cross-striations within the centromere were correlated with similar structural features detected in the chromatid arms. As the chromatid arm striations are correlated with chromatid helicity (Rattner and Lin, 1985a), by analogy the centromeric striations may reflect a similar folding pattern within the 200- to 300-nm centromeric fiber. The apparent helical folding of the extended centromere fiber can also be seen in TEM whole-mount preparations (Figure 4.15). However, the relaxation of the underlying radial loops frequently obscures this level of structure. These observations suggest that the chromatin loops of the centromere are arranged within a fiber that is helically coiled to form the native centromere. At present, we have no information to suggest that this level of helical packing is also present in the 200- to 300-nm fiber as it transverses the remainder of the chromosome. However, several recent studies of plant chromosomes suggest that such an arrangement may exist.

6. Structure of the centromere and centromeric heterochromatin during interphase

During interphase in most cells, although the remainder of the chromosome decondenses to fill the nucleus, the centromere region remains com-

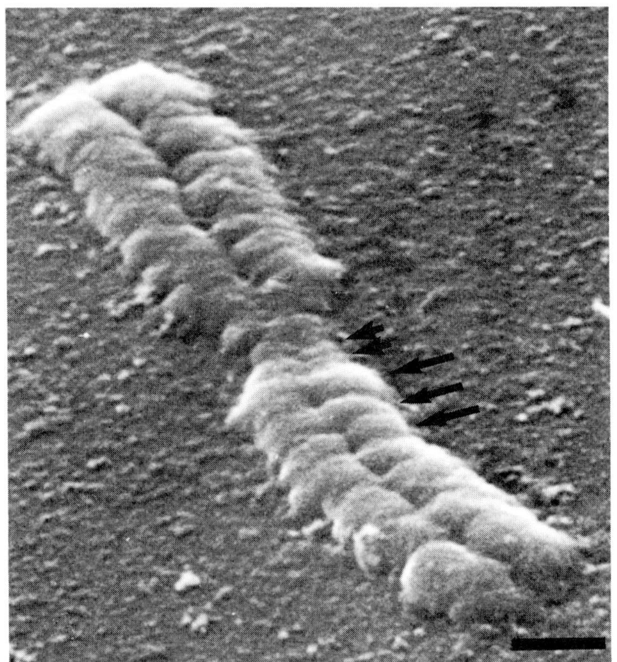

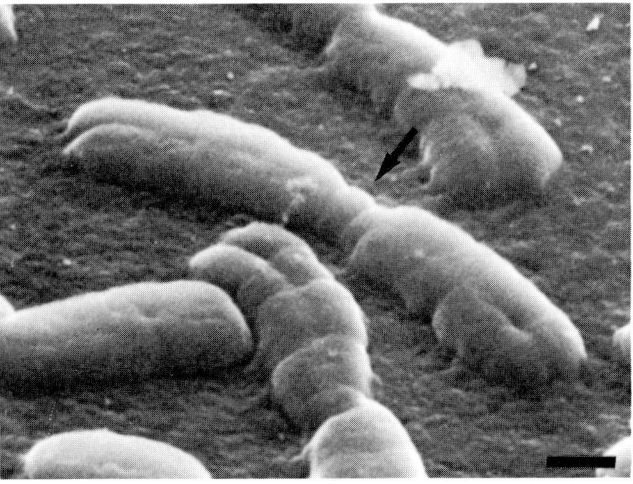

Figures 4.13–4.14. Scanning electron micrograph of chromosomes of a mouse L929 cell prepared as in Figure 4.12. The coiled nature of the centromere (small arrows) and the arms (large arrows) are clearly visible as a series of cross-striations. The close apposition of successive coils results in the loss of the demarcation between the coils (Figure 4.14, arrow). Bar = 1 μm.

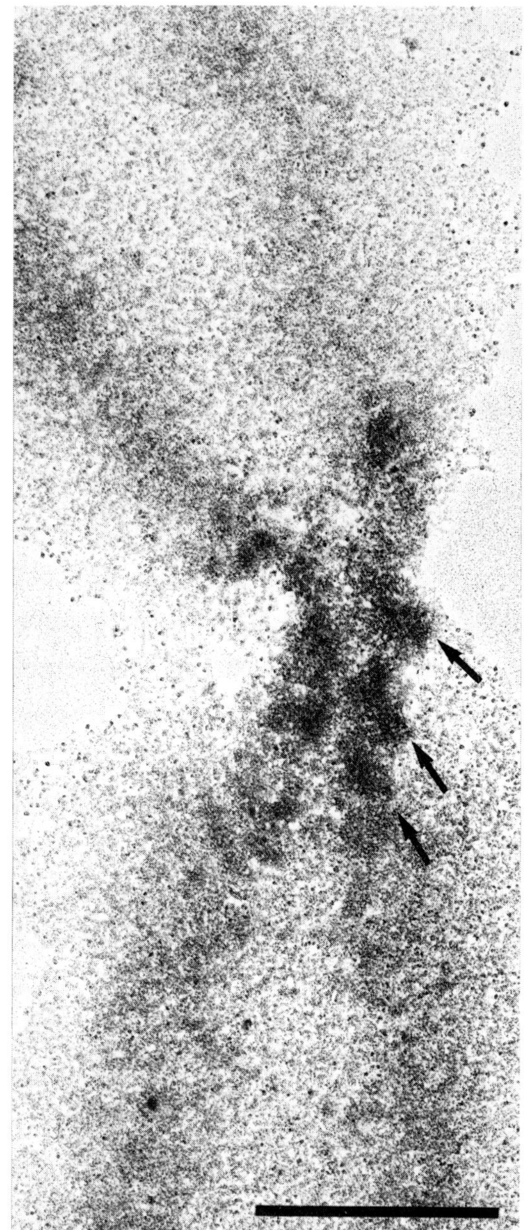

Figure 4.15. Transmission electron micrograph of a mouse chromosome from a cell grown in the presence of 5-azacytidine, illustrating the morphology of the condensing centromere. The electron-dense centromeric heterochromatin displays a pronounced coiled appearance. Successive coils are indicated by the arrow. Bar = 1 μm.

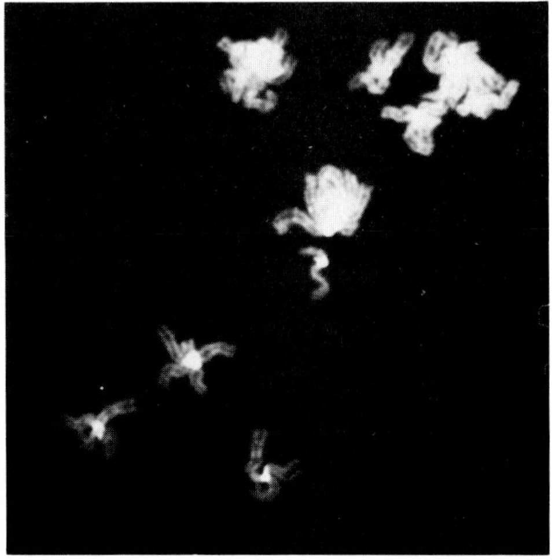

Figure 4.16. Fluorescent micrograph of chromosomal clusters released from a mouse L929 cell grown in the presence of H33258. The chromosomes are associated at the centromere regions, which are located at the center of the clusters. Bar = 1 μm.

pact so that discrete regions of centromeric heterochromatin are quite apparent. In the mouse, this arrangement is particularly prominent, and frequently several centromeric regions are associated together, forming easily identifiable chromocenters. The probable arrangement of chromatin within these regions has been partially elucidated in our study of drug-treated cells.

We have shown above, that the centromere region of mouse L929 cells grown in the presence of H33258 or 5-azacytidine during the G_2 period has an extended conformation when viewed in acid-fixed preparations. We were surprised to find that this conformation did not exist in treated chromosomes released from the cell by detergent lysis and viewed in wet mounts in the presence of the drug. In this case, the centromere displayed a more compact form. Typically, the compact centromeres from several chromosomes associate together to form chromosome clusters, with numerous clusters arising from the single cell (Figure 4.16). These clusters represent the spherical arrangement of several chromosomes, with the centromere regions associated together at the center of the cluster and the euchromatic arms extended radially. The only exception to this observation was found in cells grown in the presence of H33258 for 30 hours or longer. Detergent lysis of these cells produced chromosomes with an extended conformation analogous to that of acid-fixed preparations.

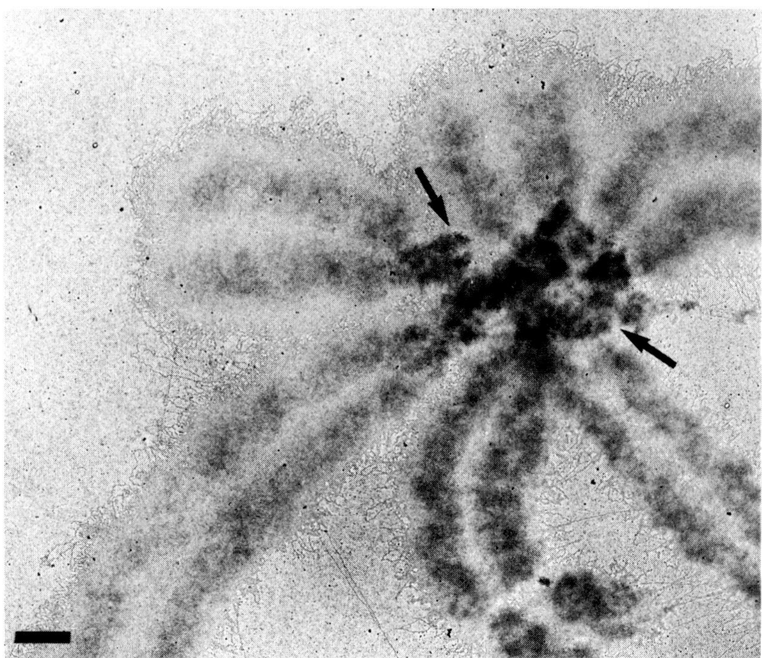

Figure 4.17. Transmission electron micrograph of a chromosomal cluster similar to those illustrated in Figure 4.16. The central region of the cluster (arrows) contains a number of individual centromere regions displaying various degrees of folding of the centromere subfiber. Bar = 1 μm.

Transmission EM preparations of the chromosomal clusters revealed that individual centromeres have a loosely coiled morphology similar to that found in 5-azacytidine-treated chromosomes that display partially condensed centromeres (Figure 4.17). Since this morphology is observed in the presence of the dye in both detergent and mechanically lysed cells, it seems unlikely that a "snap-back" of the uncondensed fiber (seen in acid-fixed chromosomes or in chromosomes from cells grown for long periods of time in the presence of H33258) has occurred. Rather, we believe that acid fixation relaxes centromere substructure by the removal of a protein or subset of proteins whose binding is affected by the drugs. Consistent with this view, conditions that generate partially condensed centromeres (e.g., the presence of 5-azacytidine) would reflect intermediate stages in the normal reassociation of these proteins.

If H33258 and 5-azacytidine inhibit the final condensation of the centromere at the onset of mitosis but do not disrupt the organization existing at the time of their addition, then the morphology of the centromere

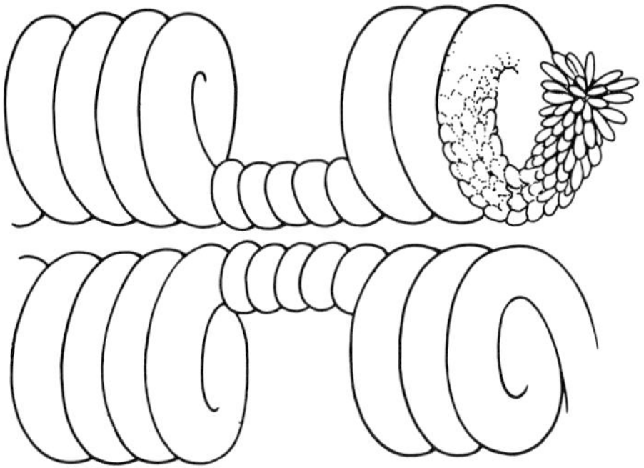

Figure 4.18. Diagram summarizing the organization of a metaphase chromosome. The 200- to 300-nm fiber composed of radial loops of the 30-nm fiber class follows a helical path within the euchromatic arms. In the centromere this fiber follows a straight path, and in turn is composed of a helically folded subfiber. The 30-nm loops of this region compose this subfiber. In the condensed state of the chromosome, the helical nature of the chromosome is obscured.

region within the clusters may reflect features of interphase centromere organization. In this view, the interphase centromere would exist as the less compact form found in the metaphase 200- to 300-nm fiber, and in the mouse similar regions from several chromosomes would associate together to form a chromocenter. Minimal reorganization and tight coiling of the extended fiber at prophase are all that would be necessary to generate final metaphase form. Thus, all the levels of chromatin packing present in the centromere at metaphase would also be present at interphase; only the organization of the highest level would seem to be different. It is assumed that this compact organization must undergo localized unfolding during replication in this region. Such an arrangement would account for the morphology of this region in cytological preparations and would reflect the apparent genetic inactivity of the centromeric heterochromatin.

7. Summary

In this chapter, we have described how the organization of chromatin changes during the cell cycle by the formation and rearrangement of discrete packing hierarchies. The distribution of these hierarchies within a chromosome accounts for the morphology of chromatin domains such as centromeric heterochromatin. In the case of the centromere, we propose that this region possesses at least five levels of high-order structure, which include (1) the 10-nm fiber, (2) the 30-nm fiber, (3) radial loops, (4) a centromere subfiber, and (5) the 200- to 300-nm fiber that represents the form of the native centromere. The euchromatic areas of the chromosome display an additional level of structure reflected in the coiling of the 200- to 300-nm fiber class. This general organization of the chromosome is illustrated in Figure 4.18.

Euchromatin and heterochromatin differ significantly in the levels of organization that are found during the interphase period. In our view, the majority of the euchromatin loses mitotic ultrastructural organization above the level of chromatin loops. On the other hand, heterochromatin retains all the hierarchies of folding present during metaphase, although in a more relaxed state. Similar variations in the distribution of chromatin hierarchies probably correlate with the appearance of facultative heterochromatin and distinct chromosomal domains such as the secondary constriction.

8. References

Adolph, K. W., Cheng, S. M., and Laemmli, U. K. (1977). Role of non-histone proteins in metaphase chromosome structure. *Cell, 12,* 805–816.

Adolph, K. W., Kreisman, L. R., and Kuehn, R. L. (1986). Assembly of chromatin fibers into metaphase chromosomes analyzed by transmission electron microscopy and scanning electron microscopy. *J. Biophys. Soc., 49,* 221–231.

Billings, P. C., Orf, J. W., and Paulmer, D. K., Talmage, D. A., Pan, C. G., and Blumenfeld, M. (1979). Anomalous electrophoretic mobility of *Drosophila* phosphorylated H_1. Is it related to the compaction of satellite DNA in heterochromatin? *Nucleic Acids Res., 6,* 2151–2165.

Blumenfeld, M., Orf, J. W., Sina, B. J., Kreber, R. A., Callahan, M. A., Mullin, J. I., and Snyder, L. A. (1978). Correlation between phosphorylated H_1 histones and satellite DNA in *Drosophila virilis. Proc. Natl. Acad. Sci. USA, 75,* 866–870.

Brutlag, D. (1980). Molecular arrangement and evolution of heterochromatic DNA. *Annu. Rev. Genet., 14,* 121–144.

Burlingame, R. W., Love, W. E., Wang, B.-C., Hamlin, R., Xuong, N.-H., and Moudrianakis, E. N. (1984). Crytallographic structure of the octomeric histone core of the nucleosome at a resolution of 3.3Å. *Science, 228,* 546–553.

Du Praw, E. J. (1965). Macromolecular organization of nuclei and chromosomes: a folded fiber model based on whole mount electron microscopy. *Nature, 206,* 338–343.

Earnshaw, W. C., and Heck, M. M. S. (1985). Localization of topoisomerase II in mitotic chromosomes. *J. Cell Biol., 100,* 1716–1723.

Felsenfeld, G., and McGhee, J. D. (1986). Structure of the 30 nm chromatin fiber. *Cell, 44,* 375–377.

Finch, J. T., and Klug, A. (1976). Solenoidal model for superstructure in chromatin. *Proc. Natl. Acad. Sci. USA, 73,* 1897–1901.

Halleck, M. S., and Gurley, R. L. (1982). Histone variants and histone modifications in chromatin fractions from heterochromatin-rich peromyscus cells. *Exp. Cell Res., 138,* 271–285.

Igo-Kemenes, T. H. W., and Zachau, H. G. (1982). Chromatin. *Annu. Rev. Biochem., 51,* 89–121.

Lica, L., and Hamkalo, B. (1983). Preparation of centromeric heterochromatin by restriction endonuclease digestion of mouse L929 cells. *Chromosoma, 88,* 42–49.

Marcus, M. R., Groitein, R., and Gropp, A. (1979). Condensation of all human chromosomes in phase G_2 and early mitosis can be drastically inhibited by 33258-Hoechst treatment. *Human Genet., 51,* 99–105.

Marsden, M. P. F., and Laemmli, U. K. (1979). Metaphase chromosome structure: evidence for a radial loop model. *Cell, 17,* 849–858.

McGhee, J. D., and Felsenfeld, G. (1980). Nucleosome structure. *Annu. Rev. Biochem., 49,* 1115–1156.

Miller, O. L., Jr., and Beatty, B. R. (1969). Visualization of nuclear genes. *Science, 164,* 955–957.

Miller, F. D., Rattner, J. B., and van de Sande, J. H. (1982). *Cold Spring Harbor Symp. Quant. Biol., 47,* 571–580.

Olins, A. L., and Olins, D. E. (1974). Spheroid chromatin units (ν bodies). *Science, 183,* 330–332.

Onnuki, Y. (1968). Structure of chromosomes. I. Morphological studies of the spiral structure of human somatic chromosomes. *Chromosoma, 25,* 402–408.

Paulson, J. R., and Laemmli, U. K. (1977). The structure of histone-depleted chromosomes. *Cell, 12,* 817–828.

Rattner, J. B., and Lin, C. C. (1985a). Radial loops and helical coils coexist in metaphase chromosomes. *Cell, 42,* 291–296.

Rattner, J. B., and Lin, C. C. (1985b). Centromere organization in chromosomes of the mouse. *Chromosoma, 92,* 325–329.

Rattner, J. B., Branch, A., and Hamkalo, B. A. (1975). Electron microscopy of whole mount metaphase chromosomes. *Chromosoma, 52,* 329–338.

Rattner, J. B., Krystal, G., and Hamkalo, B. A. (1978). Selective digestion of mouse metaphase chromosomes. *Chromosoma, 66,* 259–268.

Rattner, J. B., Goldsmith, M., and Hamkalo, B. A. (1980). Chromatin organization during meiotic prophase of *Bombyx mori. Chromosoma, 79,* 215–224.

Rattner, J. B., Goldsmith, M., and Hamkalo, B. A. (1981). Chromatin organization during male meiosis in *Bombyx mori. Chromosoma, 82,* 341–351.

Rattner, J. B., Saunders, C., Davie, J. R., and Hamkalo, B. A. (1982). *J. Cell Biol. 92,* 217–222.

Renz, M., Nehls, T., and Hozier, J. (1977). Involvement of histone H_1 in the organization of the chromosome fiber. *Proc. Natl. Acad. Sci. USA, 74,* 1879–1884.

Richmond, T. J., Finch, J. T., Rushton, B., Rhodes, D., and Klug, A. (1984). Structure of the nucleosome core particle at 7Å resolution. *Nature, 311,* 532–537.

Ris, H., and Korenberg, J. (1979). Chromosome structure and levels of organization. In *Cell Biology,* Vol. 2, ed. D. M. Prescott and L. Goldstein, pp. 268–361. New York: Academic Press.

Scheer, U., Zentgraf, H., and Sauer, H. (1981). Different chromatin structures in *Physarum polycephalen* a special form of transcriptionally active chromatin devoid of nucleosomal particles. *Chromosoma, 84,* 279–290.

Thoma, I., Koller, T., and Klug, A. (1979). Involvement of histone H_1 in the organization of the nucleosome and the self-dependent substructures of chromatin. *J. Cell Biol., 83,* 408–427.

Viegas-Pequignot, E., and Dutrillaux, B. (1981). Detection of G-C rich heterochromatin by 5-azacytidine in mammals. *Hum. Genet., 57,* 134–137.

Weith, A. (1985). The fine structure of euchromatin and centromeric heterochromatin in *Tenebria molitor* chromosomes. *Chromosoma, 91,* 287–296.
Weith, A., and Traut, W. (1980). Synaptonemal complexes with associated chromatin in a moth *Ephestia Kuehniella* Z. *Chromosoma, 78,* 275–291.
White, M. J. D. (1973). *Cytology and Evolution,* pp. 1–58. New York: Cambridge University Press.
Widom, J., and Klug, A. (1986). Structure of the 300Å chromatin filament: x-ray diffraction from Oriental samples. *Cell, 43,* 207–213.
Williams, S. P., Athey, B. D., Muglia, L. J., Scott-Schappe, R., Gough, A. H., and Langmore, J. P. (1986). Chromatin fibers are left-handed double helices with diameter and mass per unit length that depends on linker length. *Biophys J., 49,* 233–248.
Woodcock, C. L. F., Frado, L. L. Y., and Rattner, J. B. (1984). The higher order structure of chromatin: evidence for a helical ribbon arrangement. *J. Cell Biol., 99,* 42–52.
Worcel, A., Strogartz, S., and Riley, D. (1981). Structure of chromatin and the linking number of DNA. *Proc. Natl. Acad. Sci. USA, 78,* 1461–1465.
Zentgraf, H. V., Miller, V., and Franke, W. W. (1980). Supernucleosomal organization of sea urchin sperm chromatin in regularly arranged 40–50 nm large granular subunits. *Eur. J. Cell. Biol., 20,* 254–264.

V

Nuclear architecture and three-dimensional organization of chromatin: their role in the control of cell function

CLAUDIO NICOLINI

1. Introduction

Several observations from the field of genetic engineering and numerous recent biophysical characterizations of native chromatin have been accounted for, to a surprising and unforeseeable degree, by a recent model for the higher-order structure and intranuclear distribution of chromatin DNA (Nicolini, 1983). Despite significant advances, some basic questions still seem incapable of resolution: For example, what is a gene and how is it expressed? What is the three-dimensional molecular structure, and what is the function of heterochromatin? Failure to answer these questions has led to an increasing reliance on jargon and to the proliferation of hypotheses; yet, the basic questions regarding structure and function remain unanswered.

According to the 1 fibrosome = 1 gene model, each of the 75,000 fibrosomes has its own degree of supercoil locally maintained by highly heterochromatic regions of the DNA, and each has the capability locally to undergo abrupt structural transitions leading to gene expression. This equation, which does offer a coherent and concise answer, both in structural and functional terms, calls for corroborating experimental evidence. That evidence is summarized in this chapter.

Furthermore, this chapter reviews recent findings on the overall structure of the nucleus, from its pores down to its molecular constituents – findings made possible by the greater resolution achieved through application of modern, highly sophisticated, and noninvasive biophysical techniques (Nicolini, 1986). This chapter will also explore the structural and functional features that have been elucidated at the single cell level.

This work has been supported by grants from the Oncology Project of the National Research Council of Italy, and from the national program, Biophysical Control of Cell Growth, of the Ministry of Public Education.

2. Higher-order gene structure

Based on the wide range of recent experimental findings on mammalian nuclei, a three-dimensional model of DNA structure, from the nuclear level down to the gene, has been postulated (Nicolini, 1983). Parallel high-resolution image analysis, both of Feulgen-stained nuclei reconstructed three-dimensionally at the light microscopic level (Kendall et al., 1980) and of freeze-etched nuclei at the electron microscopic level (Nicolini et al., 1984a), reveals the presence of a chromosome-sized fiber 300 Å in diameter near the nuclear border and around the nucleoli (Figure 5.1). Within such a fiber, arranged like a drapery is a fundamental repeating unit, usually (but not exclusively) about 4600 Å long, and capable of unfolding to form separate nucleofilaments. This repeating unit is regularly interspersed with regions that are 2000–9000 Å long. It is highly coiled and attached to the nuclear envelope near the pores. Whatever the exact size of the repeating unit (more studies are needed to clarify this point), these findings predict (1) an upper limit to the length of soluble chromatin obtainable by the mildest nuclease digestion of nuclei; and (2) a significant decrease in chromatin solubilization obtained with the use of restriction enzymes, which cut the DNA considerably less frequently than does a nonspecific nuclease. Both predictions are indeed satisfied: A maximum mass of about 5×10^6 daltons for the nuclease-soluble segment, and a 40% limit digestion for the endonuclease *EcoRI* (Igo-Kemenes and Zachau, 1977) have been reported for rat liver nuclei, quite compatible with the observed ropelike arrangement.

It is intriguing to note that the variable length of the coiled regions (8–32 kilobase pairs [kbp]) corresponds roughly to the size of the nontranscribable DNA sequences of the human genome, which are 20–38 kbp for the ribosomal gene; these coiled regions are interspersed with corresponding transcribable sequences of approximately constant length, 14 kbp, which in turn compares well to the 17-kbp size of the basic "modulating" rope. The species dependence is evident in all genes: For example, the size of the repeating unit for the histone gene is 7 kbp in sea urchin and 15 kbp in chicken chromosomes. Electron micrographs of properly stained ribosomal DNA in the process of transcription (Miller and Beatty, 1969), when the cores of DNA are unwound from the nucleoli under conditions of low ionic strength, reveal thin axial fibers of duplex DNA covered along their length with a "Christmas tree"-like matrix of about 100 thin fibrils (i.e., RNA precursors associated with proteins) of progressively increasing length and terminating on a spherical granule (presumably RNA polymerase). Each matrix represents a transcription unit engaged in the synthesis of a 40*S* precursor and is separated by an alternating inactive stretch of axial core with a DNA packing ratio of 1.66

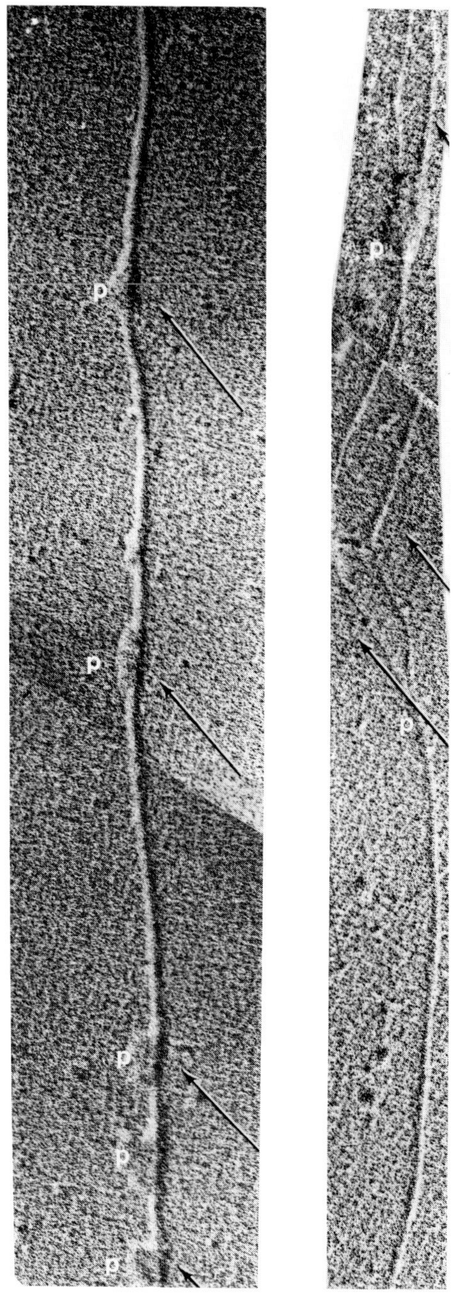

Figure 5.1. Electron micrographs of rat liver chromatin on a phospholipid monolayer. The unfolding of the native fiber into subfiber (arrows) is apparent, as is the successive refolding which occurs between pieces of biological material (P) that periodically attaches to the chromatin fiber.

bp/Å, as compared to 0.24 bp/Å of the transcription unit; curiously, the relation between these packing ratios is identical to the relation (i.e., about 7) between the packing ratios of the nucleofilament and free DNA.

All the above findings prompted me to postulate (Nicolini, 1983) that the repeating unit, which is about 17 kbp long (the length of each of the four nucleofilaments which protend from the nucleoli and the nuclear envelope and together comprise the *fibrosome*), is identical with the coding gene. When the fibrosome is unfolded completely, as occurs during metabolic activation, several mRNA of different size can be transcribed simultaneously and contiguously to yield a total of 17 kbp; alternatively, several mRNA can be coded by the same fibrosome sequence, which thus codes for different proteins of different size, depending on the initiation sites.

Strikingly, the above predictions are compatible with the most recent, still unexplained findings from genetic engineering studies: A full-length protein is translated from a complete gene, but a shorter protein, with only part of this sequence being utilized, can also be translated by use of initiation sites that lie within that same "gene."

Finally, if we consider that the average length of each human gene, when equated to an individual fibrosome (i.e., 45% transcribable plus 55% nontranscribable, as is known for chromatin overall) would be about 40 kbp, then the total number of genes for a typical mammalian cell (with a genome of 3×10^9 bp) would be approximately 75,000, which is compatible with what has been determined either directly by means of restriction enzymes or hybridization techniques (Rolton, Birnie, and Paul, 1977), or indirectly by determination of the rate of spontaneous or induced mutation (Ohno, 1971).

3. Chromatin organization

The emerging picture is confirmed by the critical role of total DNA length (Noll, Thomas, and Kornberg, 1975; Nicolini, Kendall, and Baserga, 1976) in maintaining the native quaternary (solenoid-like or ropelike) chromatin structure. Isolated unsheared chromatin of higher molecular weights were reportedly obtained by gentle, cold-water-induced swelling of isolated nuclei; even at low ionic strength, this preparation was shown to display differential scattering of circular polarized light (Nicolini and Kendall, 1977). Similar conclusions were reached in studies demonstrating the higher reactivity of specific chromatin regions toward ethidium bromide (EB) intercalation when unsheared chromatin was utilized at low ionic strength. These DNA regions thus appear to be maintained under physical constraints favorable to intercalation, namely in the presence of a number of negative superhelices (Nicolini and Kendall, 1977).

In *Drosophila,* and in the mammalian genome isolated as a compact mass of 10^9 daltons (Benyajti and Worcel, 1976), DNase I nicks must occur more frequently than every 6×10^6 daltons (or 400 nucleosomes) to cause full relaxation of the genome to a form sedimenting more slowly and without any negative superhelical turns. Since nucleosomes are preserved even in the presence of DNase I (Cook and Brazell, 1976), it would appear then that 6×10^6 daltons is the minimum molecular mass of DNA that can be considered as a closed structure with a high content of negative superhelical turns.

When a detailed analysis of mammalian nuclear DNA was carried out by continuous measurement of viscosity at 30-minute intervals over a period of more than 48 hours, three discrete unfoldings appeared evident (Nicolini et al., 1982), probably corresponding to the three steplike changes in higher-order superstructures previously described. Similar higher-order DNA foldings and topological constraints are apparent in histone-depleted interphase nuclei and *nucleoids,* cells from which substantial protein has been released by identical lysis in 2 M NaCl, detergent, and chelating agents (Cook and Brazell, 1976).

Calorimetric studies confirm these findings (Nicolini et al., 1983; Nicolini, Germano, and Diaspro, in press-b). For isolated unsheared chromatin, in the range from 300 to 400 Kelvin (K), three broad transitions (II, III, IV) are indeed clearly and reproducibly discernible at all ionic strengths; the enthalpy changes of these transitions are dramatic and are still quite different with changes in salt concentration. The data for the second transition (i.e., III), as expected from polyelectrolyte theory (Belmont et al., 1981), clearly seemed to suggest the possible assignment of transition II to the quaternary ropelike structure that is present even at 1 mM Tris–HCl. The other two transitions (i.e., II and IV), and also transition I which is consistently observed at lower temperature (331 K) only in native nuclei, were related (Nicolini et al., 1984b), respectively, to the helix–coil transition (III) and to either the quinternary drapery-like chromatin structure (I) and/or to the melting of macromolecules (i.e., membrane lipoproteins, nuclear proteins, and/or RNA) strongly bound to native chromatin (I, II). As shown in Figure 5.2 (below), thermal profiles of intact cells reveal, as compared to the corresponding isolated nuclei, the presence of at least an additional endothermic transition, centered at about 338 K, which is by far the most prominent; such a large, broad, background enthalpy in the low-melting region appears probably as a result of the melting of numerous cytoplasmic and nuclear components.

For a more comprehensive study and assignment of transitions (Nicolini et al., in press-b), calorimetry has been applied to calf thymus DNA, to yeast total RNA, and to histone chromosomal proteins in media of various ionic strengths. In all cases, both the melting temperature and the

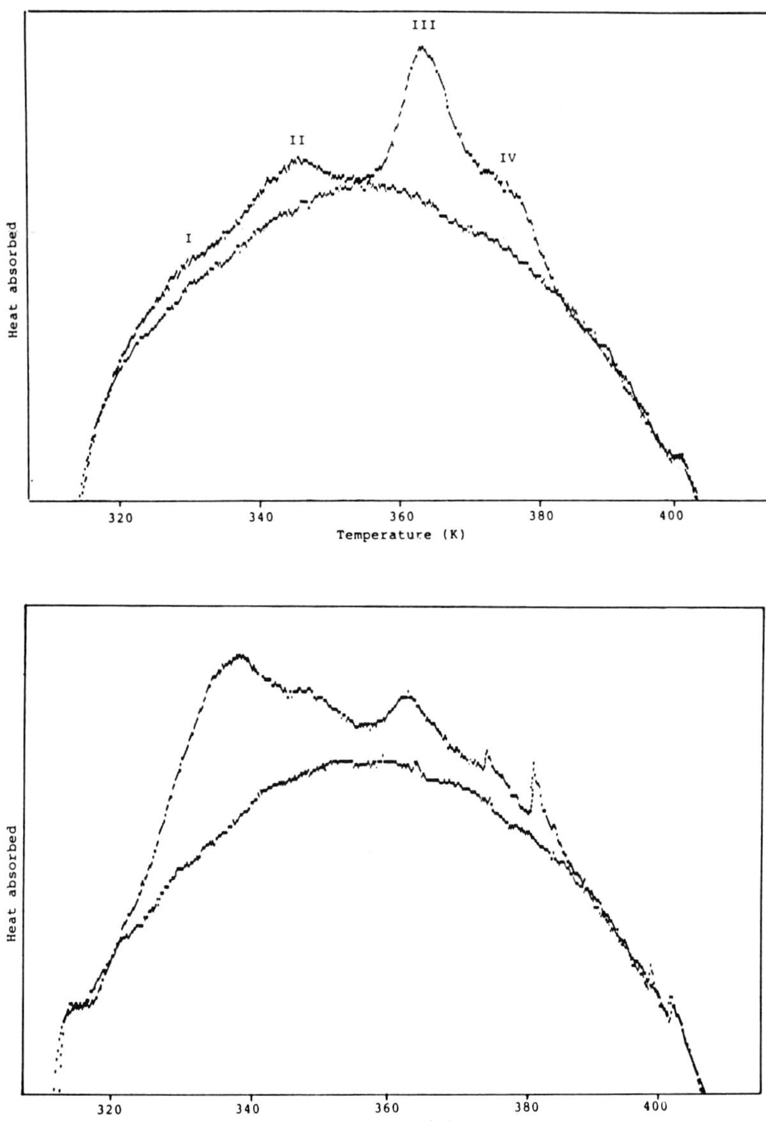

Figure 5.2. Heat capacity versus temperature (in degrees Kelvin), for resting G_0 rat liver nuclei (above) and for cells (below).

enthalpy of transition increase with the salt concentration. Furthermore DNase digestion, carried out on both cells and nuclei at different times and diverse enzyme concentrations, induces the striking disappearance of the higher-temperature transition, and conserves in the intact nuclei only two broad transitions centered at 345 K and 365 K. The thermal profiles of nuclei and cells in "heterochromatic" G_0 (Figure 5.2) and "euchromatic" G_1 states, while demonstrating the challenging complexity of any interpretation of the observed enthalphy changes, confirm the dynamics and the higher-order structure of chromatin but further clarify both the involvement of other macromolecular structures, such as ribosomal RNA and membranes and the exact nature of the various thermal transitions due to DNA (Nicolini et al., in press-b). In order to increase the reproducibility and the sensitivity of measurements of weak signals and of those barely detectable in the thermal transitions of chromatin, cells, and nuclei, we have constructed and utilized a computerized system, electronically interfaced, which has offered a unique opportunity for the acquisition, treatment, and display of the calorimetric data.

It should be noted that chromatin organization appears affected by changes in DNA secondary structure. Indeed, DNA can exist as either a right-handed or a left-handed double helix, held together by Watson–Crick hydrogen bonds. These two conformations are in equilibrium with each other. However, the left-handed (or Z conformation) has a higher energy than the right-handed (or B conformation) and hence occurs less frequently. In left-handed (Z) DNA, the nucleotides alternate in *syn* and *anti* conformations in contrast to right-handed B-DNA, in which all the conformations are *anti* (Wang et al., 1985). Z-DNA is highly immunogenic in contrast to B-DNA. In biological systems, two major factors that stabilize Z-DNA are (1) negative supercoiling and (2) proteins that bind specifically to Z-DNA but do not bind to B-DNA. An enzyme that brings about genetic recombination in lower eukaryotes has been purified and shown to be a Z-DNA binding protein (Kmiec, Angelidis, and Holloman, 1985). One of these Z-DNA binding proteins has been found attached to a segment of the control region of the viral genome known to be important in regulating gene expression in the SV40 virus (Azorin and Rich, 1985).

4. Intranuclear DNA distribution: three states

Analytical image analysis of Feulgen-stained cells has been utilized profitably to characterize nuclear DNA in situ. Current findings have been summarized at length in recent papers (Kendall et al., 1980; Nicolini, 1983) to which the reader is referred for more details. Our intent here is to present only the implications of the latest work related to chromatin structure, emerging from the acquisition of the optical density (OD) values of each of the several thousand individual pixels constituting a single nuclear image.

From such high-resolution image analysis of nuclear DNA, at both light and electron microscopic levels before and after nuclease digestion but with nuclei first fixed and then digested, chromatin DNA appears to exist in up to three resolvable states of higher-order structures: The first is a very dispersed one (0.03–0.07 OD), where "active" genes that are attacked early by DNase I appear to be located. The two other states consist of DNA not preferentially digested by DNase, and of a progressively larger degree of supercoiling: 0.14–0.20 OD (disperse "euchromatic" DNA or mononucleofilaments) and 0.30–0.51 OD (condensed "heterochromatic" DNA or tetrafilament rope).

These conclusions are supported by recent findings demonstrating (1) similar levels of condensation (average OD) of individual interphase chromosomes reconstituted from 1-μm-thick serial sections of fixed, and then Feulgen-stained melanoma nuclei; and, more significantly (2) the apparently similar levels of condensation of individual chromatin fibrils shown by uranyl acetate staining and electron microscopy. Remarkably, the relation of the OD levels of the second and third states described above (i.e., 2.8:1) is exactly what would be expected from the difference in DNA packing ratio between a ropelike tetrafilament and four adjacent nucleofilaments, both maintained at the extremity by fixed attachment sites near the nuclear envelope. Interestingly, any other quaternary folding thus far proposed would yield an increase in the packing ratio of 6 (solenoid) or higher. If the nucleofilaments were free to move within the nucleus and did not have periodic fixed attachments every few thousand ångstroms to keep them close to each other (corresponding to the structural transition of rope to nucleofilament), the increase in the packing ratio upon rope formation would have been 11.6. When nuclei are first digested with nuclease (Figure 5.3) and then fixed, the results are drastically different (Nicolini, Diaspro, and Germano, in press-a). In this instance, the intermediate state of condensation (corresponding to the mononucleofilament) also appears to be selectively attacked by DNase I quite early.

In any case, the traditional interpretation that euchromatin and heterochromatin are two separate entities (i.e., active and inactive portions of the genome, respectively) thus appears incorrect, since the degree of condensation per se of chromatin DNA is not the factor that predicts preferential DNase digestion. Other components are needed that are destroyed with fixation; only the unfolding of individual nucleosomes to form free DNA apparently allows preferential attack by DNase under any condition (i.e., before or after fixation). Such a conclusion is consistent with biomedical data now in the literature, indicating organization of 25- to 50-kb regions that are DNase I sensitive, adjacent to active genes (Stalder, Seeheck, and Borun, 1978; Weisbrod and Weintraub, 1979). Speculations concerning organization of transcriptional units into closed

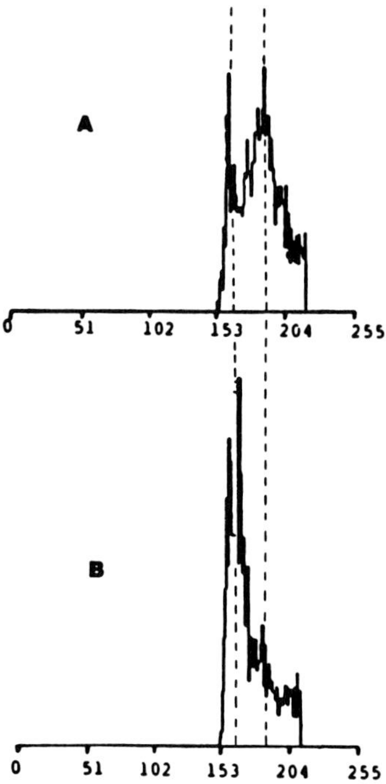

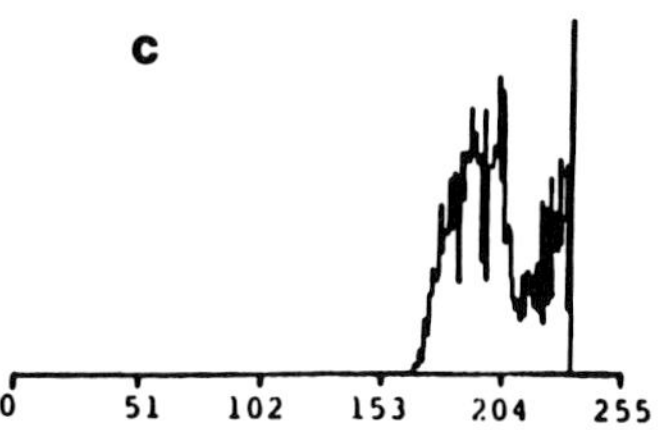

Figure 5.3. Transmittance histograms (number of pixels versus gray level) of Feulgen-stained rat liver nuclei before (A) and after (B, C) DNase digestion. The difference in results obtained by carrying out nuclease digestion before (B) versus after (C) alcohol–acetic acid fixation is apparent.

"lampbrush-like" loops (Weisbrod and Weintraub, 1979) may thus be justifiable.

Support for such a hypothesis can be found in the three types of chromatin conformation apparent in histone genes from sperm, mesenchyme blastular embryos, and early embryos (Spinelli et al., 1982). Blot hybrid-

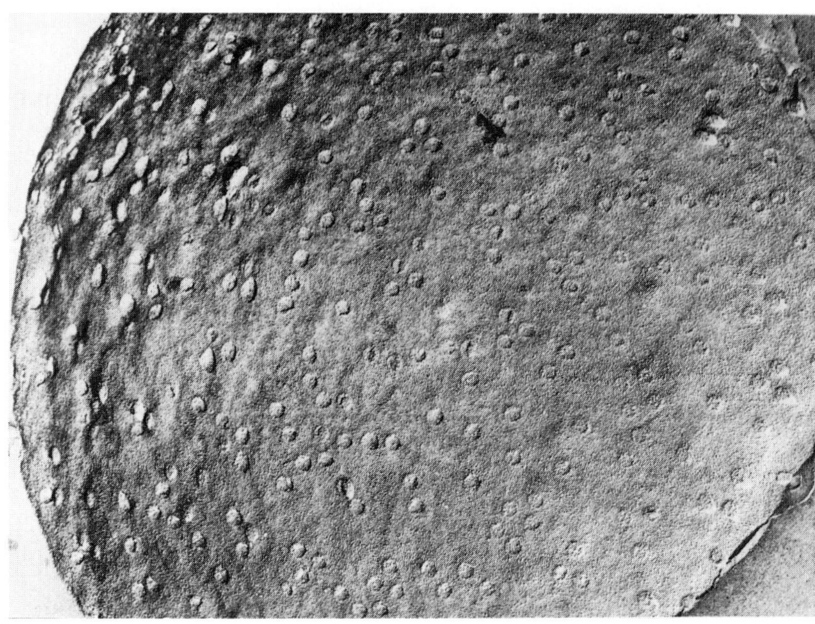

Figure 5.4. Freeze-etch electron micrograph of a rat liver nuclear membrane, showing the pores and their distribution.

ization and micrococcal nuclease digestion reveal a striking difference in the ratio of multimers to monomers, reflecting different higher-order nucleosomal structures for the chromatin from the first two nuclei (sperm and mesenchyme blastular embryos, both repressing histone genes) while the cell is in different stages of proliferation and differentiation. The same analysis suggests that actively transcribing histone genes do not exhibit the periodic nucleosomal arrangement typical of the bulk chromatin (see also Wu, Wong, and Elgin, 1979).

5. Nuclear envelope

The double membrane surrounding the nucleus contains as its most conspicuous feature nuclear pore complexes (Figure 5.4), which have been assumed to be sites of molecular and ionic exchange between nucleus and cytoplasm (Maul, 1977).

Rigorous statistical analysis carried out on computer-enhanced nuclear membranes from various types of rat liver interphase cells demonstrates that the number of pore clusters is independent of the total number of pores and the chromatin distribution, but closely conforms to the number of chromatin bodies near the nuclear border (Nicolini et al.,

1984a). The fraction of total DNA present in most chromatin clusters is about 0.9–7.0%, which is in striking agreement with the fraction of total DNA present in each metaphase chromosome from the same cells (0.8–5.0%); similarly, the proportion of the total number of pores contained in each pore cluster of the envelope also ranges between 1.2% and 9.9%. About 20 groups of nuclear pores, corresponding to the clustering of single chromatin fibers, can be objectively identified on the nuclear envelope (Figure 5.4), a number very close to the actual number (19–21) of homologous pairs of chromosomes in rat liver cells (Yoshida and Sagai, 1972).

All the above data support a possible circumstantial linkage between the highly ordered structure of interphase chromatin DNA (Sedat and Manuelidis, 1978; Kendall et al., 1980; Nicolini, 1983) and the formation of individual metaphase chromosomes, possibly resulting from the collapse of localized ropelike chromatin fibers anchored to the nuclear envelope (Comings, 1968; Olins and Olins, 1979; Kendall et al., 1980; Nicolini et al., 1984b). More direct evidence for a close association of interphase chromatin with the nuclear membrane has recently been demonstrated in unfixed nuclei (Agard and Sedat, 1983). This association, resulting in a predominantly chromosome-free central cavity, corroborates earlier findings in fixed nuclei (Nicolini, 1979, 1980; Olins and Olins, 1979; Kendall et al., 1980). Furthermore, the existence of a repeating unit of about 160 nm in the interpore distance is compatible with the observation of a repeating unit of 480 nm in the chromatin segments delimited by fragments of the nuclear lamina (as shown by Nicolini, 1983): A 480-nm contour length of the same fiber, hanging from the nuclear envelope, corresponds to a pore distance typically of 160 nm. Using optical fluorescence microscopy and three-dimensional chromosome topography of intact unfixed nuclei, Agard and Sedat (1983) have shown a complex mixture of intertwined coils and parallel chromosome segments closely packed against the inner surface of the nuclear membrane. This, along with the presence of laminar fragments within giant chromatin fibers, provides indirect justification for the nuclear skeleton concept discussed in the following section.

6. Nuclear skeleton

Within the nuclei the existence of a skeleton – that is, a scaffold consisting of a dense network of fibers – surrounded by a "halo" of loops of fibrosomal DNA anchored at both ends at the scaffold has been confirmed; these findings seem to be reproducible in both metaphase and interphase chromosomes. Still rather controversial, however, is the nuclear matrix concept, whereby the network of fibers extends *within* the nucleus (Pardoll, Vogelstein, and Coffey, 1980).

Isolation of the nuclear matrix indeed appears to depend critically upon the sequence of the isolation steps; several authors, using similar conditions of exposure of the nuclei to nuclease and Triton-X, have obtained only a peripheral lamina with a shape and size similar to those of the original nucleus. This fraction consists of 18% nucleic acid, 1% lipid, and 82% nonhistone proteins. The nonhistones reside at the nuclear pores within the lamina; their location becomes diffuse during mitosis. These are possibly the proteins involved in the attachment of chromatin fibers to the pores and in maintaining the highly packed form of the tetrafilament. The nuclear matrix could then be a consequence of the intranuclear distribution of chromatin. The location of nonhistones forming the protein network was originally determined from the location of the extracted chromatin-DNA fibers, which extend from the nuclear envelope and the nucleoli.

However, whereas the "euchromatic" chromatin basic unit appears both critically and locally to depend for its supercoil upon histone 1 (H_1) ions and other chromosomal proteins, most of the highly coiled "heterochromatic" regions – consisting of the same multifilament rope – do *not,* and therefore yield a skeleton that is unattacked by nuclease and unaffected by high salt concentration and even by detergent (Nicolini et al., 1984b); this skeleton is associated with the pores at the nuclear envelope and is possibly maintained by lipoproteins, by Ca^{2+} or Cu^{2+}, and by a fraction of nonhistone proteins (the three nuclear lamina proteins are still present at 2.0 M NaCl) that is characterized by strong binding with AT-rich regions of the genome.

The availability of purified histone mRNAs as specific hybridization probes has recently allowed this prediction to be confirmed in sea urchins, giving indirect support to the postulated role of nonhistones in maintaining the highly coiled regions of the fiber.

7. DNA replication: initiation and termination sites

Using HeLa S3 cells synchronized by selective detachment, we reported (Nicolini, Belmont, and Martelli, 1986) a parallel study of nuclear morphology and autoradiographic grain patterns between middle G_1 and middle S phases. Two distinct [^{3}H]thymidine labeling patterns were obtained. The first "peripheral" labeling pattern, which had a characteristic nuclear size distribution, in contrast to the heterogeneous and varying size distributions of Feulgen-stained nuclei, is apparently characteristic of very early S phase. The size of the second labeling pattern, characterized by homogeneous or inhomogeneous grain distribution throughout the nucleus, was equal to or larger than the first and varied with S phase progression. Together, the corresponding nuclear sizes of the

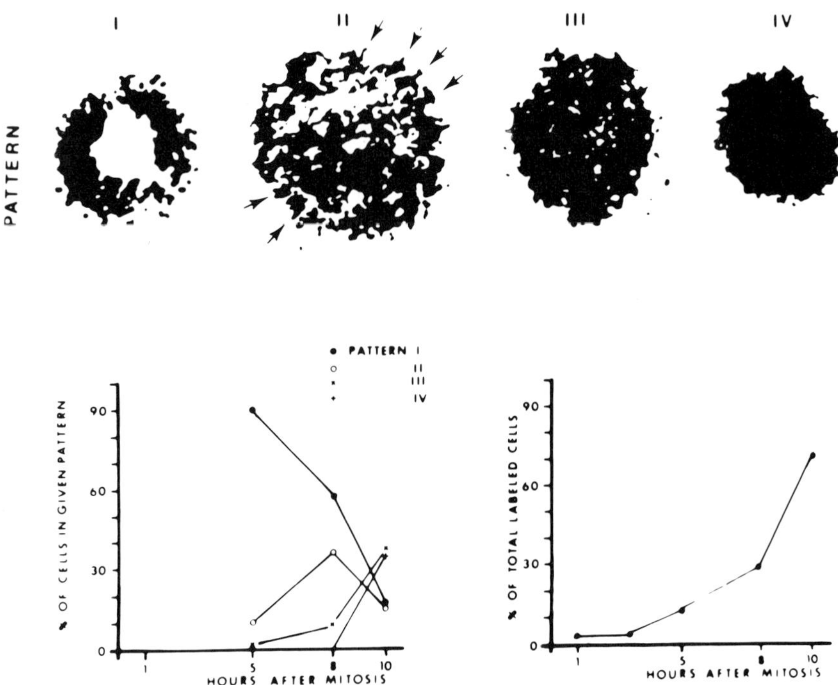

Figure 5.5. Typical labeling patterns of [^{3}H]thymidine-pulsed mammalian cells, in very early or very late S (DNA replication phase) (I) and in other parts of S phase (II–IV).

labeled nuclei represent the larger extreme of nuclear areas, and the labeling index closely parallels the fraction of nuclei having areas larger than the minimum size of the labeled nuclei. These results suggest a characteristic nuclear size (reflecting unique intranuclear DNA distribution) as a necessary, if not sufficient, requirement for S phase initiation. Parallel experimentation with rat liver cells, synchronized in vivo by partial hepatectomy and analyzed by thin-section autoradiography, confirms the existence of a peripheral labeling pattern in both the very early part and the very late part of S phase (Figure 5.5), which reconciles our data with previous results and points to the fact that both initiation and termination sites for DNA replication are near the nuclear periphery. These results obtained in two different systems, appear rather conclusive. However, in still earlier investigations, Williams and Ockey (1970), using both fluorodeoxyuridine (FUdR) and methotrexate or each separately to synchronize the cells at early S phase, obtained a peripheral labeling pattern. A homogeneous distribution was predominant at the early time points.

Moreover, in the first of these studies a peripheral pattern was found in the majority of labeled nuclei at the later time points and was identified as characteristic of late S phase. These observations indeed parallel our findings on rat liver cells.

There are, however, several important differences in our study which must be kept in mind. None of the earlier published reports analyzed the entire sequence after mitosis both in vivo and in vitro, including the very early time intervals [i.e., the only time periods with a dominant peripheral pattern and with a very low labeling index (LI) – less than 5%]; also no study was carried out using comparable high-resolution imaging techniques and comparable physiological methods of synchronization. For HeLa cells, our pattern I was predominant only at 5 hours after selective detachment, where the LI was only 5%. For the great majority of cells, this time point actually corresponds to middle G_1. By 8 hours, with the LI still only 17%, our homogeneous grain pattern II was already the predominant pattern; at this time, most S phase nuclei still have a DNA content near 2C and none has a DNA content exceeding 2.5C. The use of drugs to alter nucleotide pool sizes and thereby inhibit DNA synthesis does not necessarily prevent replicon initiation, and it is not clear whether what earlier investigations considered early S phase, actually corresponds to the *true* very early S phase, which apparently gave rise to our pattern I. Indeed, previous experiments demonstrated an altered inter-origin distance for replicons and a reduced proportion of DNA replicating at similar times in consecutive S phases when FUdR was used for synchronization. Our conclusion appeared to hold true when we viewed the entire nucleus and also when we viewed 2.0-μm-thick sections.

What our results suggested is that the origins of replication of individual replicons are fixed at the nuclear periphery and that termination of these replicons occurs also at the nuclear periphery (Nicolini, Belmont, and Martelli, 1986). Several lines of evidence are compatible with this hypothesis, indicating that the nuclear lamina and the nuclear envelope may be responsible for the topologically defined organization of the DNA. Middle replication sites, as well as RNA transcription sites, would be thus localized on the skeleton: They would comprise distal regions of the DNA loops, which appear attached to the polypeptide lamina interposed between the inner nuclear membrane and the peripheral chromatin, and which extend into the nuclear interior. However other papers are consistent with the conclusion that initiation of replication of individual replicons occurs in association with the nuclear matrix, mostly (but not totally) occurring in the inner reaches of the nucleus, rather than exclusively at the nuclear lamina membrane or periphery. If we accept as fact the largely presumptive mechanism of action of α and β DNA polymerases, initiation of less than 12% of newly replicated DNA would appear to occur in association with the nuclear envelope structure. Within such a context, our data would then suggest that very early and late S phase consist of replication of replicons associated with the nuclear periphery, whereas middle S consists of replication of replicons located in more central regions. The data presented here cannot by themselves discriminate between the two alternative possibilities. In either case, however, a peripheral pattern in early S is proven compatible with the literature (which is otherwise conflicting in most other respects); the exact time sequence and the geometric location of DNA replication sites both in vitro and in vivo are consistent with the existing data.

We would like to emphasize that the significance of our results extends beyond the simplistic view that a critical nuclear size per se is related to S phase initiation, in light of earlier work relating changes in nuclear morphology to changes in intranuclear DNA organization (Nicolini, 1983).

8. DNA transcription and DNase-sensitive regions

In recent years, much attention has been focused on the selective nuclease activities of such enzymes as micrococcal nuclease and DNase I. Many investigators, beginning with Weintraub and Groudine (1976), have reported the preferential digestion by DNase I of genes that had been activated at some time point in the epigenetic history of the cell system under study. This is in part because limited digestion by micrococcal nuclease releases histone 1 (H_1), high mobility group (HMG) proteins, and other nonhistone proteins in relatively different amounts from those in bulk chromatin; indeed it has been suggested that micrococcal nuclease also preferentially digests transcriptionally active chromatin. Furthermore, large 25- to 50-kb DNA domains with increased DNase I sensitivity have been reported to lie adjacent to potentially active genes (Stalder et al., 1978).

Recently, we studied the digestion (following fixation) by micrococcal nuclease of HeLa S3 nuclei at the G_1–S boundary through image analysis of Feulgen-stained nuclei (Kendall, Beltrame, and Nicolini, 1979; Nicolini, 1980). Within the possible experimental error for the experimental design employed, no preferential digestion of the subfraction of dispersed chromatin (i.e., "euchromatin") within the fixed nucleus was demonstrated. On the contrary, when percent digestion at given times was compared to nuclear area (for 2C DNA nuclei) and when experimental OD histograms (of picture points vs. given OD level) from individual "fixed" nuclei were compared to theoretical histograms generated by use of a hypergeometric probability distribution model for enzyme digestion (Kendall et al., 1980), preferential digestion of denser nuclei or denser regions ("heterochromatin") of chromatin was demonstrated. Questions have been frequently raised (Bonner, 1979), however, concerning the actual specificity of micrococcal nuclease for active genes, and reasonable doubts have remained also concerning the existence of a very small fraction of "active" chromatin preferentially digested at early times and at low enzyme concentration – a fraction yet undetected by image analysis in our early experiments (where digestion was carried out after fairly disruptive conditions of cell fixation).

Indeed, although the basic intranuclear DNA distribution into three orders of superpacking (Belmont, Kendell, and Nicolini, 1984) and the occurrence of discrete transitions among them during induction of cell proliferation (Nicolini, 1983) appear similar under all of the extremely diverse conditions of ionic strength and pH employed and under the many alternative procedures used for chromatin characterization (either isolated or in situ), still not clear are (1) the intranuclear distribution of active DNA regions that are DNase sensitive; (2) the role of intranuclear distribution in determining gene activation; and (3) the factors related to the preservation of overall nuclear architecture and to gene expression during cell cycle progression.

Recently, we have extended the above experiments to the study of limited digestion by enzymes (i.e., both DNase I and micrococcal) before and after fixation, in two cell systems (HeLa in vitro and rat liver in vivo), in order to analyze nuclei of different metabolic activity. We demonstrated the existence of significant cell-cycle-related changes in both (1) the intranuclear DNA distribution, which was apparent even when a crude procedure was intentionally utilized to remove partially or completely H_1 and HMG chromosomal proteins (Brasch and Setterfield, 1974); and (2) the preferential digestion (occurring only before fixation and with DNase) of the most decondensed, metabolically active chromatin subfraction, whose native intranuclear organization in nuclei is such as to allow detection even at light microscopic levels of resolution. The action of DNase I is readily apparent in G_1 nuclei that are first digested and then fixed; however, in G_1 nuclei that are first fixed and then digested, the intranuclear DNA distribution appears unaltered, with both domains being equally poorly digested at the early times points. In contrast, in metabolically inactive nuclei (G_0), only a slight redistribution occurs within the single DNA domain; in this case, only a very small fraction of the genome appears to be organized in randomly distributed domains which are highly dispersed and preferentially digested by the enzyme. Spectroscopic characterization of chromatin isolated from the same nuclei reveals that fixation causes substantial depletion of chromosomal proteins of the soluble chromatin subfraction (which is considerably larger in G_1 than in G_0 nuclei), while leaving the insoluble fraction unaltered.

All of the above evidence suggests that the DNA regions containing active genes, which are uniquely sensitive to nucleases, are significantly larger in metabolically active G_1 nuclei and are maintained by their nuclear skeleton (associated with the insoluble and inactive chromatin fraction) even after fixation; the same fixation, however, makes these active DNA regions (clearly associated with the soluble chromatin subfraction) unrecognizable to the nuclease.

It seems apparent that, while, strictly speaking, gene and chromatin cannot be probed at the resolutions possible in this study (Nicolini et al., in press-a), the changes in both nuclear architecture and intranuclear DNA distribution brought about by enzyme digestion allow us to draw several conclusions regarding chromatin-DNA structure that would be impossible to reach by other available means. Several lines of evidence on the nucleosome sensitivity of active and inactive genes have already being gleaned biochemically, and our data on the fraction of the genome digested by DNase I can no doubt be interpreted as being consistent with the conclusion of other investigators on the cell-cycle-dependent changes in chromatin sensitivity toward this nuclease (Cocco et al., 1984); however, the primary information acquired by combining light microscopy with high-resolution image analysis is of another kind. First, the two enzymes, DNase I and micrococcal nuclease, appear to be probing different aspects of the chromatin structure that remains after fixation. The differential action of these enzymes was suggested earlier for isolated chromatin (Bonner, 1979). Our data confirm this earlier speculation. Second, our results, in agreement with previous findings (Nicolini and Belmont, 1983) which demonstrated altered levels of dye intercalation into chromatin from fixed cells with the same DNA content but in different functional states, also showed that these differences were retained after Carnoy's fixation (as used in this paper). The gross higher-order structure of most chromatin DNA would thus appear to have been unaffected by the fixation. Third, whereas studies employing nick translation and autoradiography have suggested that for certain DNA regions, DNase I hypersensitivity is preserved during metaphase and remains intact after Carnoy fixation, our data do not show this. Indeed, the DNase I sensitivity of a randomly distributed small fraction of the genome and the related change in the intranuclear DNA remain evident, but are of a significantly reduced magnitude (Figure 5.3) whenever nuclei fixation is performed before the DNA is exposed to the enzyme.

The above results thus suggest that fixation-induced denaturation and chromosomal protein stripping of a selective fraction of DNA alter those regions of chromatin DNA that constitute native active genes and native potentially active genes, affecting their capacity to be recognized by the nuclease but leaving unaltered the higher-order intranuclear DNA distribution that is maintained by the nuclear skeleton, corresponding to the insoluble chromatin subfraction (for a review, see Nicolini, 1986).

9. Oncogene expression and DNA supercoiling

For the same amount of DNA, chromatin DNA is considerably less condensed in the low versus the high metastatic clone variants from B16 melanoma cells. In transformed Chinese hamster ovary–K1 (CHO-K1) cells, nuclear DNA is substantially more condensed than in the corresponding fibroblast-like cells (Nicolini and Beltrame, 1982). These observations can be explained in terms of a model for differential gene expression associated with the global increase in nuclear DNA condensation apparent during neoplastic transformation and at the onset of DNA or RNA synthesis. Within such a framework, a normal cell displays for each of its 75,000 fibrosomes ("genes") a varying number of DNA negative superhelical turns from zero (for unfolded and transcribing DNA) to a maximum "threshold" value (for a highly packed nucleofilament) beyond which the DNA undergoes an abrupt transition to a relaxed form (i.e., 0), releasing all its stored energy. This excess energy in turn triggers the transcription (or replication) of the corresponding DNA sequences – now accessible in state I – also through enhanced binding of RNA (or DNA) polymerases. The increased condensation in the overall genome during neoplastic transformation causes a similar increase in negative superhelical turns for each of the 75,000–100,000 fibrosome units, which in turn may lead either to gene repression (for those fibrosomes with superhelical density equal to zero) or to gene expression (for those with superhelical density near the initial threshold value). It is this imbalance that is assumed to lead to cancer, rather than the enhanced expression of a single oncogene, which is frequently postulated to initiate oncogenesis. Stimulation of transcriptional activity in vitro is observed at low and moderate negative superhelical densities up to the level of the natural superhelical form of the plasmid pBR322. Enhancement of transcription occurs at different levels of superhelicity, suggesting a sequence-dependent structural alteration of promoters upon changes of axial writhe, which may generate kink formation. Initiation of transcription is drastically inhibited at higher specific linking differences exceeding that of the natural superhelical form of pBR322; this inhibition is correlated with a transition from the right-handed B form to the left-handed Z form of particular DNA sequences induced by supercoiling. The enhancement and inhibition of transcription by supercoiling support the role of energetic and structural changes in topologically constrained DNA as elements of a control mechanism (Brahms, Dargouge, Brahms, and Wagner, 1985).

10. Three-dimensional structure and role of heterochromatic regions

The active genes thus appear to be organized within large domains whose level of condensation and intranuclear organization are discretely different from those of the greater proportion of chromatin. The highly dispersed chromatin observed in these metabolically active, DNase I-sensitive regions, which are nevertheless resistant to micrococcal nuclease digestion (Kendall et al., 1980; Nicolini et al., in press-a), is strikingly compatible with the recently reported nick translation of similar HeLa cell nuclei (Javaherian, Lin, and Wang, 1978; Fasman, 1979); in the latter studies, the labeled DNase I-sensitive regions appear indeed to consist of "either a high mobility-group protein-containing nucleosome in some unfolded conformation or discrete DNA fragments, wrapped around some nonhistone proteins, located in a highly DNase I-sensitive region, which is resistant to micrococcal nuclease digestion." Moreover, these DNase-sensitive domains are sufficiently large to be detectable at light microscopic resolution.

Indeed, such conclusions are consistent with the biochemical data now in the literature, indicating the organization of DNase I-sensitive regions of 25–50 kb adjacent to active genes (Stalder, Seeheck, and Borun, 1978; Weisbrod, 1982). Speculation concerning the organization of transcriptional units into closed "lampbrush-like" loops (Weisbrod, 1982) or into fibrosome units resembling a closed drapery (Nicolini, 1983) is supported by the combined results of a number of recent reports including electron microscopic studies of native chromatin, freeze-fracture scanning EM studies of nuclei, and the previously mentioned light microscopic study.

Finally, the most significant conclusion is that the nuclear skeleton appears both necessary and sufficient to maintain the higher-order structure (i.e., the intranuclear distribution) of the active DNA regions, but is not sufficient to characterize and identify active genes. Conversely, the chromosomal proteins that are removed by fixation (possibly histone 1, HMG nonhistones, and their enzymatically modified counterparts known to be present only in the soluble active chromatin fraction) represent the second level of control of gene expression, being unnecessary for mantaining the higher-order intranuclear DNA distribution, and only necessary (but not sufficient; i.e., a cofactor) for the lower-order transition between the nucleofilament and the free DNA, leading to local gene activation (Nicolini, 1986).

The first level of control could then be exercised by other factors such as (1) strongly bound nonhistones (still present after fixation or at 2 M NaCl ionic strength, and maintaining the DNA supercoil); (2) local nuclear changes in ions (those ions interacting with DNA phosphate groups and thereby affecting DNA bending and cell function); or (3) local levels of DNA methylation and hydration (critically determining secondary structure and gene activation).

The fibrosome model also implies that much of the DNA, lying in the coiled heterochromatic alternating regions, does not code for protein, and that its function is purely structural. This again appears to resolve the paradoxical relationship between DNA complexity and gene number, known as the C-paradox: Much more genome is available than would be required to code for all of the proteins that it specifies, and genome size within any species can vary up to 100 fold although the actual number of protein products remains similar within that species.

All of these observations are thus compatible with the view that most DNA heterochromatic sequences, located mostly in the nontranscribed spacers (highly AT-rich), function solely to maintain the quaternary and quinternary chromatin-DNA structure. Their specific function is apparently to maintain the integrity of the nuclear skeleton and to create fixed sites for the attachment of nucleofilaments which can then locally yield DNA with highly negative superhelical turns. As has been demonstrated previously, an increasing number of negative superhelical turns appears to lead, in and of itself, to a higher rate of transcription in protein-free DNA and to the expression of the transformed phenotype in chromatin.

11. References

Agard, D. A., and Sedat J. W. (1983). The three-dimensional architecture of a polytene nucleus. *Nature, 302,* 676–681.

Azorin, F., and Rich, A. (1985). Isolation of Z-DNA binding proteins from SV40 minichromosomes: evidence for binding to the viral control region. *Cell, 41,* 365–374.

Benyajati, C., and Worcel, A. (1976). Isolation characterization and structure of the folded interphase genoma of *Drosophila melanogaster. Cell, 9,* 393.

Belmont, A., Kendall, F., and Nicolini, C. (1984). Three-dimensional intranuclear DNA organization in situ: three states of condensation. *J. Cell Sci., 65,* 123.

Bonner, J. (1979). Expressed and nonexpressed portions of the genome: their separation and their characterization. In *Chromatin Structure and Function,* ed. C. Nicolini, pp. 15–23. New York:Plenum.

Brahms, J. G., Dargouge, S., Brahms, Y., and Wagner, V. (1985). Activation and inhibition of transcription by supercoiling. *J. Mol. Biol., 181*(4), 455–466.

Brasch, K., and Setterfield, G. (1974). Structural organization of chromosomes in interphase nuclei. *Exp. Cell. Res. 83,* 175.

Cocco, L., Gilmour, R. S., Papa, S., Capitani, S., and Manzoli, F. A. (1984). Response of isolated nuclei to phospholipid vesicles: analysis of chromatin sensitivity to DNase I and micrococcal nuclease. *Cell Biol. Int. Rep. 8,* 55–63.

Comings, D. E. (1968). The rationale for an ordered arrangement of chromatin in the interphase nucleus. *Am. J. Hum. Genet., 20,* 440.

Cook, O. R., and Brazell, I. A. (1976). Conformational constraints in nuclear DNA. *J. Cell Sci., 22,* 287.

Fasman, G. D. (1979). Circular dichroism of DNA protein and chromatin. In *Chromatin Structure and Function,* ed. C. Nicolini, pp. 67–108. New York: Plenum.

Igo-Kemenes, T., and Zachau, H. G. (1977). Domains in chromatin structure. *Cold Spring Harbor Symp. Quant. Biol., 42,* 109–114.

Javaherian, K., Lin, J. F., and Wang, J. C. (1978). Nonhistone proteins HMG_1 and HMG_2 change the DNA helical structure. *Science, 199,* 1345.

Kendall, F., Beltrame, F., and Nicolini, C. (1979). Chromatin study in situ. I. Image analysis. In *Chromatin Structure and Function,* ed. C. Nicolini, pp. 265–292. New York: Plenum.

Kendall, F., Beltrame, F., Belmont, A., Zietz, S., and Nicolini, C. (1980). The quinternary chromatin-DNA structure: three-dimensional reconstruction and functional significance. *Cell Biophys., 2,* 373–404.

Kmiec, E. B., Angelidis, K. J., and Holloman, W. K. (1985). Left-handed DNA and the synaptic pairing reaction promoted by Ustilago rec1 protein. *Cell, 40,* 139–145.

Maul, G. G. (1977). The nuclear and cytoplasmic pore complex: structure, dynamics, distribution and evolution. *Int. Rev. Cytol.* [*Suppl*], *6,* 75–186.

Miller, O. L., and Beatty, B. R. (1969). Visualization of a nucleolar gene. *Science, 164,* 955–957.

Nicolini, C., ed. (1979). *Chromatin Structure and Function.* Nato Life Sciences Series, 1–880. New York: Plenum.

Nicolini, C. (1980). Molecular and cellular approaches to cell proliferation. *Cell Biophys., 4,* 1.

Nicolini, C. (1983). Chromatin structure: from nuclei to genes. *Anticancer Res., 3,* 63–86.

Nicolini, C. (1986). *Biophysics and Cancer.* pp. 1–463. New York: Plenum.

Nicolini, C., and Belmont, A. (1983). The G1 period: two cycles of chromatin conformational changes monitored by single cell dye intercalation. *Cell Biophys., 5,* 79–94.

Nicolini, C., and Beltrame, F. (1982). Coupling of chromatin structure to cell geometry during the cell cycle: transformed versus reverse-transformed CHO. *Cell Biol. Int. Rep., 6,* 63–71.

Nicolini, C., and Kendall, F. (1977). Differential light scattering in CD spectra of chromatin: alternative corrections and inferences. *Physiol. Chem. Phys. 9,* 265.

Nicolini, C., Kendall, F., and Baserga, R. (1976). DNA structure in sheared and unsheared chromatin. *Science, 192,* 796.

Nicolini, C., Carlo, P., Martelli, A., Finollo, R., Bignone, F. A., Patrone, E., Trefiletti, V., and Brambilla, G. (1982). Viscoelastic properties of native DNA from intact nuclei of mammalian cells. *J. Mol. Biol., 161,* 155–178.

Nicolini, C., Trefiletti, V., Cavazza, B., Cuniberti, C., Patrone, E., Carlo, P., and Brambilla, G. (1983). Quaternary and quinternary structures of native chromatin DNA in liver nuclei: differential scanning calorimetry. *Science, 219,* 176.

Nicolini, C., Vernazza, G., Chiabrera, A., Maraldi, I. N., and Capitani, S. (1984a). Nuclear pores and interphase chromatin: high-resolution image analysis and freeze etching. *J. Cell Sci., 72,* 75–87.

Nicolini, C., Carlo, P., Finollo, R., Vigo, F., Cavazza, B., Ricci, E., Ledda, A., and Brambilla, G. (1984b). Phase transitions in nuclei and chromatin: is nuclear volume controlled by the chromatin or by the nuclear matrix? *Cell Biophys., 6,* 183–196.

Nicolini, C., Belmont, A. S., and Martelli, A. (1986). Critical nuclear DNA size and distribution associated with S phase initiation: peripheral location of initiation, and termination sites. *Cell Biophys., 8,* 103–117.

Nicolini, C., Diaspro, A., and Germano, P. (in press-a). Nuclear architecture, intranuclear DNA distribution and DNase sensitivity. *Cell Biophys.*

Nicolini, C., Germano, P., and Diaspro, A. (in press-b). Microcalorimetry characterization of cells and nuclei during cell proliferation. *Exp. Cell Res.*

Noll, M., Thomas, J. O., and Kornberg, R. D. (1975). Preparation of native chromatin and damage caused by shearing. *Science, 187,* 1203.

Ohno, S. (1971). Simplicity of mammalian regulatory systems inferred by single gene determination of sex phenotypes. *Nature, 234,* 134.

Olins, A. L., and Olins, D. E. (1979). Stereoelectron microscopy of the 25 nm chromatin fibers in isolated nuclei. *J. Cell Biol., 81,* 260–265.

Pardoll, D. M., Vogelstein, B., and Coffey, D. S. (1980). A fixed site of DNA replication in eukaryotic cells. *Cell, 19,* 527.

Rolton, H. A., Birnie, G. D., and Paul, J. (1977). The diversity and specificity of nuclear and polysomal poly $(A)^+$ and RNA populations in normal and MSU-transformed cells. *Cell Different., 6,* 25.

Sedat, J., and Manuelidis, L. (1978). A direct approach to the structure of eukaryotic chromosomes. *Cold Spring Harbor Symp. Quant. Biol., 42,* 331.

Spinelli, G., Albanese, I., Anello, L., Ciaccio, M., and Di Liegro, I. (1982). Chromatin structure of histones genes in sea urchin sperms and embryos. *Nucleic Acid Res. 10,* 7977–7991.

Stalder, J., Seeheck, T., Borun, R. (1978). Degradation of the DNA genes by DNase I in *Physarum polycephalum. Eur. J. Biochem., 90,* 391–396.

Wang, A. H., Gessner, R. V., van der Marel, G. A., van Boom, J. H., and Rich, A. (1985). Crystal structure of Z-DNA without an alternating purine–pyrimidine sequence. *Proc. Natl. Acad. Sci. USA, 82,* 3611–3615.

Weisbrod, S. (1982). Active chromatin. *Nature, 297,* 283.

Williams, C. A., and Ockey, C. N. (1970). Distribution of DNA replicator sites in mammalian nuclei after different methods of cell synchronization. *Exp. Cell Res., 63,* 365–372.

Wu, C., Wong, J. C., and Elgin, S. C. (1979). The chromatin structure of specific gene activity. *Cell, 16,* 807.

Weintraub, H., and Groudine, M. (1976). Chromosomal subunits in active genes have an altered conformation. *Science, 193,* 848.

Weisbrod, S., and Weintraub, H., (1979). Isolation of a subclass of nuclear proteins responsible for conferring a DNAse-I sensitive structure on globin chromatin. *Proc. Natl. Acad. Sci. USA, 76,* 630–635.

Yoshida, H., and Sagai, T. (1972). Banding pattern analysis of polymorphic karyotype in the black rat by a new differential staining technique. *Chromosoma, 37,* 387.

VI

Heterogeneity of heterochromatin of human chromosomes as demonstrated by restriction endonuclease treatment

ARVIND BABU

1. Introduction

Since various staining techniques were first applied to chromosome studies, several attempts have been made to understand chromosome banding. One of them has involved the use of various DNases to digest the chromosomes. These studies have resulted in a number of important findings. Metaphase chromosomes prepared according to conventional methanol–acetic acid fixation and air-drying techniques are susceptible to DNase digestion. When the chromosomes are exposed to micrococcal nuclease or DNase I and stained with Giemsa, the chromosomes stain faintly and uniformly (Sahasrabuddhe, Pathak, and Hsu, 1978). However, if the fixed chromosomes are treated with chromomycin A^3 (CMA), a chromomycin antibiotic that binds to GC-rich double-stranded DNA to form stable complexes, prior to digestion with pancreatic DNase and staining with Giemsa, a distinct R-banding is produced (Schweizer, 1977).

There are, however, type II restriction endonucleases, which recognize relatively short DNA base sequences and cleave at specific sites. Some of these restriction endonucleases were used almost a decade ago by Jones and his group to study chromosomal damage induced by external agents. Jones (1977) reported that human chromosomes digested by *HaeIII* showed G-like banding. Subsequently, Lima de Faria and her group (1980) used endonucleases *HaeIII* and *EcoRI* to digest the chromosomes of Indian muntjac and reported that the chromosomes treated with *HaeIII* showed a "hairy" appearance. However, not until more recent systematic studies did the importance of the restriction endonucleases become fully evident. In 1983, Miller and colleagues, using a number of restriction endonucleases, reported that treatment of human chromosomes with certain enzymes having 4- or 5-base pair (bp) recognition sequences, and subsequent staining with Giemsa, resulted in distinct differential staining patterns. The darkly stained regions of the chromosomes represent endonuclease-resistant areas, whereas the lightly stained regions correspond to areas susceptible to that particular enzyme. The staining pattern produced by each enzyme is entirely characteristic and reproducible. Interestingly, most of the endonucleases capable of inducing the differential staining produce C-like bands. These enzymes can thus be utilized to explore the nature and characteristics of the C-heterochromatic portion of the genome. Several reports have been published on the mechanism of endonuclease-induced bands and their applications. The heterogeneity in C-band regions that is obtained with the use of various restriction endonucleases is reviewed in this chapter.

2. Differential effects of restriction endonucleases on chromosomes

Among the several endonucleases used to digest metaphase chromosomes, only those enzymes with 4- or 5-bp recognition sequences induce differential Giemsa staining. It has been suggested, in general, that the reduction in Giemsa staining intensity corresponds to the extent of DNA loss from various chromosomal regions. The mechanism of the banding induced by endonucleases is discussed in section 5 of this chapter. Restriction endonucleases have at least three kinds of effects on chromosomes: Some of the enzymes have apparently no effect on metaphase chromosomes. Enzymes like *HaeIII* induce formation of G-like bands, with heavy staining of some of the C-bands. On the other hand, some endonucleases produce very characteristic C-like bands, whereas enzymes like *HinfI* not only greatly reduce the Giemsa staining of the metaphase chromosomes, but also cause negative (i.e., reduced) staining of C-band regions. The staining patterns produced by some of the enzymes have been more thoroughly studied than those of other enzymes primarily because of the distinctive nature of the bands and their potential significance for analysis of heterochromatin and the genome. For example, the bands produced by endonucleases *AluI, DdeI, HaeIII, HinfI, MboI,* and *RsaI* have been investigated more extensively and therefore are discussed in some detail in the following subsections.

2.1. *AluI* (5′ . . . AG↓CT . . . 3′)

The digestion of chromosomes with endonuclease *AluI* greatly reduces the overall Giemsa staining of human chromosomes, leaving intact only certain C-band regions (Miller et al., 1983; Mezzanotte, Ferrucci, Vanni, and Bianchi, 1983). The C-like bands produced by *AluI* are either similar in size to the corresponding C-bands or smaller. This enzyme has been used more extensively in recent chromosome studies than has any other enzyme. The *AluI*/Giemsa bands have been described in detail by Bianchi and colleagues (1985a) and Babu and Verma (1985a, 1985b, 1986a). A similar reaction producing characteristic C-like bands has been described in other mammalian species, namely mouse, muntjac, chimpanzee, gorilla, and orangutan (Kaelbling, Miller, and Miller, 1984; Babu and Verma 1985d, 1986b, 1986c; DeStefano, Romano, and Ferrucci, 1986). However, some exceptions have also been noted: In five species of *Macaca* studied by DeStefano and Ferrucci (1986), the digestion of metaphase chromosomes with *AluI* resulted in homogeneously weak staining and no differentiation of heterochromatic C-band areas.

2.2. *DdeI* (5′ . . . C↓TNAG . . . 3′)

The general effect of endonuclease *DdeI* treatment on human chromosomes resembles that of *AluI,* with extensive reduction of Giemsa staining of complete metaphase chromosomes except for some of the C-band regions (Miller et al., 1983; Bianchi et al., 1985a). A similar effect has also been reported in other primate species, namely chimpanzee, gorilla, and orangutan (Babu and Verma, 1986c).

2.3. *HaeIII* (5′ . . . GG↓CC . . . 3′)

Treatment with *HaeIII* induces the formation of G-like bands along the length of the chromosomes, and prominently stained C-bands for certain chromosomes. In the human chromosomal complement, only the major C-bands of chromosomes 1, 9, 16, and the Y are darkly stained (Miller et al., 1983). In Indian muntjac also, a small portion of C-heterochromatin located at the proximal short-arm region of the X chromosome is strongly stained; it remains endonuclease-resistant even after prolonged treatment, which induces G-like bands in the rest of the metaphase chromosomes (Babu and Verma, 1986b; Verma, Jacob, and Babu, 1986). However, typical C-bands together with G-like banding have been reported in mouse chromosomes (Kaelbling et al., 1984). In all species studied, more complete digestion of chromosomes with *HaeIII* resulted in further reduction of G-like bands, whereas the prominent C-bands were apparently unaffected.

2.4. *HinfI* (5′ . . . G↓ANTC . . . 3′)

The endonuclease *HinfI* is quite distinct in its reaction on human chromosomes. The major C-bands of human chromosomes 1, 9, 16, and the Y are either indifferently stained or even negatively stained (Miller et al., 1983; Bianchi et al., 1985a). However, certain heteromorphic regions of chromosomes 3 and 4, and some of the satellite regions of acrocentric chromosomes are darkly stained (see later section). Among other primate species studied with the use of *HinfI,* gorilla and chimpanzee show a large amount of resistant chromatin, which is mostly located at the terminal regions of short- and long-arm regions of several chromosomes. In contrast, the orangutan has very little *HinfI*-resistant chromatin (Babu and Verma, 1986c). A characteristic C-banding is induced by *HinfI* in mouse chromosomes (Kaelbling et al., 1984). In muntjac chromosomes, only a small portion of C-heterochromatin located in the proximal region of the long arm of X and Y_1, in the pericentric region of Y_1, and in a very small region of Y_2 are differentially stained (Babu and Verma, 1986b).

2.5. *MboI* (5′ . . . ↓GATC . . . 3′)

The restriction enzyme *MboI* is similar to *AluI* and *DdeI* in its reaction; it produces modified characteristic C-like bands with overall reduction of Giemsa staining in the remaining regions of the human chromosomal complement (Miller et al., 1983; Bianchi et al., 1985a). A similar reaction has been reported in other species, including chimpanzee, gorilla, orangutan, muntjac, and mouse (Babu and Verma, 1986c; Kaelbling et al., 1984).

2.6. *RsaI* (5′ . . . GT↓AC . . . 3′)

The endonuclease *RsaI* also brings about the reduction of Giemsa staining in human chromosomal regions except for certain C-bands (Miller et al., 1983; Bianchi et al., 1985a). The C-like bands are characteristic and differ from the others. Also it is interesting that the residual bands after digestion with *RsaI* – in contrast to those obtained with other C-band-inducing enzymes *AluI, DdeI,* and *MboI* – do not quite correspond to the G-band pattern. This makes chromosome identification rather confusing. A similar effect is reported in other mammalian species, such as chimpanzee, gorilla, and orangutan (Babu and Verma, 1986c).

2.7. Other endonucleases

A number of other endonucleases have been used to digest the human chromosomes; however, because they do not induce any significant reaction or staining pattern, they have very little importance for chromosome cytology. Among the endonucleases thus far reported, *CfoI, HhaI,* and *HpaII,* having 4-bp recognition sequences, and *AvaII,* having a 5-bp recognition sequence, had no detectable effect. Enzymes with recognition

Table 6.1. Banding patterns of metaphase chromosomes produced by restriction endonuclease treatment

Enzyme	Recognition sequence	Species	Type of banding	Reference
AluI	AG↓CT	human	C	1,2,3,4
		mouse	C	5
		muntjac	C	6
		orangutan	C	8,10
		gorilla	C	10
		chimpanzee	C	10
		macaca (5 species)	no C	9
AvaII	G↓G$^{A}_{T}$CC	human	no effect	1
BstNI	CC↓$^{A}_{T}$GG	human	G, C	5
		mouse	G, C	5
DdeI	C↓TNAG	human	C	1,3
		chimpanzee	C	10
		gorilla	C	10
		orangutan	C	10
EcoRI[a]	G↓AATTC	human	no effect	1
		human	G	2
EcoRII	↓CC$^{A}_{T}$GG	human	C	1
		mouse	G, C	5
HaeIII	GG↓CC	human	G, C	1,2,3
		mouse	G, C	5
		muntjac	G, C	6
		chimpanzee	G, C	10
		gorilla	G, C	10
		orangutan	G, C	10
HhaI	GCG↓C	human	no effect	1
HindIII[a]	A↓AGCTT	human	no effect	1
		human	G	2
HinfI	G↓ANTC	human	neg. C, except for Nos. 3, 4, and some satellites	1,3,7
		mouse	C	5
		muntjac	neg. C, except for Y1 and Y2	6
		chimpanzee	neg. centromeric C-bands, but many terminal C-bands	10
		gorilla	neg. centromeric C-bands, but	10

Table 6.1. *(cont.)*

Enzyme	Recognition sequence	Species	Type of banding	Reference
			many terminal C-bands	
		orangutan	neg. C-bands, except for a few regions	10
HpaII	C↓CGG	human	no effect	1
MboI	↓GATC	human	C	1,3
		mouse	C	5
		muntjac	C	6
		chimpanzee	C	10
		gorilla	C	10
		orangutan	C	10
MspI	C↓CGG	human	no effect, except for neg. staining of amplified rDNA	1
RsaI	GT↓AC	human	C	3
		chimpanzee	C	10
		gorilla	C	10
		orangutan	C	10

AvaI, BalI, BamHI, BglI, BstEII, CfoI, HaeII, HincI, KpnI, PvuI, PuvII, SacI, SmaI, SstI, and *XbaI* have recognition sequences that are at least 6 bp long and thus did not produce any banding pattern (Miller et al., 1983).
[a]Discrepancy between two reports.
Sources: (1) Miller et al., 1983; (2) Mezzanotte et al., 1983a: (3) Bianchi et al., 1985a; (4) Babu and Verma, 1985a, 1985b, 1986a; (5) Kaelbling et al., 1984; (6) Babu and Verma, 1985c, 1986b; (7) Agarwal et al., 1986; (8) DeStefano et al., 1986; (9) DeStefano and Ferrucci, 1986; (10) Babu and Verma, 1986c.

sequences of 6 bp long or longer have very little effect or may produce mild G-like bands (Table 6.1; Miller et al., 1983; Mezzanotte, Bianchi, Vanni, and Ferrucci, 1983).

3. Restriction endonuclease/Giemsa bands of human chromosomes

Those endonucleases, particularly those mentioned earlier, that produce selective staining of different C-band regions are of special interest in clinical cytogenetics. The wide spectrum of patterns can be useful in studies of human heterochromatin and associated heteromorphisms and thus deserves a separate section on applications. In investigations of heteromorphic regions, proper selection of the enzyme depends upon the chro-

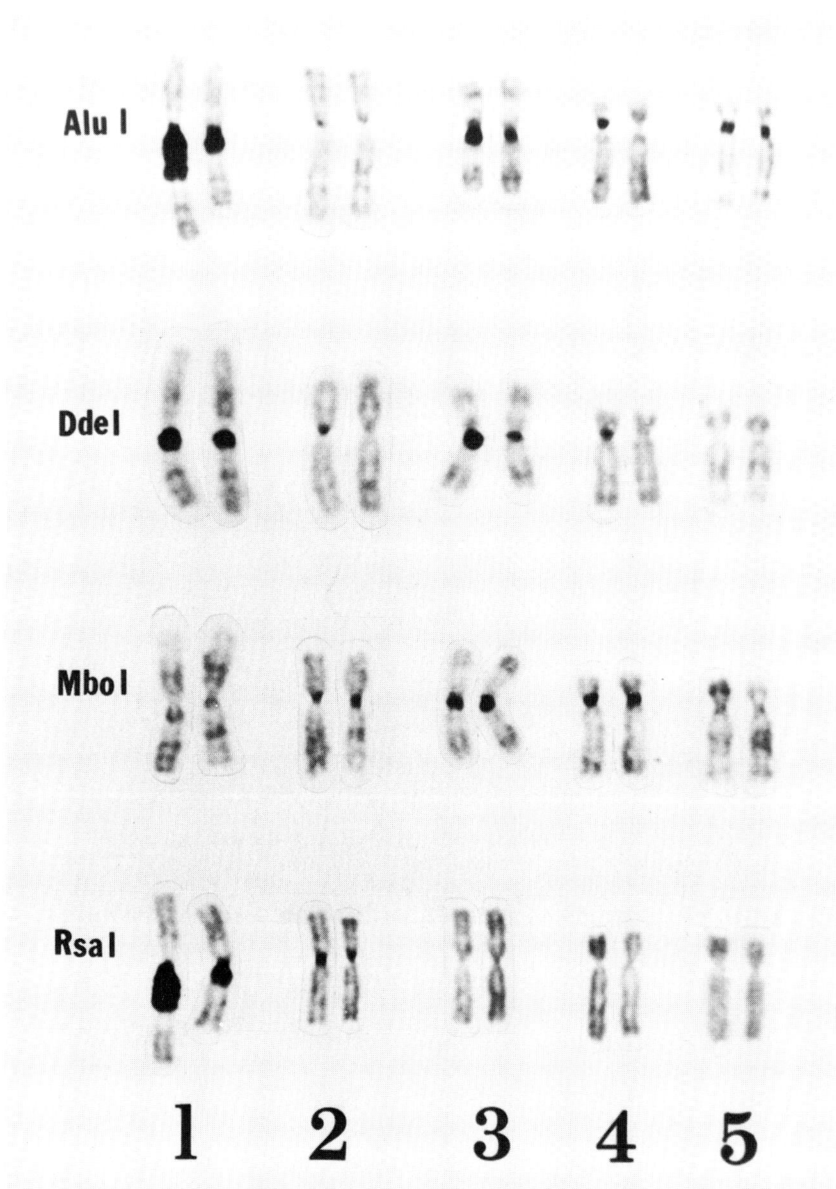

Figure 6.1. Giemsa staining patterns induced by restriction endonucleases *AluI*, *DdeI*, *MboI*, and *RsaI* in human chromosomes 1–5.

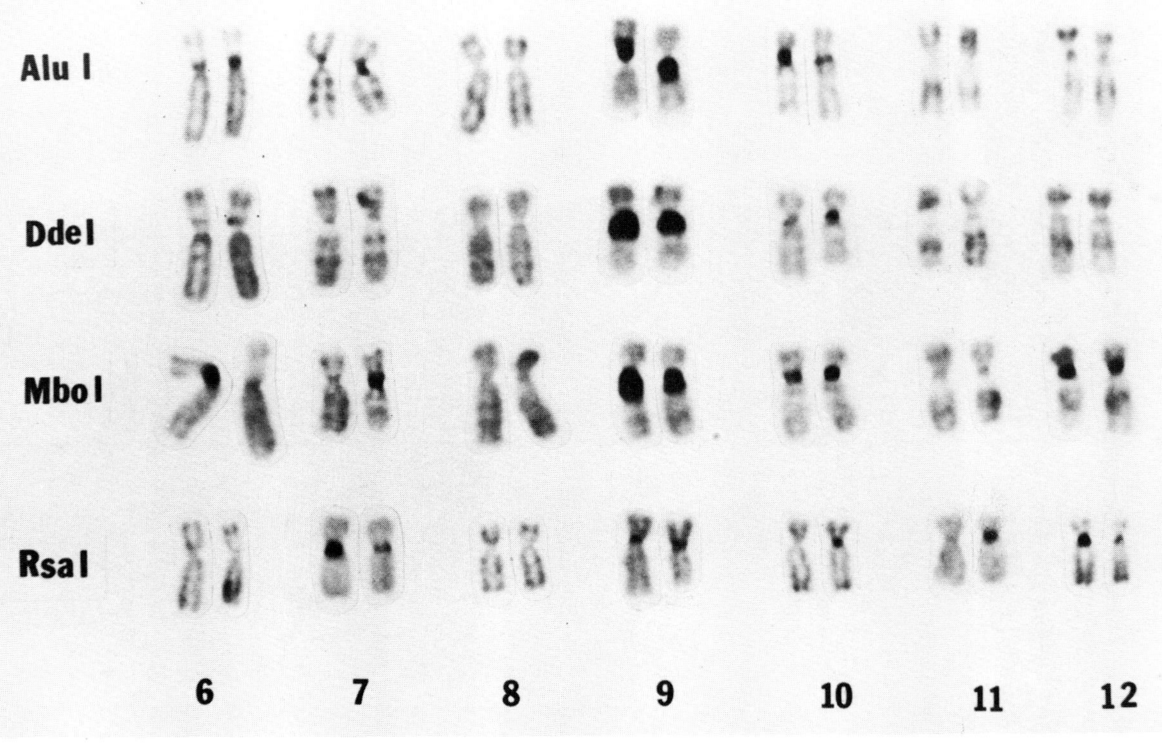

Figure 6.2. Staining profiles induced by various endonucleases in chromosomes 6–12.

mosome being analyzed. This section reviews the basic staining patterns of human chromosomes produced by these enzymes and serves as a quick reference. The staining patterns induced by endonucleases *AluI, DdeI, MboI,* and *RsaI* in each human chromosome are illustrated in composite Figures 6.1 to 6.4. The bands produced by *HaeIII* and *HinfI* are shown in Figures 6.5 and 6.6, respectively. The staining patterns induced by these endonucleases are described in Table 6.2. There are some minor discrepancies in staining profiles of some of the chromosomes [compare Bianchi et al. (1985a) and Babu and Verma (1986a; in press-a)]. These differences could probably be due to the inherent heteromorphisms and also to the subtle nature of the bands in those chromosomes.

1. *Chromosome 1.* The pericentric heterochromatin of chromosome 1 is resistant and darkly stained following treatment with endonucleases *AluI, DdeI, HaeIII,* and *RsaI* but not with *MboI* and *HinfI.* The bands produced by *AluI* and *RsaI* treatments are similar to CBG bands, whereas the bands induced by *DdeI* and *HaeIII* are slightly smaller than the C-bands because the region at the primary constriction is not stained (Miller et al., 1983; Mezzanotte et al., 1983a; Babu and Verma, 1984, 1986a, 1986d; Bianchi et al., 1985a).
2. *Chromosome 2.* The centromeric heterochromatin of chromosome 2 is stained by *MboI* and *RsaI;* the bands produced are apparently comparable to CBG bands. There is no remarkable staining by *HaeIII* and *HinfI.* In some of the chromosomes, the enzymes *AluI* and *DdeI* induced very fine bands which were much smaller than the CBG bands or *MboI*/Giemsa and *RsaI*/Giemsa bands. These bands most likely represent the heterogeneous region of chromatin and are frequently heteromorphic (Figure 6.1).
3. *Chromosome 3.* Most endonuclease – that is, *AluI, DdeI, HinfI,* and *MboI,* but not *HaeIII* and *RsaI* – induce characteristic staining of the pericentric region of chromosome 3. The bands produced by *AluI, DdeI,* and *HinfI* are smaller than the corresponding CBG bands and are significantly heteromorphic. They are comparable to the brightly fluorescent region obtained with QFQ (Bianchi et al., 1985a; Babu and Verma, 1986a). However, the relation between the *MboI*/Giemsa bands and other bands has not been adequately reported.
4. *Chromosome 4.* The pericentric heterochromatin, in general, is consistently resistant to *MboI,* and the resultant bands are similar to CBG bands. The enzymes *AluI, DdeI,* and *HinfI* induce small bands only in certain chromosomes; these bands are smaller than CBG bands and are usually located at the proximal short arm (Figures 6.1 and 6.6). These bands represent a heterogeneous subset of chroma-

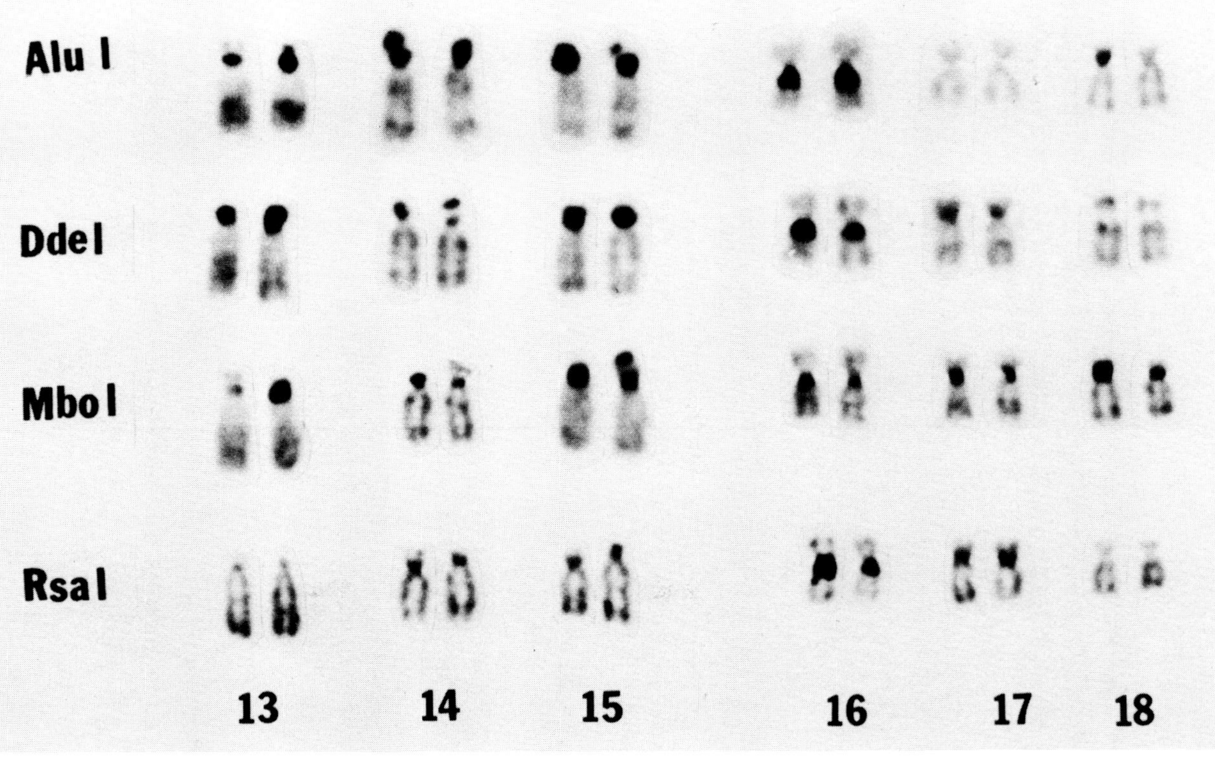

Figure 6.3. Restriction endonuclease/Giemsa bands in chromosomes 13–18.

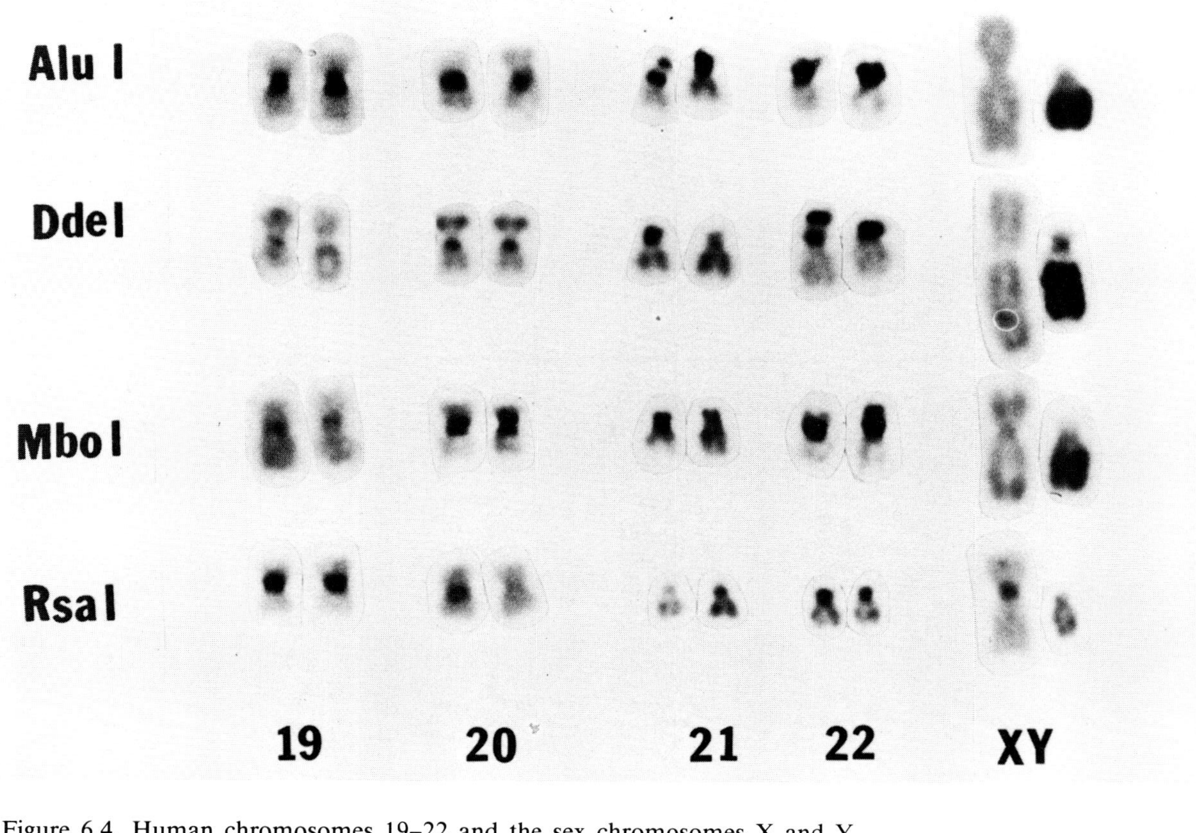

Figure 6.4. Human chromosomes 19–22 and the sex chromosomes X and Y, showing the characteristic Giemsa bands induced by *AluI, DdeI, MboI,* and *RsaI.*

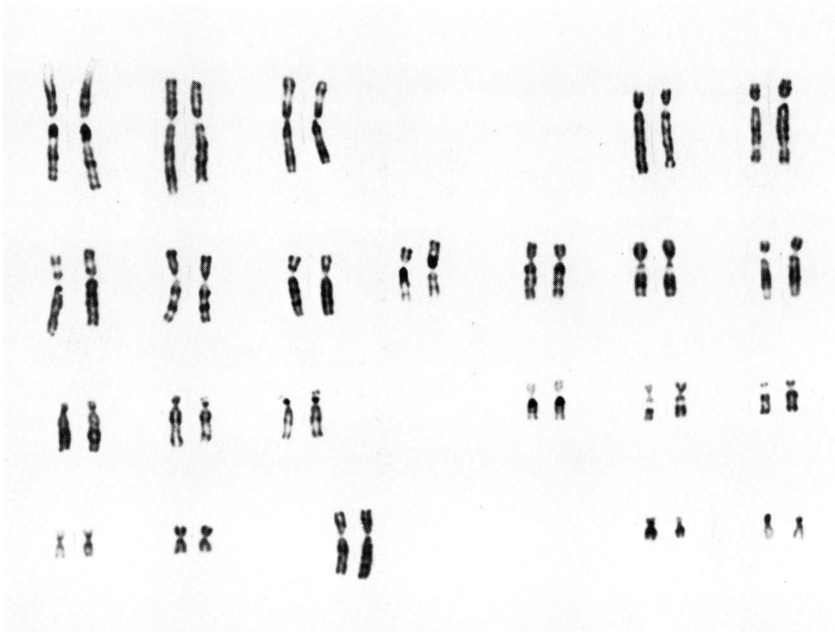

Figure 6.5. Karyotype of human chromosomes treated with restriction endonuclease *HaeIII* and stained by Giemsa.

tin. There is significant correspondence between heteromorphisms marked by QFQ and *AluI* treatments (Babu and Verma, 1986e).

5. *Chromosome 5.* The pericentric heterochromatin of chromosome 5 is stained only after treating with *AluI* and not with any other enzyme. The *AluI*/Giemsa bands are comparable to CBG bands (Babu and Verma, 1986a, in press-a).
6. *Chromosome 6.* Chromosome 6 shows no bands following treatment with *DdeI, HaeIII, HinfI,* or *RsaI.* The endonucleases *AluI* and *MboI* produce small bands in some of the chromosomes, usually in the proximal short-arm region. These bands are heteromorphic and differ between homologous chromosomes (Figure 6.2).
7. *Chromosome 7.* There are consistent small bands induced by *AluI* at the centromeric region. The enzymes *MboI* and *RsaI* induce bands of varying sizes in some chromosomes 7. These bands seem to be heteromorphic, but no definite relationship between these bands and CBG bands has been established (Figure 6.2). No bands are detected after *DdeI* and *HaeIII* treatments.
8. *Chromosome 8.* Bianchi et al. (1985a) have reported that for all individuals, chromosome 8 was shown to have an *MboI*-resistant peri-

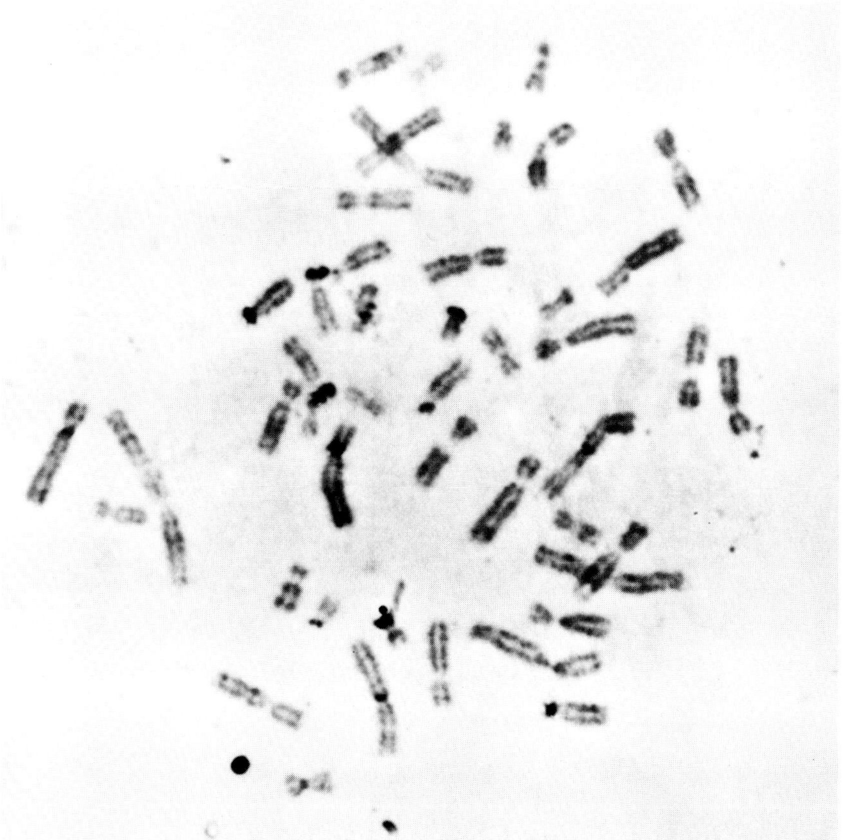

Figure 6.6. Human metaphase chromosomes treated with restriction enzyme *HinfI* and stained by Giemsa.

centric region; only in one individual were bands obtained with *RsaI* treatment. However, our data consistently showed that chromosome 8 gave no differential staining with any of the enzymes used (see Figure 6.2; Agarwal, Babu, and Verma, 1986; Babu and Verma, 1987a).

9. *Chromosome 9.* Distinct pericentric bands have been obtained with *AluI, DdeI, HaeIII,* and *MboI* treatments (Bianchi et al., 1985a). However, the bands produced by *AluI* and *HaeIII* are usually smaller than the corresponding CBG bands because a small portion of C-heterochromatin at the primary constriction is found to be endonuclease-susceptible and negatively stained (Babu and Verma, 1986a). On the other hand, the bands induced by *DdeI* and *MboI*

Table 6.2. The banding patterns induced by endonucleases in human chromosomes

Chromosome No.	*AluI* AG↓CT	*DdeI* C↓TNAG	*HaeIII* GG↓CC	*HinfI* G↓ANTC	*MboI* ↓GATC	*RsaI* GT↓AC
1	+	+	+	−	−	+
2	−	−	−	−	+[a]	+
3	+[b]	+	−	+[b]	+	−
4	+[b]	−	−	+[b]	+[a]	−
5	+	−	−	−	+	+[a]
6	+[a]	−	−	−	+[a]	−
7	+	−	−	−	+	+
8	−	−	−	−	−[c]	−
9	+	+	+	−	+	+
10	+[c]	−	−	−	+	+[a]
11	−	−	−	−	−	+
12	−	−	−	−	+	+[c]
13	+[a]	+	+[a]	−	+	+[a]
14	+[a]	+	+[a]	−	+	+[a]
15	+	+	+	−	+	+
16	+	+	+	−	+	+
17	+	−	−	−	+[a]	+
18	+[b]	−	−	−	+	−
19	+[c]	−	−	−	+[a]	+
20	+[a]	−	−	−	+	+
21	+[a]	+	+[a]	−	+	+[a]
22	+[a]	−	+[a]	−	+	+[a]
X	−	−	−	−	−	−[a]
Y	+	+	+	−	+	−[a]

Note: Plus signs (+) indicate endonuclease-resistant bands stained darkly by Giemsa; minus signs (−) indicate no band.
[a]Bands varying in their intensity between homologues.
[b]Bands are present only in heteromorphic chromosomes.
[c]Disparity exists in the literature.
Sources: Bianchi et al., 1985a,b; Babu and Verma, 1986a,b.

seem to be generally similar to CBG bands. Interestingly, with *RsaI*/Giemsa, chromosome 9 yields only a small band which is much smaller than the CBG band at the primary constriction (see Figure 6.2; Bianchi et al., 1985a; Babu and Verma, 1987b). No bands, or even negative staining, is seen at the pericentric regions after *HinfI* treatment.

10. *Chromosome 10.* Consistent pericentric bands are found with *AluI* and *MboI* treatments, whereas small bands are seen in some chromosomes after *DdeI* and *RsaI* treatments. The bands produced by the two latter enzymes may represent heteromorphisms.

11. *Chromosome 11.* Chromosome 11 showed no bands with any of the enzymes used except for *RsaI* (Bianchi et al., 1985a). However, the *RsaI*-resistant bands are absent in some individual chromosomes; these may represent heteromorphisms (Babu and Verma, unpublished results).
12. *Chromosome 12.* The pericentric heterochromatin of chromosome 12 is stained following *MboI* and *RsaI* but not after other enzymes. Among these two enzymes, the bands produced by *RsaI* seem to show greater heteromorphism (Figure 6.2).
13. *Chromosomes 13–15.* With all endonucleases studied, the acrocentric chromosomes of the D group show great variation in the staining of the centromeric heterochromatin, the short arm, and the satellites (Figures 6.3, 6.5, and 6.6). Although it is difficult to generalize a single pattern from the results obtained with *AluI, DdeI, MboI,* and *HaeIII* treatments, the observations may be summarized as follows: The pericentric regions and the satellites, where present, of all the chromosomes tend to be heavily stained following *AluI* treatment. The least amount of staining is obtained after *RsaI.* There is extensive heteromorphism, even within one individual. The bands in chromosome 14 are usually less intensely stained than those in the other chromosomes of this group. Interestingly, with *HinfI,* the pericentric heterochromatin of these chromosomes is negatively stained, whereas some of the satellite regions are prominently displayed (Figure 6.6; see also Bianchi et al., 1985a).
14. *Chromosome 16.* The pericentric region of chromosome 16 is positively stained following all endonuclease treatments except for *HinfI.* Bands produced by *AluI* and *RsaI* treatments are comparable to the corresponding CBG bands, whereas bands obtained with *DdeI* and *HaeIII* are slightly smaller, owing to the negative staining of the primary constriction region (Babu and Verma, 1986a; Agarwal et al., 1986). It is remarkable that a very small band at the primary constriction region is obtained with *MboI. HinfI* produces either indifferent or negative staining (Bianchi et al., 1985a).
15. *Chromosome 17.* Distinct pericentric bands are observed after *MboI* and *RsaI.* The staining detected after treatments with other enzymes is unremarkable.
16. *Chromosome 18.* The pericentric heterochromatin is consistently stained by *MboI.* The bands produced by *MboI,* although heteromorphic, are larger than those produced by *AluI* and probably compare well with CBG bands. On the other hand, the bands induced by *AluI* are present only in certain chromosomes. They are much smaller and clearly represent a heterogeneous subject of the CBG band. The size and position of the *AluI*-resistant band vary among different chromosomes. These bands, when present, are frequently

located toward the proximal short-arm region and rarely situated at the proximal long-arm region (Babu and Verma, 1985a, 1986c, 1986f; Verma, Babu, and Patil, 1986b).

17. *Chromosome 19.* Treatment with *AluI* and *RsaI* produced clear pericentric bands in chromosome 19. The enzymes *DdeI* and *MboI* induced slight differential staining in some preparations, but these were considered unremarkable. Some authors do not concur with our assessment (e.g., see Bianchi et al., 1985a).
18. *Chromosome 20.* There are distinct centromeric bands following *MboI* and slightly smaller bands after *AluI.* However, after *DdeI* treatment, there are very fine, small bands which are difficult to document in some of the preparations. Similarly, with *RsaI,* only minor differential staining is noted in some chromosomes (see also Bianchi et al., 1985a).
19. *Chromosomes 21–22.* The G-group chromosomes, like those of the D group, show considerable variation in staining obtained with different enzymes. The general comments described for the D-group chromosomes also apply to those of the G group. Chromosome 22 tends to stain more intensely than chromosome 21 following all enzyme treatments (Figure 6.4).
20. *X chromosome.* No differential staining at the pericentric region of the X chromosome has been reported to follow any of the enzyme treatments (Bianchi et al., 1985a; Babu and Verma, 1987a). In our unpublished findings, however, the X chromosome in some individuals showed some differential staining at the centromeric region after *RsaI* (Figure 6.4).
21. *Y chromosome.* In the Y chromosome, the distal long-arm heterochromatic region is heavily stained whereas the centromeric region shows only slight differential staining following treatments with *AluI, DdeI, HaeIII,* and *MboI.* In contrast, the long arm does not show any dark staining after *HinfI* and *RsaI* treatments.

4. Heterogeneity of human C-heterochromatin

The pericentric constitutive heterochromatin of human chromosomes is uniformly stained by the C-banding (CBG) technique (Arrighi and Hsu, 1971). A variation of Giemsa staining at alkaline pH 11.0 (i.e., the *G-11 technique*) stains only certain chromosomal C-band regions. The differences in the bands produced by the two techniques reflect cytochemical differences in the chromatin of C-bands (Bobrow, Madan, and Pearson, 1972). Significant heterogeneity has been reported in chromosomes 1 and 9 with the use of G-11 and distamycin A plus 4,6-diamidino-2-phenylindole (DA/DAPI) techniques in conjunction with C-banding (Magenis,

Donlon, and Wyandt, 1978; Donlon and Magenis, 1981; Buys et al., 1981). An even higher degree of heterogeneity has been reported in the C-bands of certain chromosomes in comparative analyses of the bands produced by restriction endonucleases and previously known techniques.

Sequential staining by G-11 and CBG has demonstrated two subsets of heterochromatin, G-11-negative and G-11-positive C-band regions, in chromosome 1 (Magenis et al., 1978). However, several studies of Babu and Verma (1984, 1986a, 1986d), using various selective staining methods (i.e., C, G-11, and DA/DAPI) together with restriction endonucleases *AluI* and *HaeIII,* have revealed that the C-band heterochromatin of human chromosome 1 consists of at least three kinds of cytochemically different heterochromatin. They are (1) the primary constriction region of the C-band, which is positively stained by *AluI*/Giemsa but negatively stained by others; (2) the proximal region of the long arm, which is stained by *AluI*/Giemsa, *HaeIII*/Giemsa, and DA/DAPI, but not by G-11; and (3) the C-band region adjacent to the euchromatin of the long arm, which is positively stained by all of the methods. Although this pattern seems the most prevalent for chromosome 1, there exist individual variations of each of these chromatin types with respect to their relative positions or the sequence of arrangement (Babu and Verma, 1986a, 1986d).

Similar heterogeneity in major C-bands of chromosomes 9 and 16, and certain minor C-bands has been described with the use of multiple banding techniques (Babu and Verma, 1986a). The heterogeneity observed in chromosome 9 with restriction endonuclease treatment is compatible with that reported earlier with the use of sequential DA/DAPI and C-banding techniques (Buys et al., 1981). The primary constriction region of the C-band in chromosome 9 is negative after *AluI*/Giemsa, *HaeIII*/Giemsa, and G-11 treatments, and negative or faintly fluorescent after DA/DAPI, whereas the proximal long-arm region of C-band is positive by all these methods (Donlon and Magenis, 1981; Buys et al., 1981; Babu and Verma, 1985a). The C-band region of chromosome 16 also exhibits two distinct subsets of chromatin (Babu and Verma, 1986a).

A highly significant contribution of the endonuclease/Giemsa technique has been its ability to demonstrate the heterogeneity of several minor C-bands in, for example, chromosome 3, 4, 5, 7, 17, and 18 (Babu and Verma, 1985b, 1985c, 1986d, 1986e, 1986f). Some of these have been more elaborately studied than others because of their clinical implications. One such heterogeneity is found in chromosome 4. The pericentric heterochromatin of chromosome 4 has been shown to be variable by Q-banding but is rarely so in C-banded preparations (McKenzie and Lubs, 1975). Another heteromorphic variant of chromosome 4 was reported; an additional G-band-positive segment in the short arm proximal to the centromere was noted by Docherty and Bowser-Riley (1984). This heteromorphism may extend to the proximal region of the short arm (Bardhan, Singh, and Davis, 1981). These variants of chromosome 4 could be better evaluated with the use of the *AluI*/Giemsa technique. Usually, chromosome 4 does not exhibit a positive band at the pericentromeric region. However, heteromorphic chromosomes with noticeably larger C-bands have a chromatin fraction that fluoresces brightly when the Q-banding technique is used. The intensity of quinacrine fluorescence varies, depending on the amount of this particular kind of chromatin; in some instances, where the amount is small, the fluorescence is obliterated. This variant C-band fraction is more discretely stained with Giemsa after *AluI* treatment (Babu and Verma, 1986e). The study indicates that although the C-band of chromosome 4 usually consists of a single kind of chromatin, it may, more frequently than previously thought, also contain a distinctly heterogeneous chromatin toward the proximal short arm.

A very interesting heteromorphism due to the presence of *AluI*/Giemsa-positive heterogeneous chromatin is reported in the C-band of chromosome 18 (Babu and Verma, 1985c, 1986f; Verma, Babu, and Patil, 1986b; Babu et al., in press). In chromosome 18, in general, the *AluI*/Giemsa bands are much smaller than the corresponding C-bands. Thus the *AluI*/Giemsa band represents a small portion of the C-band and is frequently located at the proximal short-arm region. However, other variants such as no *AluI*/Giemsa band, or band location toward the proximal long-arm region, are not infrequent. Therefore, it is evident that the C-band of chromosome 18 usually contains two heterogeneous types of chromatin. Furthermore, it is worth noting that the pericentric band in chromosome 18 produced by application of the *MboI*/Giemsa technique is consistently larger than that obtained with *AluI*/Giemsa and is more comparable to the CBG-bands (Bianchi et al., 1985a). Results suggest that the bands induced by *AluI*/Giemsa can be used as chromosomal markers to designate the parental origin in cases of trisomy 18, which is discussed at greater length in section 7.

5. Mechanism of restriction endonuclease/Giemsa banding

Metaphase chromosomes fixed in methanol–acetic acid and spread on a glass surface are faintly stained with Giemsa when they are treated with DNases prior to staining. This suggests that the DNA in metaphase chromosomes prepared according to the conventional staining procedure is susceptible to nucleases. In fact, the DNA of chromosomes treated with DNase I or micrococcal nuclease is extensively cleaved, yielding fragments of a few base pairs which are readily extracted. The result is an overall reduction of Giemsa staining (Alfi, Donnel, and Derencsenyi,

1973; Sahasrabuddhe et al., 1978). However, if the chromosomes are exposed to the chromomycin antibiotic, chromomycin A_3 (which stably binds GC-rich double-stranded DNA), prior to the digestion by DNase, the GC-rich DNA, predominantly located in G-negative regions, is protected whereas the remaining DNA is cleaved and removed, resulting in the appearance of R-bands on subsequent staining (Schweizer, Ambros, and Andrele, 1978). A similar mechanism has been suggested by Miller et al. (1983) for the production of various patterns of staining by restriction endonucleases. The basic difference between DNase I and restriction endonuclease action is that the latter enzymes have short recognition sequences and cleave the DNA at specific sites, producing relatively larger fragments of DNA as compared to DNase I. However, the chromosomal regions containing DNA with a large number of recognition sites for a particular endonuclease, upon digestion with that enzyme, would be cut into short extractable DNA fragments, resulting in the loss of Giemsa staining. For example, the human DNA has many *AluI* sites; thus when human chromosomes are exposed to this enzyme, extensive loss of Giemsa staining is observed, except in certain C-bands. The endonuclease *HaeIII,* which cuts at GGCC, produces a G-like banding pattern, reducing Giemsa staining of R-band regions, which are likely to have abundant *HaeIII* sites. Moreover, it is believed that the DNA in the entire human chromosome complement is available to enzyme digestion because every region is susceptible and is subjected to the reduction of Giemsa stain by at least one of the enzymes (Miller et al., 1983).

Furthermore, it has been estimated that DNA fragments longer than 1000 bp (1 kbp) remain in the chromatin whereas smaller fragments of about 100 bp long are extracted from the chromosomes. These rough estimates are based on the following observations: Double digestion with *HaeIII* plus *MspI,* which reduces the Y-specific 2.1-kbp DNA into fragments of 0.95 and 1.15 kbp in length, did not extract DNA from the distal long arm of Y (Miller et al., 1983). Nevertheless, the endonuclease *HinfI,* which cuts satellites II, III, and IV DNAs into smaller fragments usually about 80 bp long results in extensive reduction of the Giemsa staining of most C-bands (Miller et al., 1983). Studies on mouse chromosomes have provided further evidence of the appropriateness of these estimates. The major-density satellite DNA of *Mus musculus* consists of approximately 1 million copies of 234-bp-long tandem repeats, most of which contain a recognition sequence for endonuclease *BstNI* (CCA↓TGG). Mouse chromosomes treated with *BstNI* show G-banding and reduced C-bands. When *BstNI*-treated mouse chromosomes are hybridized to the radiolabeled cloned 200-bp satellite segment PMR 196, there is a corresponding reduction in hybridization to the C-band regions. These results suggest that short DNA fragments up to about 200 bp in length may be extracted (Miller, Gosden, Hastie, and Evans, 1984).

The study of Bianchi et al. (1985a) on the extraction of radioactivity from labeled chromosomes by restriction endonuclease digestion has also demonstrated that enzymes with a high frequency of recognition sites in the DNA produce shorter DNA fragments and a greater loss of DNA; the loss of DNA is directly associated with decreased staining. Moreover, if endonuclease-treated chromosomes are subsequently exposed to proteinase K, there is further loss of DNA, which probably corresponds to longer fragments normally retained in the matrix of the chromosomal proteins but now released upon protein degradation. Similar findings have been reported by Mezzanotte and colleagues (1985), who compared endonuclease-induced losses from treated metaphase chromosomes and nuclei. The endonuclease *AluI,* which induces C-like bands, removed more than half the DNA, whereas *HaeIII,* which produces G- and C-like bands, removed one-quarter to less than one-half of the DNA. These studies provide substantial evidence that DNA fragments of several hundred base pairs are readily extracted from fixed metaphase chromosomes and differential extraction of the DNA is the major cause of endonuclease-induced bands.

There is, however, an alternative suggestion regarding the mechanism of banding induced by restriction endonucleases. Mezzanotte et al. (1983b, 1985) have reported that the endonucleases *EcoRI, HpaII, HindIII,* and *BstNI,* which removed negligible amounts of DNA, also caused G-like banding. This led these authors to suggest that the structural organization of chromatin, as well as the frequency of restriction sites and the amount of DNA extracted, may play an important role in chromosome banding. Importantly, the findings reported by Mezzanotte et al. (1983b, 1985) and Miller et al. (1983) are inconsistent and thus need to be explored further.

It is patently clear that chromatin condensation is *not* a factor in the endonuclease-induced bands. Prematurely condensed G_1 chromosomes treated with *AluI, HaeIII,* or *MboI* showed clear bands, and the major bands that could be recognized compared well with the patterns seen in metaphase chromosomes. Also, treated interphase nuclei showed clear chromocenters, although they varied from cell to cell in phytohemagglutinin-stimulated lymphocytes (Bianchi et al., 1985a). Furthermore, the C-band regions of chromosomes 1, 9, 16, and Y, which remain uncondensed by experimental treatment with 5-azacytidine, showed staining similar to that seen in untreated metaphase chromosomes when restriction endonucleases *AluI, DdeI, HaeIII, MboI,* and *RsaI* were employed (Babu and Verma, 1986g). These observations suggest that endonuclease banding is independent of chromosome condensation and cell-cycle stage.

6. Comments and technical notes

Staining methods using restriction endonucleases are relatively simpler and more rapid, as compared to many of the other available techniques. The restriction enzymes are usually supplied as concentrated enzyme units in small quantities of their respective buffer solutions. These enzymes are diluted to the necessary working concentration (usually 100–200 U/ml or as required) with standard assay buffer specific for each buffer. The composition of assay buffer for each enzyme is also suggested by the enzyme manufacturers; this is useful as a ready reference. The buffers are made to provide the required pH at 37°C. The descriptive methodology for use of two of the enzymes, *AluI* and *HaeIII,* has been described by Babu and Verma (1985a). Conventionally prepared air-dried chromosomal material, preferably not older than one week or 10 days, is treated with the enzyme at 37°C. The working solution is placed on a slide, overlaid with a coverglass, and incubated in a humidified chamber for the required time, which usually ranges from two to four hours, depending on the concentration of enzyme in the assay buffer. Following the treatment, the slides are rinsed briefly in distilled water and stained with Giemsa.

Treatment of chromosomal material with endonucleases between the second and fifth day after preparation yields the best differentiation of staining and preserves ideal chromosome morphology. Even freshly prepared material can be used without further delay. However, finer details of the bands may not be of acceptable quality; furthermore, the chromosomes may acquire a ghostlike appearance when treatment is only slightly excessive. Material treated with endonucleases after the second day following preparation have a much higher resistance to endonuclease treatment and tolerate a wider range of treatment durations with excellent retention of morphology.

The residual Giemsa bands obtained after endonuclease digestion are usually adequate for chromosome identification. However, if necessary, one can photograph the metaphase chromosomes by the QFQ technique prior to treatment with enzyme, but exposure to the light source should be kept to a minimum. Prolonged exposure may interfere with or even block endonuclease action, probably as a result of radiation-induced DNA–protein cross-links (Bianchi, Bianchi, and Cleaver, 1984; Bianchi, Morgan, and Cleaver, 1985). It should be mentioned that prior QFQ banding affects each enzyme treatment differently. For example, the distal long-arm region of the Y chromosome and the pericentric region of the X chromosome tend to remain darker than control slides treated without prior QFQ (Babu and Verma, unpublished). Preparations that are treated for prolonged periods, if not stained to the required intensity by Giemsa, can, in most cases, be observed and photographed by use of phase-contrast optics. Furthermore, different batches of enzyme may vary in their activity; thus it is advisable to run several samples, subjecting them to various periods of treatment or various concentrations of enzyme.

7. Applications of restriction endonuclease/Giemsa banding in clinical cytogenetics

A number of selective staining techniques such as G-11, DA/DAPI, Hoechst/netropsin, and D287/170 (a chemical variant of DAPI) have recently been developed to stain specifically a variety of chromosomal regions (Bobrow et al., 1972; Schweizer et al., 1978; Sahar and Latt, 1980; Schnedl et al., 1981). These methods have had unequivocally impressive impact in the field of cytogenetics and have greatly contributed chromosome studies in both clinical and basic research. Newer techniques utilizing restriction endonucleases and Giemsa undoubtedly have provided additional advantages over those already in practice, and will, as probing tools, facilitate further exploration of chromosomes.

The potential importance of restriction endonucleases for this field has already been demonstrated in our previous reports. The restriction endonuclease/Giemsa techniques are not only powerful; they are also much simpler to use and are less time consuming than many other techniques. These methods can be utilized for fresh slides with no requirement for aging or storing, and slides are usually ready for microscopic observation in a few hours. This feature has stimulated interest in the use of these techniques, especially in prenatal diagnosis, for rapid identification of unusual heteromorphisms (Rodriguez, Babu, and Verma, 1986). Furthermore, preparations made by applying these techniques are permanent, and better resolution is achieved with very little distortion in chromosome morphology.

The heteromorphic sites of human chromosomes have been studied extensively with the use of C- and Q-banding techniques. Some of the heteromorphisms, especially those of smaller magnitude, could not be delineated by C-banding, but could be demonstrated by Q-banding in rare instances. One such example is human chromosome 4, which is variable in Q-banded preparations (McKenzie and Lubs, 1975; Bardhan et al., 1981). G-banding has been useful only when the heteromorphic segment is large enough to be discerned as a distinct G-positive band (Docherty and Bowser-Riley, 1984). However, application of the *AluI*/Giemsa technique has highlighted even very minor heteromorphisms which could easily be missed by Q-banding (Babu and Verma, 1986e). Although no systematic study is available, other endonucleases such as *DdeI* and *HinfI* may also be useful in revealing the heteromorphisms of chromosome 4 (Figure 6.1). Another significant heteromorphism demonstrated by using *Alui*/Giemsa has been reported by Babu and Verma (1985c).

Chromosome 18 generally consists of a relatively small pericentric C-band, and thus CBG staining is apparently inadequate to reveal heteromorphisms. Nonetheless, the bands obtained when this chromosome is digested with *AluI* are much smaller than the corresponding C-bands, indicating that only a small portion of the C-band is resistant. This particular band is remarkably variable in size and position, or may even be totally absent. This heteromorphic subset of C-band chromatin can be adopted as a chromosomal marker to identify the parental origin of the additional chromosome in Edwards' syndrome (Babu and Verma, 1986f, Babu and Verma, in press-b).

Restriction endonucleases that produce C-like bands have revealed a number of heterogeneous regions that demonstrate the distinct cytological profiles of heterochromatin. Some of these characteristics have been used to correlate chromatin of different chromosomes that may contain similar DNA (Babu and Verma, 1986a). Enzymes that induce specific bands for regions containing repetitive DNA can be utilized to recognize such areas of chromosomes, in preference to elaborate methods like in situ hybridization (Miller et al., 1983). Furthermore, the characteristic staining patterns are useful in identifying the origin of some of the rearranged marker chromosomes in cancer cytogenetics, and the frequent heteromorphisms are utilized to discriminate donor from recipient cells in bone marrow transplants (Babu, Verma, and Macera, 1987).

8. References

Agarwal, A. K., Babu, A., and Verma, R. S. (1986). Expression of 'rare' heteromorphisms of the human genome by restriction endonucleases. *Am. J. Hum. Genet., 39,* A185.

Alfi, O. S., Donnel, G. N., and Derencsenyi, A. (1973). C-banding of human chromosomes produced by DNase. *Lancet, 2,* 505.

Arrighi, F. E., and Hsu, T. C. (1971). Localization of heterochromatin in human chromosomes. *Cytogenetics, 10,* 81–86.

Babu, K. A., and Verma, R. S. (1984). Heterogeneity of heterochromatin in human chromosome 1. *J. Cell Biol., 99* (No. 4, Pt. 2), 248a.

Babu, K. A., and Verma, R. S. (1985a). The application of restriction endonucleases *AluI* and *HaeIII* in evaluation of heterochromatic regions in human genome. *Karyogram, 11,* 23–27.

Babu, K. A., and Verma, R. S. (1985b). The staining profiles of constitutive heterochromatin in human chromosomes. *Genetics, 110* (No. 3, Pt. 2), S3.

Babu, K. A., and Verma, R. S. (1985c). Identification of additional chromosome 18 in Edwards' syndrome using restriction endonuclease *AluI*. *Am. J. Hum. Genet., 37*(4), A43.
Babu, K. A., and Verma, R. S. (1985d). Restriction endonuclease resistant fractions of Indian muntjac *(Muntiacus muntjak)* chromosomes. *J. Cell Biol., 101*(5), 74a.
Babu, A., and Verma R. S. (1986a). Characterization of human chromosomal constitutive heterochromatin. *Can. J. Genet., Cytol. 28,* 631–645.
Babu, A., and Verma, R. S. (1986b). Cytochemical nature of constitutive heterochromatin of Indian muntjac *(Muntiacus muntjak). Cytogenet. Cell Genet., 41,* 96–100.
Babu, A., and Verma, R. S. (1986c). Characterization of heterochromatin in the chromosomes of Hominidae by restriction endonucleases *AluI, DdeI, HaeIII, HinfI, MboI,* and *RsaI. Am. J. Hum. Genet., 39,* A186.
Babu, A., and Verma, R. S. (1986d). Cytochemical heterogeneity of C-band in human chromosome 1. *Histochem. J., 18,* 329–333.
Babu, A., and Verma, R. S. (1986e). Heteromorphic variants of chromosome 4. *Cytogenet. Cell Genet., 41,* 60–61.
Babu, A., and Verma, R. S. (1986f). The heteromorphic marker on chromosome 18. *Am. J. Hum. Genet., 38,* 549–554.
Babu, A., and Verma, R. S. (1986g). The role of chromatin condensation in restriction endonuclease induced bands in human metaphase chromosomes. *Genetics, 113* (No. 1, Pt. 2), S17.
Babu, A., and Verma, R. S. (1986h). Characterization of banding patterns induced by restriction endonucleases and their application in clinical cytogenetics, Berlin, West Germany. *Abstracts,* Pt. 1, p. 190.
Babu, A., and Verma, R. S. (1987a). Chromosome structure: Euchromatin and heterochromatin. *Int. Rev. Cytol. 108,* 1-60.
Babu, A., and Verma, R. S. (1987b). Classification and incidence of pericentric heterochromatin of human chromosome 18. *Chromosoma, 95,* 163–166.
Babu, A., Verma, R. S., and Macera, M. J. (1987). Identification of marker chromosomes by restriction endonucleases/Giemsa-technique in neoplastic cells. *Cancer Genet. Cytogenet., 24,* 367–369.
Bardhan, S., Singh, D. N., and Davis, K. (1981). Polymorphism in chromosome 4. *Clin. Genet., 20,* 44–47.
Bianchi, N. O., Bianchi, M. S., and Cleaver, J. E. (1984). The action of ultraviolet light on the patterns of banding induced by restriction endonucleases in human chromosomes. *Chromosoma, 90,* 133–138.
Bianchi, M. S., Bianchi, N. O., Pantelias, G. E., and Wolff, S. (1985a). The mechanism and pattern of banding induced by restriction endonucleases in human chromosomes. *Chromosoma, 91,* 131–136.
Bianchi, N. O., Morgan, W. F., and Cleaver, J. E. (1985b). Relationship of ultraviolet light-induced DNA-protein cross-linkage to chromatin structure. *Exp. Cell Res., 156,* 405–418.
Bobrow, M., Madan, K., and Pearson, P. L. (1972). Staining of some specific regions of human chromosomes, particularly the secondary constriction of No. 9. *Nature, 238,* 122–124.
Buys, C. H. C. M., Gouw, W. L., Blenkers, J. A. M., and van Dalen, C. H. (1981). Heterogeneity of human chromosome 9 constitutive heterochromatin as revealed by sequential distamycin A/DAPI staining and C-banding. *Hum. Genet., 57,* 28–30.

DeStefano, G. F., and Ferrucci, L. (1986). New cytogenetic techniques in the study of primate genome evolution. *Hum. Genet., 72,* 98–100.
DeStefano, G. F., Romano, E., and Ferrucci, L. (1986). The *AluI*-induced bands in metaphase chromosomes of orangutan *(Pongo pygmaeus). Hum. Genet., 72,* 268–271.
Docherty, Z., and Bowser-Riley, S. M. (1984). A rare heterochromatic variant of chromosome 4. *J. Med. Genet., 21,* 470–472.
Donlon, T. A., and Magenis, R. E. (1981). Structural organization of the heterochromatic region of human chromosome 9. *Chromosoma, 84,* 353–363.
Jones, K. W. (1977). Repetitive DNA and primate evolution. In *Molecular Structure of Human Chromosomes,* ed. J. J. Yunis, pp. 295–326. New York: Academic Press.
Kaelbling, M., Miller, D. A., and Miller, O. J. (1984). Restriction enzyme banding of mouse metaphase chromosomes. *Chromosoma, 90,* 128–132.
Lima de Faria, A., Isaksson, M., and Olsson, E. (1980). Action of restriction endonucleases on the DNA and chromosomes of *Muntiacus muntjak. Hereditas, 92,* 267–273.
Magenis, R. E., Donlon, T. A., and Wyandt, H. E. (1978). Giemsa-11 staining of chromosome 1: a newly described hetermorphism. *Science, 202,* 64–65.
McKenzie, W. H., and Lubs, H. A. (1975). Human Q and C chromosomal variants: distribution and incidence. *Cytogenet. Cell Genet., 14,* 97–115.
Mezzanotte, R., Ferrucci, L., Vanni, R., and Bianchi, U. (1983a). Selective digestion of human metaphase chromosome by *AluI* restriction endonuclease. *J. Histochem. Cytochem., 31,* 553–556.
Mezzanotte, R., Bianchi, U., Vanni, R., and Ferrucci, L. (1983b). Chromatin organization and restriction nuclease activity on human metaphase chromosomes. *Cytogenet. Cell Genet., 36,* 562–566.
Mezzanotte, R., Ferrucci, L., Vanni, R., and Summner, A. T. (1985). Some factors affecting the action of restriction endonucleases on human metaphase chromosomes. *Exp. Cell Res., 161,* 247–253.
Miller, D. A., Choi, Y-C., and Miller, O. J. (1983). Chromosome localization of highly repetitive human DNA's and amplified ribosomal DNA with restriction enzymes. *Science, 219,* 395–397.
Miller, D. A., Gosden, J. R., Hastie, N. D., and Evans, H. J. (1984). Mechanism of endonuclease banding of chromosomes. *Exp. Cell Res., 155,* 294–298.
Rodriguez, J. G., Babu, A., and Verma, R. S. (1986). A rapid method of culturing bloody amniotic fluid for chromosome analysis. *Am. J. Obstet. Gynecol., 154,* 969–970.
Sahar, E., and Latt, S. A. (1980). Energy transfer and binding competition between dyes used to enhance staining differentiation in metaphase chromosomes. *Chromosoma, 79,* 1–28.
Sahasrabuddhe, C. G., Pathak, S., and Hsu, T. C. (1978). Responses of mammalian metaphase chromosomes to endonuclease digestion. *Chromosoma, 69,* 331–338.
Schnedl, W., Abraham, R., Dann, O., Geber, G., and Schweizer, D. (1981). Preferential fluorescent staining of heterochromatic regions in human chromosomes 9, 15, and the Y by D287/170. *Hum. Genet., 59,* 10–13.
Schweizer, D. (1977). R-banding produced by DNase I digestion of chromomycin-stained chromosomes. *Chromosoma, 64,* 117–124.
Schweizer, D., Ambros, P., and Andrele, M. (1978). Modification of DAPI banding on human chromosomes by prestaining with DNA binding oligopeptide antibiotic, distamycin A. *Exp. Cell Res., 111,* 327–332.

Verma, R. S., Jacob, J. P., and Babu, A. (1986a). Heterochromatin organization in the nucleus of Indian muntjac *(Muntiacus muntjak). Can. J. Genet. Cytol., 28,* 628–630.
Verma, R. S., Babu, A., and Patil, S. R. (1986b). Characterization of *AluI* resistant chromatin of human chromosome 18. *Am. J. Hum. Genet., 39,* A136.

VII
Heteromorphisms of heterochromatin

RAM S. VERMA

1. Introduction

In the broader sense, polymorphisms can be classified into two groups: chromosomal polymorphisms and genetic polymorphisms. This chapter is intended to describe only human chromosomal polymorphisms. At the Paris Conference on Standardization in Human Cytogenetics (1971), it was recommended that the term *variant* be used to describe morphological deviations of chromosome morphology. Subsequently, in the Paris Conference Supplement (1975), the term *polymorphism* was designated to replace variant. The term *heteromorphism* has been employed in the literature published since that time.

Prior to the advent of various banding techniques, it was recognized that certain homologues of human chromosomes exhibit consistent morphological differences. As a result of recent advances in staining techniques, it has become possible to recognize heteromorphic regions with great precision. Each year an increasing number of heteromorphic sites are being recognized in the genome. It is now an accepted fact that variability of constitutive heterochromatin appears to be an integral characteristic of eukaryotes. White (1954) first recognized the heteromorphic nature of heterochromatin in grasshopper chromosomes. Later, Lubs and Ruddle (1970) drew attention to the considerable range of variation in heterochromatic regions of the human genome. It is remarkable to note that every human chromosome within the genome is variable (McKenzie and Lubs, 1975). Heterochromatin from heteromorphic chromosomes can be classified by size, position, staining properties, or any combination of these criteria through the use of selective staining techniques. It is possible to subdivide further a heteromorphic segment into many subunits (Verma, Rodriguez, and Dosik, 1984). Several questions regarding human chromosomal heteromorphisms have now been resolved:

1. Heteromorphisms are consistent within a person, but are variable from person to person.
2. Variations in heteromorphisms are continuous, rather than discrete.

3. Variation in the morphology of chromosomes is due to the presence of heterochromatin, and every chromosome is variable.
4. The mode of inheritance of heteromorphisms is Mendelian, but a few exceptions have been recorded.
5. Often there is no direct relationship between a heteromorphism identified by one technique and that identified by another (Verma and Lubs, 1975a).
6. Constitutive heterochromatin contains highly redundant DNA and is apparently not transcribed.
7. Meiotic crossing over in heterochromatic regions is rare or virtually absent (Kurnit, 1979).
8. De novo rearrangements of constitutive heterochromatin occur rarely, if ever, in meiotic cells but are well documented in somatic cells.
9. The biological significance of heteromorphisms remains obscure (Jacobs, 1977; Lubs, 1977; Verma and Dosik, 1980; Erdtmann, 1982).
10. Heteromorphisms are used extensively as reliable markers in clinical medicine (Olson, Magenis, and Lovrien, 1986).
11. The molecular and structural organization of heterochromatic markers is currently being disclosed (Verma and Dosik, in press).

2. Heteromorphisms by various staining techniques

Following the inception of various banding techniques, the most interesting phenomena observed in the human genome were stain-specific heteromorphisms. These heteromorphic segments could be classified by size, position, staining intensity, or any combination of those criteria, using various staining methods. Heteromorphic classification has been based on estimation rather than actual measurements. Furthermore, despite the fact that variation is continuous rather than discrete, several attempts at classification have been made on the basis of arbitrary scales, and different levels have been assigned. Each banding technique provides a unique type of staining pattern, thus deserving a separate description.

2.1. QFQ heteromorphisms

It was the Q-banding technique using quinacrine mustard that revolutionized the field of human cytogenetics; it not only led to the correct identification of individual chromosomes, but also revealed that certain regions stain with variable intensities. These different fluorescent intensities can be classified into at least five levels, as established at the Paris Conference (1971). The brightest level was assigned a code of 5, and the

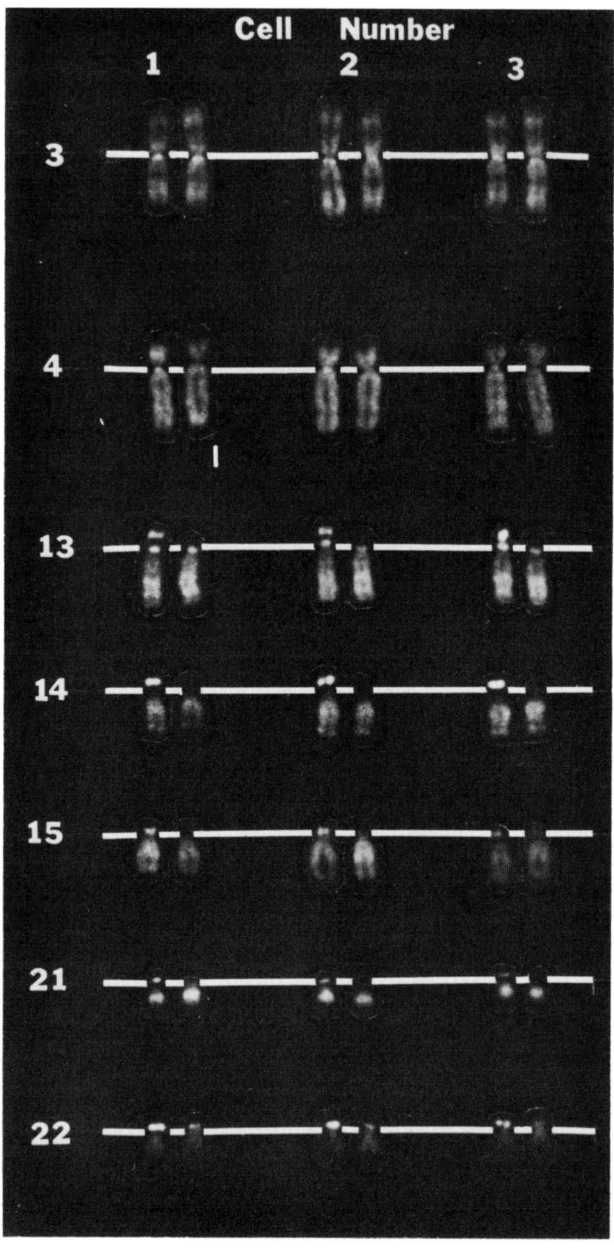

Figure 7.1. Heteromorphic regions as demonstrated by the QFQ technique.

least fluorescent level a code of 1. The bands in the centromeric regions of chromosomes 3 and 4; the short arm of acrocentric chromosomes 13–15, 21, and 22; and the long arm of Y were found to be heteromorphic by the QFQ technique (Figure 7.1).

Intensity heteromorphisms observed in the short arm of acrocentric chromosomes among various ethnic groups are significantly different (Lubs et al., 1977; Al-Nassar, Palmer, Conneally, and Yu, 1981). The centromeric heteromorphism of chromosome 3 ranges in frequency from 40% to 70% in a normal population. Nevertheless, there are wide variations among different studies reporting the frequency of QFQ heteromorphisms. Using the fluorescence intensity of the centromeric region of chromosome 3, it has become possible to identify inversions of the pericentromeric heterochromatin. Inversion heteromorphisms of chromosome 3 are familial in nature and do not have clinical consequences. The incidence of bright variable bands in the centromeric region of chromosome 4 is different from that in chromosome 3. Furthermore, the range of variation seen in chromosome 4 is broader than that in chromosome 3. Inversion heteromorphisms are not as frequently observed by QFQ technique, but the *AluI*/Giemsa procedure has revealed the existence of inversion heteromorphisms on chromosome 4 (Babu and Verma, 1986).

The distal one- to two-thirds of the long arm of the human Y chromosome is brightly fluorescent when the QFQ technique is employed and has been found to be highly variable; the genetically active material is located in the nonfluorescent segment and is considered nonvariable (Bobrow, Pearson, Pike, and El-Alfir, 1971). There is current evidence that the length of the nonfluorescent segment also accounts for racial variation. (For a review, see Verma, Dosik, Scharf, and Lubs, 1978b; Verma, Huq, and Dosik, 1983.)

2.2. CBG heteromorphisms

Another important banding technique that has been intensively used to investigate heteromorphisms involving the centromere is the C-banding technique. This technique, described by Arrighi and Hsu (1971) has allowed constitutive heterochromatin to be identified. C-banding stains the secondary constriction regions (h) of chromosomes 1, 9, and 16, the distal one- to two-thirds of chromosome Y, the centromeres of all chromosomes, and satellites of the acrocentric chromosomes. Heteromorphisms of these regions are well documented (McKenzie and Lubs, 1975), and, although these variations are continuous, they can be classified according to two categories, size and position. Patil and Lubs (1977), using the short arm of chromosome 16 (16p) as a reference, classified the

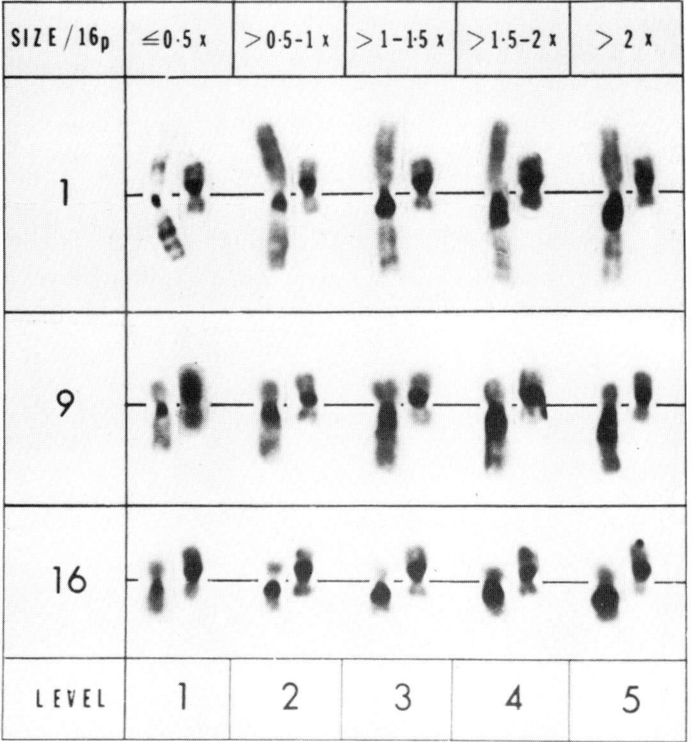

Figure 7.2. Application of the CBG technique to size classification of secondary constriction regions with 16p as a reference. (Courtesy of Dr. S. R. Patil.)

size of the h region into five groups, ranging in size from very small to very large (Figure 7.2). In addition, Verma, Dosik, and Lubs (1978) classified inversion heteromorphisms into five categories: no inversion (NI); partial inversion, minor (MIN); half inversion (HI); partial inversion, major (MAJ); and complete inversion. Recently, I compared the incidence of size heteromorphism in the h regions of chromosome 1, 9, and 16 among Caucasians, Indians, and blacks. A higher incidence of heteromorphisms for chromosome 1 was noted in Indians, whereas Caucasians and blacks exhibited similar but lower frequencies. Chromosome 9 showed a significantly higher incidence of heteromorphisms in Caucasians, whereas Indians and blacks had a similar incidence. Chromosome 16 was the least heteromorphic in all groups, and no racial variation was noted (Verma, Rodriguez, and Dosik, 1982).

East Indians and blacks showed similar frequencies of inversion heteromorphisms for chromosomes 1 and 9, whereas Caucasians demonstrated a significantly lower incidence of this type of heteromorphism. Inversions of the h region in chromosome 16 are very rare; they were not found in black or Caucasian populations, but 1.5% of the East Indian population showed chromosome 16 inversions. Many clinical parameters have been correlated with various types of heteromorphisms revealed by CBG techniques. However, no definite conclusions have been reached regarding the significance of these heteromorphisms.

Heterochromatin of the centromere and of the secondary constriction regions contains highly repetitive DNA and is heterogeneous in nature. Our knowledge of the organization of human centromeric heterochromatin is derived from studies of satellite DNA. The h regions of chromosomes 1, 9, and 16 contain large amounts of constitutive heterochromatin. The h regions of chromosome 1 contain predominantly satellite DNA II and III; chromosome 9 contains all four types (I–IV), whereas chromosome 16 contains predominantly satellite II DNA (John, 1980). Large heterochromatic regions frequently appear segmented when stained by various selective staining techniques (Donlon and Magenis, 1981; Verma, Rodriguez, and Dosik, 1984). Heterochromatic regions always loop from the major chromosome axis and appear not to be involved in pairing during meiosis (Page, 1973). The exact cause(s) of inversions of the h regions is unknown; however, an excessive amount of satellite DNA may precipitate an unusual condition that leads to inversion heteromorphisms.

2.3. RFA heteromorphisms

Heteromorphisms of the short arms of human acrocentric chromosomes have been noted through the use of acridine orange reverse banding (RFA; Verma and Lubs, 1975a). Six different color-heteromorphisms could be classified (Figure 7.3), and five separate size categories ranging from very small to very large were established (Verma and Lubs, 1975b). In one of my earlier investigations, I tried to detect any interrelationship between the heteromorphisms on human acrocentric chromosomes obtained with the QFQ techniques and those obtained with the RFA technique. Results disclosed no consistent relationship between negative or brilliant QFQ variants and the various colors observed with the RFA technique (Verma, Dosik, and Lubs, 1977). Furthermore, more variation was detected in human acrocentric chromosomes by application of the RFA technique than by any other previously described method (Verma and Dosik, 1981). These findings added a new dimension to the heteromorphisms observed among different ethnic groups. Furthermore, the numerous heteromorphic markers that have been obtained by this approach are being used extensively to examine aneuploidy in humans.

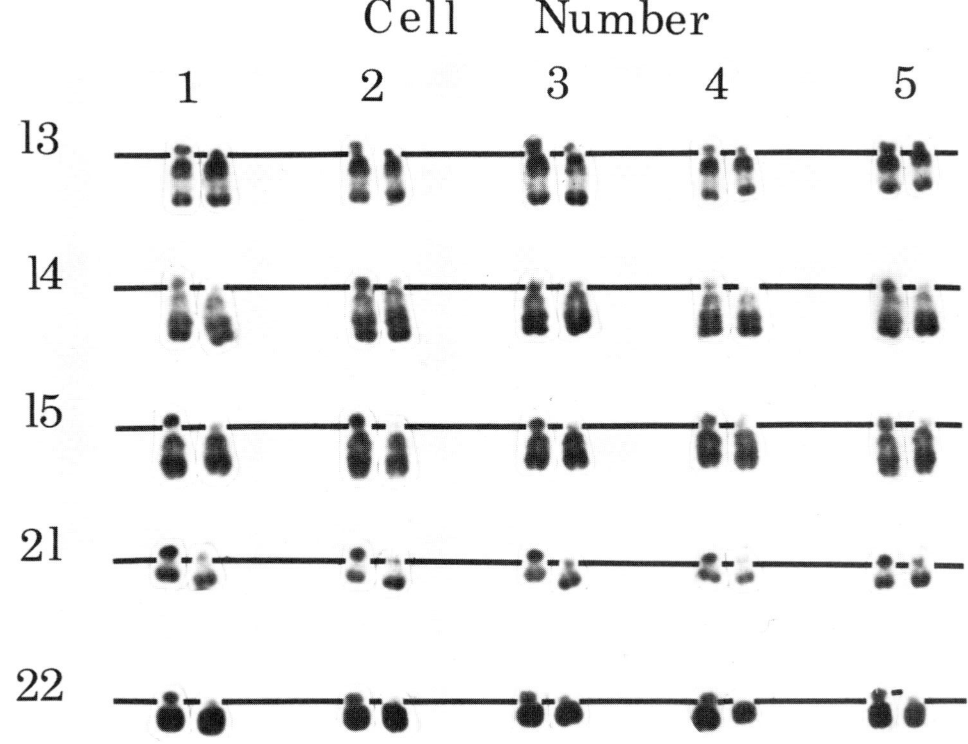

Figure 7.3. Size variation of acrocentric chromosomes demonstrated by the RFA technique.

The nature and the molecular mechanism of the different colors revealed by this technique remain unknown. However, heteromorphic regions containing various types of satellite DNA may denature differently during heat treatment, causing the acridine orange molecules to react to altered binding sites (Verma and Lubs, 1976).

2.4. NOR heteromorphisms

The silver staining technique, which is most commonly known as the NOR technique, stains only the nucleolar organizer regions (NOR) of human acrocentric chromosomes that were active in the previous cell cycle (Figure 7.4).

Morphological size variability of the NOR regions is well documented by this approach. However, not all acrocentric chromosomes show Ag-positive NORs (Varley, 1977). Furthermore, the heteromorphisms are so extreme that they can be classified into at least five sizes ranging from very small to very large (Verma, Benjamin, Rodriguez, and Dosik, 1981). A higher "incidence" of NOR heteromorphisms was obtained with silver staining than with the RFA technique. Presumably, the small or very small NORs could not be detected by RFA at the microscopic level, but because of the impregnation of silver, the appearance was clearer. Furthermore, the silver staining technique may have many other advantageous staining properties. The findings suggest that for studying racial and population heteromorphisms of NORs, the silver staining technique is far more useful than the RFA technique. Polymorphic forms of different acrocentric chromosomes have also been evaluated using base-specific fluorochromes, antibiotics, and fluorescence antibodies in conjunction with silver staining (Okamoto, Miller, Erlanger, and Miller, 1981). Furthermore, the inactive ribosomal gene may be undetected by one technique but be demonstrated by another. It is suggested that there is no relationship between the amount of rDNA per NOR and the intensity of silver staining of the NOR. Consequently, activation of rDNA at various stages occurs independently of the amount of rDNA present at the NOR site, as determined by silver staining.

Silver-stained areas of acrocentric chromosomes represent the sites of ribosomal cistrons (Babu and Verma, 1985). Presumably, the large NORs contain multiple copies of the 18S and 28S ribosomal cistrons. All 10 acrocentric chromosomes have NORs, but either because there are very few copies of rDNA present in small NORs or because NORs are very small, they may not impregnate with silver. Alternatively, not every chromosome has active NORs sites, leading to the different frequencies of heteromorphisms observed by various techniques (Babu, Macera, and Verma, 1986).

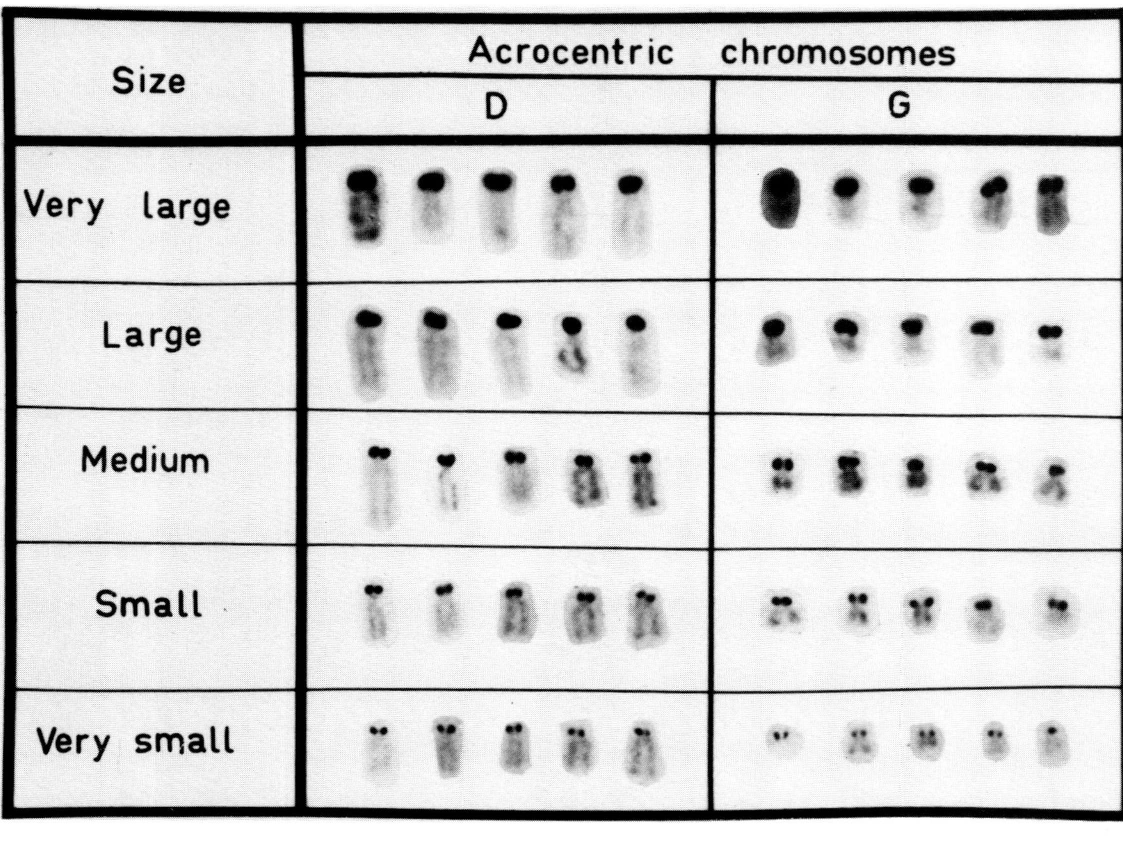

Figure 7.4. Size variation of NOR regions shown by the silver staining technique

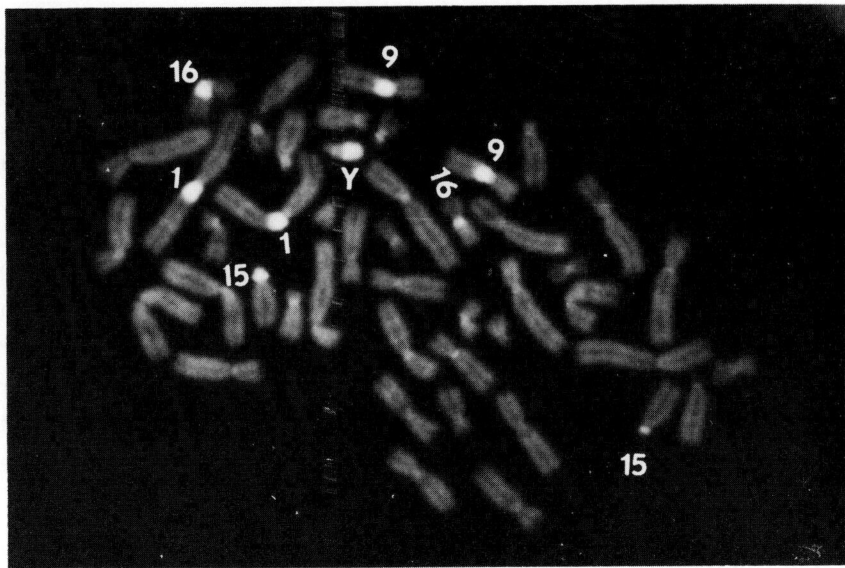

Figure 7.5. Application of the DA/DAPI technique. Intensely stained chromosomes are labeled. (Courtesy of Dr. D. Schweizer.)

2.5. Heteromorphisms by other selective staining methods

In recent years, a battery of other selective staining techniques have been used to explore the heteromorphic regions of human chromosomes. These selective techniques stain highly specific regions of the chromosomes. For example, distamycin A plus 4′,6-diamidino-2-phenylindole (DA/DAPI) (Schweizer, Ambros, and Andrle, 1978; see Figure 7.5), Hoechst/netropsin (Sahar and Latt, 1980; Schnedl et al., 1981), and methylcytosine antibodies (Miller, Schnedl, Allen, and Erlanger, 1974) produce relatively comparable bands, with the fluorescence being particularly intense for paracentromeric regions of chromosomes 1, 9, and 16; the proximal short arm of chromosome 15; and the distal two-thirds of the long arm of the Y chromosome. Even more selective are the G-11 technique (Bobrow and Madan, 1973) and fluorochrome D287/170 (Schnedl et al., 1981). The combinations of various selective staining procedures with other techniques has been useful for studying the heterogeneity of heterochromatin (Buys, Gouw, Blenkers, and van Dalen, 1981; Donlon and Magenis, 1981; Magenis, Donlon, and Wyandt, 1978). For example, sequential staining with G-11 and CBG revealed at least two subsets of heterochromatin, the proximal G-11-negative and distal G-11 positive, in both chromosomes 1 and 9 (Magenis et al., 1978). The presence of similar subregions in chromosome 9 was also described by Buys et al. (1981): After sequential C-banding, the more brightly fluorescent region was identified with DA/DAPI, whereas the other stained intensely with Giemsa. Furthermore, comparative studies using in situ hybridization and DA/DAPI staining of the heterochromatin of chromosome 9 indicated that the fluorescence intensity was positively correlated with the location of satellite DNAs (Gosden, Lawrie and Cooke, 1981; Gosden, Spowart, and Lawrie, 1981). Similarly, the fluorescence intensity obtained with G-11 staining correlated with the location of satellite DNA III (Bobrow and Madan, 1973) and also with satellite DNA IV (Gosden, Buckland, Clayton, and Evans, 1975). The heterogeneity of human constitutive heterochromatin observed by sequential banding using these selective staining techniques has been better characterized in terms of sequences or familial sequences localized in the chromosomes (Miller, Choi, and Miller, 1983; Gosden et al., 1981a, 1981b).

3. Inherent properties

The existence of heteromorphisms can be tested by familial studies. This has been demonstrated by Van Dyke and colleagues (1977) through their studies on twins. A large number of familial studies have indicated that heteromorphisms are inherited in Mendelian fashion (Verma and Lubs, 1976). Nevertheless, a few studies have implied that apparent noninherited variation between parent and offspring occurs as somatic mosaicism and preferential segregation. However, noninherited variation may have arisen from a mismatching of repetitive DNA sequences with subsequent unequal crossing over. Furthermore, the observed mosaic patterns provide suggestive evidence that such an event occurs in somatic cells as well as during meiosis.

4. Structural variability and lateral asymmetry

It has been observed that when 5-bromodeoxyuridine (BrdU)-substituted chromosomal preparations are stained under appropriate conditions, the staining properties of the substituted chromatin are modified. Lin and his colleagues (1974) observed lateral asymmetry in mouse centromeric heterochromatin after a single replication in the presence of BrdU. This lateral asymmetry is presumably associated with an unequal distribution of the thymidine-rich chain of satellite DNA in C-banded regions, since one chain of mouse satellite DNA contains 45% thymidine and the other chain 22% (Flamm, McCallum, and Walker, 1967). Lateral asymmetry of heterochromatin of the h region of human chromosomes 1, 9, and 16 has

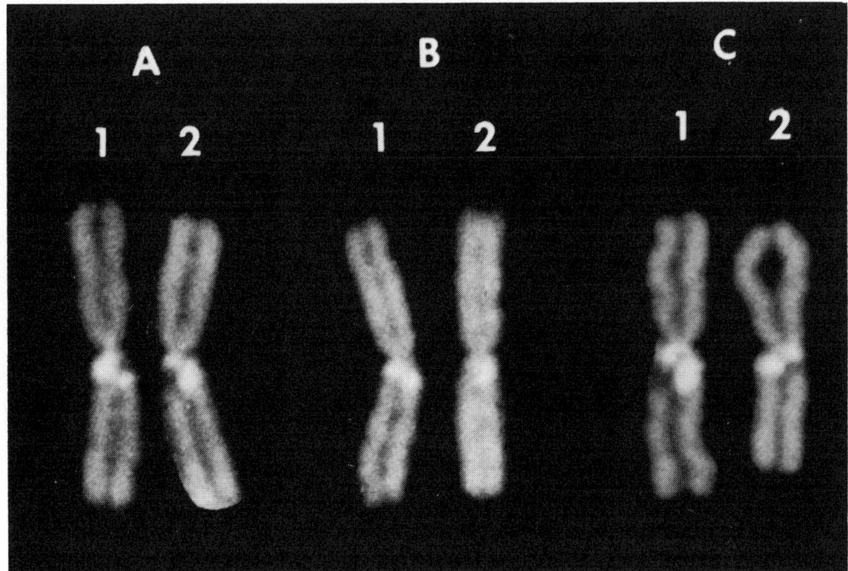

Figure 7.6. Lateral asymmetry of chromosome 1 from three different individuals. (Courtesy of M. S. Lin.)

been demonstrated. The distal two-thirds of the long arm of the Y chromosome has also been shown to exhibit asymmetrical heterochromatin (Lin and Alfi, 1978). An example of lateral asymmetry in chromosome 1 is shown in Figure 7.6. Furthermore, lateral asymmetry has also been suggested for the centromeric heterochromatin of chromosomes 2, 3, 4, 5, 7, 13, 14, 17, 20, 21, and 22 (Ghosh, Rani, and Nand, 1979).

The areas of asymmetrical staining in the human genome all contain satellite DNAs, and all are located within regions that stain positively by C-banding techniques. However, the asymmetry seen in certain chromosomal regions may not be due solely to variation in satellite DNA content, but may also be due to other repetitive types of DNA (Emanuel, 1978). The results of several investigators suggest that the high frequency of asymmetry of heterochromatin in populations is consistent and also heritable (Angell and Jacobs, 1978).

5. Applications of heteromorphic markers

Chromosomal heteromorphisms found in the human genome provide a useful tool for clinical studies since they are inherited in a Mendelian fashion, are stable, and are presumed to have a low mutation rate. Selective staining techniques have been used to determine the origin of extra chromosomes in Down's and Edwards' syndromes. The mechanisms producing an extra chromosome in aneuploid fetuses have been well documented (Verma, Babu, Chemitiganti, and Dosik, 1986). The genetic basis of human triploids has been well explored with the use of these markers. Maternal cell contamination during amniotic fluid cell cultures can now be detected. Heteromorphic markers have also been used to establish paternity. Cytogenetic markers are being used extensively to detect graft cells during bone marrow transplantation. Heteromorphic markers are currently being used to elucidate the chromosomal mechanisms involved in the production of mosaics and to study chimeras. Their use in zygosity in twins and other multiple births has great potential. Furthermore, heteromorphic variants may be used more extensively for gene mapping in somatic cell hybridization.

6. Clinical significance

The biological and clinical significance of chromosomal heteromorphisms is poorly understood. Nevertheless, a series of clinical parameters have been studied. An association between one or more particular heteromorphisms and clinical abnormalities has been suggested. A significant increase in giant satellites was seen in G-group chromosomes from children with a major congenital anomaly detected at birth as compared to children with a normal genome (Lubs and Ruddle, 1970). A number of studies have suggested that carriers of extreme heteromorphisms have a reduced fertility and may produce children with chromosomal abnormalities (Blumberg, Shulkin, Mohandas, and Kaback, 1982). Some investigators have found longer Y chromosomes in criminals, whereas others have not found any length differences between criminals and controls. Similar controversy has been noted for the longer Y chromosome (Verma, Shah, and Dosik, 1983; Rodriguez-Gomez, Martin-Sempere, and Abrisqueta, 1987). A recent review of Atkin and Brito-Babapulle (1981) on the role of heterochromatin in cancer suggests that a particular heterochromatic variant may influence the risk of cancer. It is thus apparent that the clinical significance of chromosomal heteromorphisms requires further investigation in carefully controlled studies.

7. Conclusions

Human chromosomal heteromorphisms identified by various staining techniques have led to a greater understanding of the structural organization of chromosomes. Many individuals previously thought to be abnormal have had only a normal heteromorphic variant. Current research on the mechanisms of the origin of heteromorphic markers is gaining momentum. Though they are inherited from generation to generation and have valuable applications in clinical medicine, their biological significance remains obscure.

8. References

Al-Nassar, K. E., Palmer, C. G., Conneally, P. M., and Yu, P.-L. (1981). The genetic structure of the Kuwaite population II: the distribution of Q-band chromosomal heteromorphisms. *Hum. Genet., 57,* 423–427.

Angell, R. R., and Jacobs, P. A. (1978). Lateral asymmetry in human constitutive heterochromatin: frequency and inheritance. *Am. J. Hum. Genet., 30,* 144–152.

Arrighi, F. E., and Hsu, T. C. (1971). Localization of heterochromatin in human chromosomes. *Cytogenetics, 10,* 81–86.

Atkin, N. B., and Brito-Babapulle, V. (1981). Heterochromatin polymorphism and human cancer. *Cancer Genet. Cytogenet., 3,* 261–272.

Babu, K. A., and Verma, R. S. (1985). Structural and functional aspects of nucleolar organizer regions (NORs) of human chromosomes. *Int. Rev. Cytol., 94,* 151–176.

Babu, A., and Verma, R. S. (1986). Heteromorphic variants of human chromosome 4. *Cytogenet. Cell Genet., 41,* 60–61.

Babu, K. A., Macera, M. J., and Verma, R. S. (1986). Intensity heteromorphisms of human chromosome 15p by DA/DAPI technique. *Hum. Genet. 73,* 298–300.

Blumberg, B. D., Shulkin, J. D., Mohandas, T., and Kaback, M. M. (1982). Minor chromosomal variants and major chromosomal anomalies in couples with recurrent abortions. *Am. J. Genet., 34,* 948–960.

Bobrow, M., and Madan, K. (1973): A comparison of chimpanzee and human chromosomes using the Giemsa-II and other chromosome banding techniques. *Cytogenet. Cell Genet., 12,* 107–116.

Bobrow, M. P. L., Pearson, P. L., Pike, M. C., El-Alfi, O. S., (1971). Length variation in quinacrine banding segment of different size. *Cytogenetics, 10,* 190–198.

Buys, C. H. C. M., Gouw, W. L., Blenkers, J. A. M., and van Dalen, C. H. (1981). Heterogeneity of human chromosome 9 constitutive heterochromatin as revealed by sequential distamycin A/DAPI staining and C-banding. *Hum. Genet., 57,* 28–30.

Donlon, R. A., and Magenis, R. E. (1981). Structural organization of the heterochromatic regions of human chromosome 9. *Chromosoma, 84,* 353–363.

Emanuel, B. S. (1978). Compound lateral asymmetry in human chromosome 6: BrdU-dye studies of 6q12–6q14. *Am. J. Hum. Genet., 30,* 153–159.

Erdtmann, B. (1982). Aspects of evaluation significance and evolution of human C-band heteromorphism. *Hum. Genet., 61,* 281–294.

Flamm, W. G., McCallum, M., and Walker, P. M. B. (1967). The isolation of complementary strands from a mouse DNA fraction. *Proc. Natl. Acad. Sci. USA., 57,* 1729–1734.

Ghosh, P. K., Rani, R., and Nand, R. (1979). Lateral asymmetry of constitutive heterochromatin in human chromosomes. *Hum. Genet., 52,* 79–84.

Gosden, P. R., Buckland, R. A., Clayton, R. P., and Evans, H. J. (1975). Chromosomal localization of DNA sequences in condensed and dispersed human chromatin. *Exp. Cell Res., 92,* 138–147.

Gosden, J. R., Lawrie, S. S., and Cooke, H. J. (1981a). A cloned repeated DNA sequence in human chromosome heteromorphisms. *Cytogenet. Cell Genet., 29,* 32–39.

Gosden, J. R., Spowart, G., and Lawrie, S. S. (1981b). Satellite DNA and cytological staining patterns in heterochromatic inversions of human chromosoma 9. *Hum. Genet., 58,* 276–278.

Jacobs, P. A. (1977). Human chromosome heteromorphisms (variants). *Prog. Med. Genet., 2,* 251–274.

John, B. (1980). Heterochromatin variation in natural populations. In *Chromosomes Today,* Vol. 7, ed. M. D. Bennett, M. Bobrow, and G. M. Hewitt, pp. 128–137. London: Allen and Unwin.

Kurnit, D. M. C. (1979). Satellite DNA and heterochromatin variants: the case for unequal mitotic crossing over. *Hum. Genet., 47,* 169–186.

Lin, M. S., and Alfi, O. S. (1978). Detection of sister chromatid exchanges by 4′, 6-diamidino-2-phenylindole fluorescence. *Chromosoma, 57,* 219–225.

Lin, M. S., Latt, S. A., and Davidson, R. L. (1974). Microfluorometric detection of asymmetry in the centromeric region of mouse chromosomes. *Exp. Cell Res., 86,* 392–395.

Lubs, H. A. (1977). Occurrence and significance of chromosome variants. In *Molecular Human Cytogenetics,* ed. R. S. Sparks, D. E. Comings, and C. F. Fox, pp. 443–446. New York: Academic Press.

Lubs, H. A., and Ruddle, F. H. (1970). Application of quantitative karyotype of chromosome variation in 4400 consecutive newborns. In *Human Population Genetics,* ed. P. A. Jacobs, W. H. Price, and P. Law, pp. 120–142. Edinburgh: Edinburgh University Press.

Lubs, H. A., Kimberling, W. J., Hecht, F., Patil, S. R., Brown, J., Gerald, P., and Summitt, R. L. (1977). Racial differences in the frequency of Q and C chromosomal heteromorphisms. *Nature, 268,* 631–632.

Magenis, R. E., Donlon, T. A., and Wyandt, H. E. (1978). Giemsa-II staining of chromosome 1: a newly described heteromorphism. *Science, 202,* 64–65.

McKenzie, W. H., and Lubs, H. A. (1975). Human Q and C chromosomal variations: distribution and incidence. *Cytogenet. Cell Genet., 14,* 97–115.

Miller, D. A., Choi, Y. C., and Miller, O. J. (1983). Chromosome localization of highly repetitive human DNAs and amplified ribosomal DNA with restriction enzymes. *Science, 219,* 295–397.

Miller, O. J., Schnedl, W., Allen, J., and Erlanger, B. F. (1974). 5-Methylcytosine localized in mammalian constitutive heterochromatin. *Nature, 251,* 636–637.

Okamoto, E., Miller, D. A., Erlanger, B. F., and Miller, O. J. (1981). Polymorphism of 5-methylcytosine rich DNA in human acrocentric chromosomes. *Hum. Genet., 58,* 255–259.

Olson, S. B., Magenis, R. E., and Lovrien, E. W. (1986). Human chromosome variation: the discriminatory power of Q-band heteromorphism (variant) analysis in distinguishing between individuals, with specific application to case of questionable paternity. *Am. J. Hum. Genet., 38,* 235–252.

Page, B. (1973). Identification of chromosome 9 in human male meiosis. *Cytogenet. Cell. Genet., 12,* 254–258.

Paris Conference (1971). *Standardization in Human Cytogenetics: Birth Defects.* Original Article Series No. VIII. New York: The National Foundation.

Paris Conference (1975). *Supplement to Standardization in Human Cytogenetics: Birth Defects.* Original Article Series No. XI. New York: The National Foundation.

Patil, S. R., and H. A. Lubs (1977). Classification of qh regions in human chromosomes 1, 9 and 16 by C-banding. *Hum. Genet., 38,* 35–38.

Rodriguez-Gomez, M. T., Martin-Sempere, M. J., and Abrisqueta, J. A. (1987). C-band length variability and reproductive wastage. *Hum. Genet., 75,* 56–61.

Sahar, E., and Latt, S. A. (1980). Energy transfer and binding completion between dyes used to enhance staining differentiation in metaphase chromosomes. *Chromosoma, 79,* 1–28.

Schnedl, W., Abraham, R., Dann, O., Geber, G., and Schweizer, D. (1981). Preferential fluorescent staining of heterochromatic regions in human chromosomes 9, 15 and the Y by D287/170. *Hum. Genet., 59,* 10–13.

Schweizer, D., Ambros, P., and Andrle, M. (1978). Modification of DAPI banding on human chromosomes with a DNA binding oligopeptide antibiotic: dystamycin A. *Exp. Cell Res., 111,* 327–332.

Van Dyke, D. L., Palmer, C. G., Nance, W. E., and Yu, P. L. (1977). Chromosome polymorphisms and twin zygosity. *Am. J. Hum. Genet., 29,* 431–447.

Varley, J. M. (1977). Patterns of silver staining of human chromosomes. *Chromosoma, 61,* 207–214.

Verma, R. S., and Dosik, H. (1980). Human chromosomal heteromorphisms: nature and clinical significance. *Int. Rev. Cytol, 62,* 361–383.

Verma, R. S., and Dosik, H. (1981). Human chromosomal heteromorphisms in American blacks III. Evidence for racial difference in RFA and QFQ intensity heteromorphisms. *Hum. Genet., 56,* 329–337.

Verma, R. S., and Dosik, H. (in press). Structural organization of heterochromatin in the human genome. In *Cytology and Cell Physiology,* ed. G. H. Bourne. New York: Academic Press.

Verma, R. S., and Lubs, H. A. (1975a). Variation in human acrocentric chromosomes with acridine orange reverse banding. *Hum. Genet., 31,* 225–235.

Verma, R. S. and Lubs, H. A. (1975b). A simple R-banding technique. *Am. J. Hum. Genet., 27,* 110–117.

Verma, R. S., and Lubs, H. A. (1976). Inheritance of acridine orange R-variants in human acrocentric chromosomes. *Hum. Hered., 26,* 315–318.

Verma, R. S., Dosik, H., and Lubs, H. A. (1977). Demonstration of color and size polymorphisms in human acrocentric chromosomes by acridine orange reverse banding. *J. Hered., 68,* 262–263.

Verma, R. S., Dosik, H., and Lubs, H. A. (1978a). Size and pericentric inversion heteromorphisms of secondary constriction regions (h) of chromosomes 1, 9 and 16 as detected by CBG technique in Caucasians: classification, frequencies, and incidence. *Am. J. Med. Genet., 2,* 331–339.

Verma, R. S., Dosik, H., Scharf, T., and Lubs, H. A. (1978b). Length heteromorphisms of the fluorescent (f) and non-fluorescent (nf) segment of human Y chromosomes: classification, frequencies, and incidence in Caucasians. *J. Med. Genet., 15,* 277–281.

Verma, R. S., Benjamin, C., Rodriguez, J., and Dosik, H. (1981). Population heteromorphisms of Ag-stained nucleolus organizer regions (NORs) in the acrocentric chromosomes of East Indians. *Hum. Genet., 59,* 412–415.

Verma, R. S., Rodriguez, J., and Dosik, H. (1982). The quantitative analysis of constitutive heterochromatic regions of human chromosomes 1, 9, and 16 in relation to size and inversion heteromorphisms in East Indians. *Experientia, 38,* 324–326.

Verma, R. S., Huq, A., and Dosik, H. (1983a). Racial variation of nonfluorescent segment of the Y chromosome in East Indians. *J. Med. Genet., 20,* 102–106.

Verma, R. S., Shah, J. V., and Dosik, H. (1983). Size of Y chromosome not associated with abortion risk. *Obstet. Gynecol., 61,* 633–634.
Verma, R. S., Rodriguez, J., and Dosik, H. (1984). Chromatin of h regions of human chromosomes at high resolution. *Experientia, 40,* 878–881.
Verma, R. S., Babu, A., Chemitiganti, S., and Dosik, H. (1986). A possible cause of nondisjunction of additional chromosome 21 in Down syndrome. *Mol. Genet. 202,* 339–341.
White, M. J. D. (1954). *Animal Cytology and Evolution.* Cambridge: Cambridge University Press.

VIII

Heterochromatin: early results and future prospects

RAM S. VERMA

1. Commentary

It has been proven that the hypothesis proposed by Heitz over a half century ago – that condensed intermitotic chromatin belongs to a unitary class – is incorrect. DNA–protein complexes which are solely responsible for the organization of chromosomes, exist in both diffuse and condensed forms. The active and inactive states of heterochromatin are now well recognized, even at the light microscopic level. In recent years, a number of banding techniques have been developed, which produce a differential response in the isopycnotic euchromatic and heterochromatic regions of chromosomes. The causal relationship with respect to tissue specificity has been worked out. The fundamental structures of heterochromatin and euchromatin are clearly different in behavior. Constitutive heterochromatin is composed of DNA sequences with distinctive characteristics. It is a documented fact that chromocenters do not form in all types of heterochromatin; this was the original basis for defining heterochromatin. With a few exceptions, it is a widely accepted fact that heterochromatin is, in general, late replicating. One of the most frequently cited examples of heterochromatin is the X chromosome in normal karyotypes of human females with two X chromosomes. Heterochromatinization of one of the X chromosomes in female mammals is quite intriguing. The molecular mechanism(s) of X inactivation is (are) not fully understood, but several investigations have linked inactivation to DNA methylation. Heterochromatin plays a major role in the evolution of chromosomes. Its addition has led to an increased genomic size. In general, heterochromatin is accumulated at the expense of euchromatin. Heterochromatin can also be lost from the genome, leading to extreme polymorphisms. The critical question remains whether heterochromatin is genetically inert. However, the possibility remains that heterochromatin serves a specific germ-line function without any significance for somatic development.

One of the most intriguing findings has been that a substantial amount of DNA is composed of repetitive sequences. Analysis of the sequence organization and nucleotide sequence of highly tandemly repeated DNA has further led to understanding of the structural organization of heterochromatic DNA. In addition, the use of cloned segments of heterochromatic DNA in combination with classical genetic and cytogenetic methods has further allowed us to understand the function of heterochromatin during development. In humans, there are at least eight different human satellite DNAs: four classified chronologically I, II, III, and IV by Corneo; and four designated A, B, C, and D by Chung and Saunders. These satellite DNA sequences occur as tandem repeat units of various lengths. Analysis of satellite DNAs in mammals has clearly shown that they are divergent sequences. Highly repetitive DNAs consist of related sequences of families that have resulted from the amplification of a simple sequence and the divergent sequences through mutation.

The highly repeated DNA sequences are well programed. Even a minor change in a very short repetitive segment would change the physical property of heterochromatin. Restriction endonuclease digestion techniques have further revealed that sequence variation in these tandemly repeated regions are random. It has been further noted that most satellite DNAs are completely resistant to specific enzymes or are digested into a very regular pattern of DNA fragment sizes.

The remarkable variation (heterogeneity) observed between highly repeated DNA sequences within and among different eukaryotes has been an enigma ever since its discovery. The preferential localization of satellite DNA in various regions of human chromosomes is well documented (Chapter II). The four types of satellite DNA do not correlate directly with the size of the C-band; furthermore, their proportions in each C-band differ. There is good evidence that the structural organization of heterochromatin is complex and highly heterogeneous. Satellite DNAs themselves are not simple, highly repeated sequences, as was once thought. Moreover, satellite DNAs with different buoyant densities may contain sequences that cross-react with one another, obscuring further any understanding of their nature with respect to heterochromatin. Additional DNA-specific probes are being identified which are free of highly repetitive sequences and will be particularly interesting in further studies of the molecular organization of heterochromatin. Following the advent of various technical innovations, the knowledge gained through molecular cytogenetics is quite gratifying. Several mechanisms have been proposed to explain the evolution of repeated sequences. By use of recombinant DNA technology, some of the hypotheses described in this monograph have now been confirmed.

Because the structural organization of heterochromatin is so obscure, the functions of heterochromatin have been a major topic of controversy and reviews in the literature. Some functions have been well explained by Bernard John (Chapter I). Interracial and interchromosomal comparisons of heterochromatin strongly suggest an important biological function.

It is believed that chromosomes are simply a linear sequence of linked gene loci. Molecular genetics has dictated that many eukaryotes carry far more DNA than is required. The extra DNA, termed "junk" DNA in recent literature, is considered genetically inert. The heterogeneity of heterochromatic DNA is complex and obscure. At this time, there is no evidence to suggest that the heteromorphic regions constitute any special risk of abnormality (Chapters VI and VII). Much anxiety has been aroused by the confusion between a chromosomal abnormality and some harmless normal heteromorphic variant. There are interesting and eminently readable reviews of the discoveries of the structural and functional accounts of heterochromatin. Some of the most valuable references of particular relevance, dealing with the meticulous details of this very exciting subject are given in the final section of this chapter.

2. Selected references on heterochromatin

Babu, K. A., and Verma, R. S. (1987). Chromosome structure: Euchromatin and heterochromatin. *Int. Rev. Cytol., 108,* 1–60.

Bradbury, E. M., MacLean, N., and Matthews, H. R. (1981). *DNA, Chromatin and Chromosomes.* Boston: Blackwell Scientific.

Eissenberg, J. C., Cartwright, I. L., Thomas, G. H., and Elgin, S. C. R. (1985). Selected topics in chromatin structure. *Annu. Rev. Genet., 19,* 485–536.

Hamlin, J. L., Milbrandt, J. D., Heintz, N. H., and Azizkhan, J. C. (1984). DNA sequence amplification in mammalian cells. *Int. Rev. Cytol., 90,* 31–82.

Igo-Kemenes, T., Hortz, W., and Zachau, H. G. (1982). Chromatin. *Annu. Rev. Biochem., 51,* 89–121.

John, B., and Miklos, G. L. G. (1979). Functional aspects of satellite DNA and heterochromatin. *Int. Rev. Cytol., 58,* 1–114.

Kornberg, R. D. (1977). Structure of chromatin. *Annu. Rev. Biochem., 46,* 931–954.

Seuanez, H. N. (1979). *The Phylogeny of the Human Chromosome.* New York: Springer-Verlag.

Singer, M. F. (1982). Highly repeated sequences in mammalian genome. *Int. Rev. Cytol., 76,* 67–112.

Subirana, J. A., Munoz-Guerra, S., Aymami, J., Radermacher, M., and Frank, J. (1985). The layered organization of nucleosomes in 30-nm chromatin fibers. *Chromosoma, 91,* 377–390.

Widom, J., Finch, J. T., and Thomas, J. O. (1985). Higher order structure of long repeat chromatin. *EMBO J., 4,* 3189–3194.

Index

(Pages numbers in italics indicate material in tables or figures.)